The Readable Darwin

The Readable Darwin

The Origin of Species Edited for Modern Readers

Second Edition

Jan A. Pechenik

OXFORD
UNIVERSITY PRESS

Oxford University Press is a department of the University of Oxford. It furthers
the University's objective of excellence in research, scholarship, and education
by publishing worldwide. Oxford is a registered trade mark of Oxford University
Press in the UK and certain other countries.

Published in the United States of America by Oxford University Press
198 Madison Avenue, New York, NY 10016, United States of America.

Library of Congress Cataloging-in-Publication Data
Names: Darwin, Charles, 1809–1882, author. | Pechenik, Jan A., editor.
Title: The readable Darwin : the origin of species /
edited for modern readers by Jan Pechenik.
Other titles: On the origin of species
Description: New York, NY : Oxford University Press, [2023] | Includes index.
Identifiers: LCCN 2022040692 (print) | LCCN 2022040693 (ebook) |
ISBN 9780197575260 (hardback) | ISBN 9780197575291 (epub)
Subjects: LCSH: Evolution (Biology) | Natural selection.
Classification: LCC QH365.O25 P434 2023 (print) |
LCC QH365.O25 (ebook) | DDC 576.8/2—dc23/eng/20220920
LC record available at https://lccn.loc.gov/2022040692
LC ebook record available at https://lccn.loc.gov/2022040693

DOI: 10.1093/oso/9780197575260.001.0001

1 3 5 7 9 8 6 4 2

Printed by Sheridan Books, Inc., United States of America

Contents

Preface by Jan A. Pechenik

> You care for nothing but shooting, dogs, and rat-catching, and you will be
> a disgrace to yourself and all your family.
> —From *Charles and Emma* (2009, by Deborah Heiligman)

Charles Darwin received this rant from his father (Robert Darwin, a successful medical doctor and investor) as a teenager while studying medicine at the university in Edinburgh, in Scotland. He had just decided to abandon a career in medicine after seeing several people being operated on without anesthesia.

The father's outburst is perhaps understandable, but how wrong he was! Here are some comments about the importance of Charles Darwin's first edition of *The Origin of Species*:

> It's safe to say that The Origin of Species . . . is one of the most influential books ever written. (David Quammen, *The Reluctant Mr. Darwin: An Intimate Portrait* of *Charles Darwin and the Making of His Theory of Evolution*; 2006, W. W. Norton & Company)

> [The *Origin*] is a book that makes the whole world vibrate. (Adam Gopnik, *Angels and Ages: Lincoln, Darwin, and the Birth of the Modern Age*; 2009, Vintage Press)

> The Darwinian revolution is widely considered the most important event in the entire history of the human intellect. (Michael Ghiselin, 2008)

Charles Darwin published *The Origin of Species* in 1859, one year after he and Alfred Russel Wallace had their remarkably similar papers on natural selection presented at a meeting of a major scientific group, the Linnean Society, in London. I think it's fair to say that the world hasn't been the same since. Here is the crux of his argument:

- Organisms vary in a great many characteristics (traits), both anatomical and behavioral.
- These characteristics are typically passed on to offspring—they are inherited.
- Organisms compete for space, food, and mates and are subject to predation, so that there is a constant struggle to stay alive and successfully reproduce.
- Those pressures give an advantage to individuals with certain characteristics.
- Individuals with those beneficial traits are most likely to survive and leave offspring, which will then inherit those useful traits.

- Over very long periods of time, those characteristics will come to be found in a great many individuals, and a new species will be thereby created.
- Diversification and extinction will then, over incredibly long periods of time, create species that differ more and more from each other, creating new classes, families, genera, and even new phyla.

It's a brilliant argument, and the argument is brilliantly and exhaustively made.

The first printing of *The Origin* sold out on the first day. Remarkably though, the book is rarely read today, not even in high school or college biology courses. That's a shame. Not only is it the basis for so much of our modern biological research in so many areas, but it's also a wonderful reminder of the incredible diversity of life on this planet and a wonderful example of a fair and honest argument based on evidence and logical thinking. Darwin explains his supporting evidence thoroughly but is also very up front about the things he doesn't yet understand, even those that might pose problems for his theory.

Reading Darwin's book *is* a bit of a slog though. The style of writing is very much of the 1800s, and although some of his prose was delightfully memorable, many of his paragraphs are long and unwieldy, as are many of the sentences, making the reading rather difficult for many modern readers. He also refers to many people without ever saying who they are, and he mentions the names of a great many animals and plants that most people today are not familiar with.

To make it easier for teachers and students to read this wonderful and incredibly important book, and to make it more accessible as well for interested adults, I have edited the entire book—all 15 chapters—into more readable prose. In doing so, I have tried to keep as much of Darwin's original wording as possible and to preserve all of the sense of what he is saying.

I chose to work with Darwin's final edition (the Sixth), which was published in 1872. The first edition was published in 1859, and it was revised as new information was collected (by Darwin and by many others around the world) and as new arguments were advanced against Darwin's proposition. In the sixth edition we can see how Darwin responded to all of his critics. For the sixth edition, Darwin also added an entirely new and wonderful chapter (Chapter 7), which includes fascinating information on the evolution of baleen whales from toothed ancestors, the evolution of climbing in plants, and the evolution of breasts in mammals. And it was in the sixth edition of *The Origin* that Darwin introduced the word "evolution" (he had first used it in *The Descent of Man*, published the previous year).

One disadvantage of reading the sixth edition rather than the first, however, is that because Darwin was under increasing pressure to develop a complete theory of natural selection, he did more grasping at straws than he did in the first edition in dealing with the possible causes of variability among individuals and the inheritance of those variations by offspring. Even by the time of Darwin's death, in 1882, nobody knew what caused variation among individuals or how those variations were transmitted to offspring. Mendel published what eventually turned out to be the beginnings of modern genetics in 1866, but the paper was cited only three times in the next 35 years. It was published in an obscure journal, and those who read the paper seem not to have understood its implications, particularly that the findings applied to all sorts of traits, not just the especially distinctive traits that Mendel had been working with in his pea plant studies. Intriguingly, Mendel read a German translation of Darwin's work, but, as far as we know, he never tried to contact Darwin, even when he visited England in 1862. Apparently Mendel himself didn't see the connection between Darwin's ideas and his own work. In any event, it was many decades later that the implications of Mendel's work were realized and several decades after that that the basic mechanisms of inheritance, including the role of DNA, were finally worked out and the mechanism of evolution through the process of natural selection, as Darwin had proposed, was finally well accepted. Indeed, some of that work is still ongoing.

Darwin didn't get everything right. I have included a number of footnotes to indicate where Darwin was correct and where he was wrong. But his ideas about the role of variation in shaping morphological and behavioral diversity through natural selection were on target, as were his ideas about the role of natural selection in gradually creating new species and new larger taxonomic categories. As a result, natural selection is now widely accepted as the major explanation for present and past organismal diversity and for the appearance of major new groups of organisms over time.

My Edits

I have spent much time during the past 10 years or so editing all 15 of Darwin's chapters, making multiple passes over every paragraph. But this is *not* my book: it's Darwin's, and I have tried to maintain his voice as much as possible. I have based much of my editing on the various rules from my *Short Guide to Writing About Biology* (ninth edition). Indeed, I like to think that this is how Darwin would have written *The Origin* had he read my *Short Guide* first! Here are the main rules that I have followed while editing this material, along with some examples:

1. Omit unnecessary words, especially unnecessary prepositions.
    Original: I suspect that the chief use of the nutriment in the seed is to favor the growth of the young seedling.... (Chapter 3)
    My revision: I suspect that the seed's nutrients are chiefly used to promote the growth of the young seedling....
2. Eliminate weak verbs (what I call "wimpy verb syndrome"), in part by making organisms the agent of the action.
    Original: Many plants are known which regularly produce....
    My revision: Many plants regularly produce....
3. Warn readers of what lies ahead, using words like "For example," "Similarly," and "On the other hand," and remind readers from time to time of what they have just read.

4. Incorporate definitions into sentences.

   Original: One of the most serious is that of **neuter insects**, which are often differently constructed from either the males or fertile females (Chapter 6, p. 174)

   My revision: One of the most serious is that of **neuter, non-reproductive insects**, which are often differently constructed from either the males or the fertile females of the species.

5. Use repetition, summary, and appropriate punctuation to link thoughts and improve the flow of ideas: Never make readers back up!

   Original: Although many statements may be found in works on natural history **to this effect**, I cannot find even one that seems to me of any weight.

   My revision: Although many works on natural history **claim that some structures in one species do indeed serve for the exclusive benefit of a different species**, I cannot find even one example that seems to me to hold any weight.

   Here is one more example:

   Original: Yet it cannot be said that small islands will not support at least some small mammals, **for they occur** in many parts of the world on very small islands ...

   My Revision: Yet it cannot be said that small islands will not support at least some small mammals, **for small mammals** occur in many parts of the world on very small islands ...

I have made a variety of other changes as well. One would-be reader commented on Amazon.com that "Another thing which made the book a little harder for me is that Darwin mentions a lot of people and animals I've never heard of." I now clarify what the various animals and plants are that Darwin refers to. For example, when Darwin mentions cirripedes or *Balanus*, I help readers by including the term "barnacles" somewhere in the sentence. I also have added a little more information to some sentences, to help make them more meaningful to modern readers. For example:

Original: Thus, we can hardly believe that the webbed feet of the upland goose or of the frigate-bird are of special use to these birds.

My revision: Thus, we can hardly believe that the webbed feet of the upland goose of South American grasslands, or of the frigate-bird, which cannot swim or even walk well, and which takes most of its food in flight, are of special use to these birds.

I have also broken up overly long sentences and overly long paragraphs and altered sentence structure when necessary to make the sentences easier to read.

Original: A trailing palm tree in the Malay Archipelago climbs the loftiest trees by the aid of exquisitely constructed hook clusters around the ends of its branches, and this contrivance, no doubt, is of the highest service to the plant; but as we see nearly similar hooks on many trees which are not climbers, and which, as there is reason to believe from the distribution of the thorn-bearing species in Africa and South America, serve as a defense against browsing quadrupeds, so the spikes on the palm may at first have been developed for this function, and subsequently have been improved and taken advantage of by the plant, as it underwent further modification and became a climber.

My revision: In my version, that single sentence has now become three sentences.

Original: Seeing how important **an organ of locomotion the tail is** in most aquatic animals. . . .

My revision: Seeing how important **the tail is as an organ of locomotion** in most aquatic animals. . . .

Finally, I have written short previews for the start of each chapter, have boldfaced particularly important sentences, and added many illustrations. Darwin's sixth edition originally had only a single illustration, a schematic drawing of evolutionary branching patterns Figure 4.11, in Chapter 4. My version includes numerous drawings and photographs of the various plants and animals that Darwin talks about, along with some other figures that clarify some of his major points.

Acknowledgments

Preparing this "translation" gave me a wonderful opportunity to read a variety of books about *The Origin* and about Darwin's life. In particular, Janet Browne's *Charles Darwin: A Biography, Volume 1: Voyaging* (1996, Princeton University Press) and *Volume 2: The Power of Place* (2003, Princeton University Press); James Costa's *The Annotated Origin* (2009, Harvard University Press) and David Reznick's *The Origin Then and Now: An Interpretative Guide to the Origin of Species* (2010, Princeton University Press) were invaluable in helping me to fully understand Darwin's thinking.

I am very grateful to my geological colleagues Jack Ridge (Tufts University) and Alan Young (Salem State University) for carefully reading my versions of Darwin's Chapters 10 and 11, about difficulties in interpreting the geological record. I also wish to thank my friend and colleague Gordon Hendler at the Natural History Museum of Los Angeles County for critical feedback on Chapter 3 and for moral support, advice, and encouragement throughout the project. My Tufts colleague George Ellmore also provided helpful information about plant biology.

Thanks to the following reviewers for their critiques of the original book proposal and sample chapter:

Stevan Arnold, Oregon State University
John Avise, University of California, Irvine
Alex Badyev, University of Arizona
David Begun, University of California, Davis
Douglas Futuyma, The State University of New York at Stony Brook
Richard Harrison, Cornell University
Mark Kirkpatrick, University of Texas at Austin
Carol Lee, University of Wisconsin
Michael Lynch, Indiana University
Mohamed Noor, Duke University
Patrick Phillips, University of Oregon
Adam Porter, University of Massachusetts
David Rand, Brown University
Michael Wade, Indiana University
Bruce Walsh, University of Arizona

I am also grateful to the following biologists who read and commented on complete texts of the first eight chapters of the manuscript:

Greg Bole, The University of British Columbia

Becky Fuller, University of Illinois at Urbana-Champaign

Frid-a Jóhannesdóttir, Cornell University

Adi Livnat, Virginia Polytechnic Institute and State University

Joel W. McGlothlin, Virginia Polytechnic Institute and State

Benjamin Normark, University of Massachusetts Amherst

Jeremy Searle, Cornell University

Christopher S. Willett, University of North Carolina at Chapel Hill

Pamela Yeh, University of California, Los Angeles

Rebecca Zufall, University of Houston University

It's also a pleasure to thank Andy Sinauer for his enthusiastic reception of this idea and his willingness to see it through and his entire team for their professionalism and expertise in bringing the project to completion so smoothly: Dean Scudder, Christopher Small, Chelsea Holabird, Stephanie Bonner, David McIntyre, Joan Gemme, Jefferson Johnson, and Tom Friedmann. I am also grateful to those at Oxford University Press for seeing this project through to completion with the full volume of Darwin's wonderful work, *The Origin of Species*.

My son Oliver and his wife Ardea Thurston-Shaine helped me find many appropriate photographs, answered many of my questions about birds, and offered helpful suggestions on various drafts. Ardea also contributed an excellent drawing of various pigeon breeds for Chapter 1 and developed the concept that resulted in the fabulous image that adorns the cover of this book. One of my graduate students (Casey Diederich) and several undergraduates (Elizabeth Card and Chinami Michaels) also contributed wonderful figures for the first part of the project, and another of my graduate students, Daria Clark, helped find appropriate figures for the second part of the project. And finally I thank my entire family—especially my wife, Regina—for their love, support, and patience throughout the adventure. J. S. Bach, Franz Schubert, Johannes Brahms, and Carl Nielsen have also played major roles in keeping me sane throughout the project.

—Jan A. Pechenik, 2022

Introduction

For nearly five years, from December 27, 1831, until October 2, 1836, I served as naturalist aboard the *HMS Beagle*, exploring. During that voyage I was much amazed by how the various types of organisms were distributed around South America, and how the animals and plants presently living on that continent are related to those found only as fossils in the geological record elsewhere. These facts, as will be seen in later chapters, seemed to me to throw some light on the origin of species—that "mystery of mysteries," as it has been called by one of our greatest scientists, John Herschel.[1] After I returned home, it occurred to me, in 1837, that I might be able to help address this great question by patiently accumulating and reflecting on all sorts of facts that might have any bearing on it. Finally, after five years of work, I allowed myself to speculate on the subject and wrote up some brief notes. I enlarged these in 1844, into a sketch of the conclusions that seemed to be most probable from the evidence I had collected. Over the subsequent 15 years I have steadily pursued the same object: trying to understand how new species come about. I hope you will excuse me for entering these personal details of my work, as I give them only to show that I have not been hasty in coming to a decision.

Now, in 1859, my work is nearly finished.[2] Still, it will take me many more years to complete it, and as my health is not strong, I have been urged to publish this brief version of my findings. I have more especially been urged to do this as Mr. Alfred Russel Wallace, who is now studying the natural history of the Malay Archipelago north of Australia, has reached almost exactly the same conclusions that I have reached concerning the origin of species. In 1858, he sent me an account on the subject, requesting that I pass it along to the geologist Sir Charles Lyell, who in turn sent it for presentation at a meeting of the Linnean Society. Mr. Wallace's paper has now been published, in the third volume of that society's journal. Sir Lyell and my colleague the excellent botanist Dr. Joseph Hooker, who both knew of my work—Dr. Hooker having read my sketch of it in 1844—honored me by suggesting that I also publish a summary of my ideas in that journal (some brief extracts from my own manuscripts) along with Mr. Wallace's excellent paper.

---

[1] John Herschel was a chemist, astronomer, mathematician, botanist, inventor, and the son of the brilliant astronomer William Herschel.

[2] The first edition of *The Origin of Species* was published in 1859. Subsequent editions were published in 1860, 1861, 1866, 1869, and 1872, as Darwin collected more information and refined and expanded his ideas.

The Readable Darwin. Second Edition. Jan A. Pechenik, Oxford University Press. © Oxford University Press 2023.
DOI: 10.1093/oso/9780197575260.003.0001

The brief summary of my ideas that I record here must necessarily be imperfect. I do not have sufficient space in this volume to give references and authorities for many of my statements; I must trust the reader to have some confidence in my accuracy. No doubt errors will have crept in, though I hope I have always been cautious in trusting to good authorities alone. I can here give only the general conclusions at which I have arrived, with a few facts in illustration, which I hope, in most cases, will suffice. No one can feel more clearly than I do the necessity of eventually publishing in detail all the facts, with references, on which I base my conclusions, as I am well aware that some evidence can be found apparently leading to conclusions directly opposite to those that I have arrived at, for nearly every point I make. A fair result can be obtained only by fully stating and balancing the facts and arguments on both sides of each question, and it is impossible to do so here.

I also much regret that lack of space prevents my having the satisfaction of acknowledging the generous assistance I have received from many naturalists, some of whom I have never actually met. I cannot, however, let this opportunity pass without expressing my deep obligations to Dr. Hooker, who, for the past 15 years, has aided me in every possible way with his large stores of knowledge and his excellent judgment.

In considering the origin of species, it is certainly conceivable that some naturalist, just by reflecting on the mutual interactions of all living organisms, on their embryological similarities and differences, their geographical distribution, their geological succession over time, and other such facts, might reach the conclusion that individual species had not been independently created but had in fact descended from other species. But such a conclusion, even if well founded and well argued, would be unsatisfactory until it could be shown *how* the innumerable species inhabiting this world had been modified so as to acquire that perfection of structure and coadaptation that now justly excites our admiration. What is the mechanism by which this could come about?

Naturalists continually refer to external conditions, such as climate and food, as the only possible source of variation. In one limited sense, as I will discuss in detail later, this may be true; but it is preposterous to attribute to mere external conditions the structure of any organism—for instance that of the woodpecker, with its feet, tail, beak, and tongue so wonderfully adapted to catch insects under the bark of trees. And consider, too, the case of the mistletoe plant, which draws its nourishment from particular trees, and which has seeds that must be transported by certain birds, and which has flowers with separate sexes absolutely requiring the help of certain insects to bring pollen from one flower to the other. It is equally preposterous to account for the structure of this parasitic plant, and its intimate and essential relationship with such a range of other organisms, by the effects of external conditions, or of habit, or of the wishes and desires of the plant itself.

Thus it is of the highest importance to gain a clear insight into the means through which organisms become modified and coadapted to interact with other organisms. When I began my observations it seemed to me that a careful study of domesticated animals and of cultivated plants would offer me the best chance of resolving this difficult problem. Nor have I been disappointed: in this and in all other perplexing cases, I have invariably found that our knowledge of variation under domestication, imperfect as it is, provided the best and most satisfying clues to how the process works in the wild. I am fully convinced of the value of such studies, even though they have so far been typically neglected by naturalists.

For this reason I devote the first chapter of this summary of my ideas and findings to Variation under Domestication. We shall see that a large amount of hereditary modification is at least possible, and perhaps even more importantly, we shall see how great is our power to accumulate slight variations in traits over time by simply choosing which animals to breed together.

In Chapter 2, I will talk about the variability of species in the wild, although I'll not have space here to present the long catalogs of facts that the topic really requires. I will, however, be able to discuss the circumstances that are most favorable to causing variation. In the next chapter (Chapter 3), I will consider the struggle for existence that takes place among all living things throughout the world and show how it is an inevitable consequence of the exponential rates[3] at which the populations of all living beings tend to grow (see Figure 3.3[4]). This is the basic doctrine of the English cleric and demographer Robert Malthus, applied now to the whole animal and vegetable kingdoms: **As many more individuals of each species are born than can possibly survive, and as, consequently, there is a frequently recurring struggle for existence, it follows that any organism will have a better chance of surviving if it varies even just slightly in any way that is helpful to itself under the complex and sometimes varying conditions of life.** Thus, that individual will be *naturally selected for*. As so many variations are transmitted through inheritance, any selected variety will then tend to propagate its new and modified form among its offspring.

I will discuss this fundamental topic of "natural selection" in some detail in Chapter 4 and will show how natural selection almost always causes extinction of the less capable forms of life and leads to what I have called "divergence of character" within a population. In the next chapter (Chapter 5), I will discuss the complex and little-known laws of variation. In the succeeding three chapters, I will carefully consider the most apparent and gravest difficulties presented by my theory. These include (1) the difficulty in understanding how a simple organism or a simple organ can eventually be changed and perfected into a highly developed being or into an elaborately constructed organ (Chapters 6 and 7) and (2) the difficulty in understanding how complex animal behaviors and mental capacity may be shaped by the same forces that shape morphology (Chapter 8).[5] Then, in Chapter 9, I use examples of successful crosses between what are considered to be separate species (i.e., hybridism) to tighten the link between varieties and species, and, in the following two chapters

---

[3] With exponential growth, a population continues to grow by the same percentage each year, which means that the population size will increase faster and faster over time.

[4] Figure 3.3 was not included in Darwin's *The Origin of Species*.

[5] *The Origin of Species*, sixth edition, contains 15 chapters; *The Readable Darwin* covers Chapters 1–8.

(Chapters 10 and 11), I discuss the imperfection of the fossil record and how that imperfection explains why we don't see all of the intermediate stages of evolutionary change preserved as fossils. Chapters 12 and 13 explain the fascinating patterns of how various species are distributed around the world and why, for example, we see some of the same species currently living in very different, often widely separated locations. In Chapter 14, I discuss how similarities in particular developmental characteristics and the presence of rudimentary organs in adults provide evidence that all of today's living species have descended from ancient, common ancestors. In the final chapter (Chapter 15), I give a brief recapitulation of the entire subject and a few concluding remarks.

No one should be surprised that there is as yet much that is unexplained regarding the origin of species and varieties considering how profoundly ignorant we are about the mutual interrelations of all the animals and plants that live around us. Who can explain why one species ranges widely and is abundant, while a related species has a narrow range and is rare? Yet these relationships are of the greatest importance for they determine the present welfare and, I believe, the future success and modification of every inhabitant of this world. We know even less of the mutual relations of the innumerable inhabitants of the world during the many past geological epochs in Earth's history. **Although much remains obscure, and will long remain obscure, I can entertain no doubt, after the most careful study and dispassionate judgment of which I am capable, that the view which most naturalists entertain, and which I previously entertained myself—that each species was independently created—is erroneous.** I am fully convinced that species are not unchangeable, but rather that all those species that now belong to what are called the same "genera" are direct descendants of some other species, a species that is probably now extinct. In the same way, the acknowledged varieties of any one species are the direct descendants of that species. Furthermore, I am convinced that natural selection has been the main engine, although not the exclusive one, of modification over long periods of time.

1
Variation Under Domestication

In this chapter, Darwin talks about how domesticated animals and plants have gradually come to look as they do through our powers of selecting for the particular traits
that we value, over the course of many generations. The keys to this process are individual variability and the passing of particular traits to offspring.

Variability

For any given species, when we compare members belonging to the same variety or
subvariety of our older cultivated animals and plants—there are at least 7,500 varieties of apples, for example, all members of the species *Malus domestica*—one of the
first things we notice is that they generally differ more from each other than do the
individuals of any one species or variety in nature. It seems clear that organisms must
be exposed during several generations to new conditions to cause any great amount
of variation, but, once the characteristics have begun to vary, they generally continue
to vary for many generations. We know of no case in which any variable organism
has ceased to vary under domestic cultivation. Our oldest cultivated plants, such as
wheat, still yield new varieties, and our oldest domesticated animals can still be rapidly modified or improved because of the variations that they continue to exhibit.[1]

Effects of Habit and of the Use or Disuse
of Parts; Correlated Variation; Inheritance

Changed habits produce an inherited effect, as, for example, the time during the year
at which a plant begins to flower after being transported from one climate to another.
With animals, the increased use or disuse of parts has had a more marked influence.
Thus I find that in the domestic duck, the bones of the wing weigh less in proportion
to the weight of the entire skeleton than do the same bones in the wild duck, while

---

[1] I have shortened this section considerably. As you will see throughout this book, Darwin, having no
knowledge of chromosomes, DNA, or even the basics of Mendelian genetics (first put forward by Gregor
Mendel, in an obscure paper published in 1866), was at a loss to explain the causes of variation, or how
those variations were transmitted to offspring. He returns to this topic in Chapters 2 and 5.

The Readable Darwin. Second Edition. Jan A. Pechenik, Oxford University Press. © Oxford University Press 2023.
DOI: 10.1093/oso/9780197575260.003.0002

the bones of the leg weigh more; this change may be safely attributed to the domestic duck flying much less and walking more than its wild parents in nature.[2]

Many laws regulate variation, some few of which can be dimly seen. Let me first say something about what may be called "correlated variation." Major changes in embryonic or larval development, for example, will probably cause some changes in the mature animal. Some instances of correlation are quite whimsical: for example, cats that are entirely white and have blue eyes are generally deaf, while the work of Dr. Karl Friedrich Heusinger von Waldegg suggests that while white sheep and pigs are poisoned by certain plants, dark-colored individuals are not. The American biologist Professor Jeffries Wyman has recently given a good illustration of this phenomenon: when he asked some Virginia farmers why all of their pigs were black, they told him that the pigs ate the paint-root plant (genus *Lachnanthes*), which colored their bones pink and caused the hoofs of all but the black individuals to fall off, so that the farmers now select only the black members of each litter for raising and breeding since those have the best chance of surviving. Hairless dogs have imperfect teeth; long-haired and coarse-haired animals have long horns or many horns; pigeons with feathered feet also have skin between their outer toes; pigeons with short beaks have small feet, and those with long beaks have large feet; and animal breeders believe that long limbs are almost always accompanied by an elongated head. Thus if humans go on selecting, and thus augmenting, any peculiarity in a species, we will almost certainly modify other parts of the structure—completely unintentionally—owing to the mysterious laws of correlation.

The causes of variation are unknown, or at best only dimly understood, and the results of those laws are infinitely complex and diversified. Similarly, the laws governing the inheritance of those variations are for the most part unknown: no one can say why the same peculiarity in different individuals of the same species, or in different species, is sometimes inherited and sometimes not; or why a child often reverts in certain characteristics to its grandfather or grandmother, or even to a more remote ancestor; or why a peculiarity is often transmitted from one sex to both sexes, or to one sex alone, more commonly but not exclusively to a member of the same sex.

But the key issues here are that both plants and animals do vary in many characteristics within a species and that many of those variations are passed along to offspring. Indeed, the number and diversity of inheritable variations in structure, of both slight and great physiological importance, are endless. It is those inherited variations, regardless of what the causes of variation are or what the mechanisms of inheritance turn out to be, that are important to my argument.

---

[2] In the absence of any convincing data to the contrary, Darwin thought that the ideas of Jean-Baptiste Lamarck (1744–1829), a prominent French naturalist, about the inheritance of changes resulting from the increased use or disuse of various body parts seemed reasonable. As mentioned earlier, Darwin knew nothing about the causes of variation or the laws governing inheritance.

Character of Domestic Varieties; Difficulties of Distinguishing Between Varieties and Species; Origin of Domestic Varieties from One or More Species

Members of the various domesticated races often differ from each other, and from other species within the same genus, to an extreme degree in some particular part, especially when compared with the species in nature to which they are most closely allied. With these exceptions, domesticated races belonging to a single species differ from each other to about the same degree that closely allied species of the same genus differ from each other in nature. Thus, the distinction between species and the varieties found within a single species is not easily made. Indeed, the domesticated races of many animals and plants have been ranked by some competent judges as the descendants of what were originally several distinct species and by other competent judges as mere varieties of a single species. If the distinction between species and domesticated races within species was markedly clear, this source of doubt would not be so common.

When trying to estimate the amount of structural difference between allied domestic races, we are soon placed in doubt by not knowing for sure whether the various races are descended from one or from several different parent species. This point, if it could be cleared up, would be quite interesting: if, for instance, it could be shown that the greyhound, bloodhound, terrier, spaniel, and bull-dog, which we all know propagate their kind very accurately, were in fact all offspring of a single ancestral species,[3] then we would surely doubt the immutability of the many closely allied natural species—for instance, of the many fox species—inhabiting different parts of the world. Unfortunately, in the case of our most anciently domesticated animals and plants, it is not possible to come to any definite conclusion about whether they are descended from one or from several wild species.

It has often been assumed that man has chosen to domesticate only animals and plants that have an extraordinarily developed natural tendency to vary and likewise to withstand diverse climates. I do not dispute that these capacities have increased the value of most of our domesticated organisms. But how could a savage possibly know, when he first tamed some particular animal, whether it would vary in succeeding generations and whether it would be able to tolerate other climates? Has the small amount of natural variability that we see in the donkey and the goose, or the limited ability of reindeer to withstand warmth, or the limited ability of the common camel to withstand cold prevented their domestication? No, it has not. I cannot doubt that if other animals and plants belonging to as great a diversity of classes and countries as our domesticated productions belong were taken from nature and bred for an

---

[3] As shown by modern molecular work, this turns out to be the case, contrary to what Darwin believed: all dogs did in fact evolve from a single ancestral species, the gray wolf, *Canis lupis*.

equal number of generations under domestication, they would on average vary just as much as the parent species of our existing domesticated productions have varied.

The origin of most of our domestic animals will probably remain vague, but some authors hold the absurd belief that every race that breeds true has had a separate prototype species in nature, even if the distinctive characteristics between the races are ever so slight. If so, then there must have existed at least 20 species of wild cattle and at least 20 species of wild sheep in Europe alone. Indeed, one author believes that there must formerly have been at least 11 wild species of sheep unique to Great Britain!

Although those beliefs seem extremely unlikely, I do believe that a small part of the difference between the various domestic dog breeds is due to their being descended from several distinct species. But in the case of strongly marked races of some other domesticated species, there is presumptive or even strong evidence that all are truly descended from a single wild stock.

For example, having kept nearly all the English breeds of fowl alive myself over some years, having bred and crossed them, and having examined their skeletons very carefully, it seems to me almost certain that all are descended from the wild Indian species *Gallus bankiva* (Figure 1.1). Indeed, this is exactly the conclusion reached by Mr. Edward Blyth and others who have studied this bird in India. Similarly for ducks and rabbits, some breeds of which differ a great deal from each other, the evidence is clear that they are all descended from the common wild duck and rabbit.

I would like to talk now about how we have made good use of inherited variations over the years in domesticating our animals and cultivating our crops.

Figure 1.1 *Gallus bankiva* now goes by *Gallus gallus bankiva*, the red junglefowl.

Figure 1.2 A sampling of pigeon diversity. From left to right: line 1, fantail, runt, English carrier; line 2, barb, rock pigeon, turbit; line 3, pouter, short-faced tumbler, Jacobin.

Breeds of the Domestic Pigeon, Their Differences and Origin

Believing that it is always best to study some special group, I have, after much careful thought and investigation, taken up the breeding of domestic pigeons. One advantage of working with pigeons is that males and females can be easily mated for life: thus different breeds can be kept together in the same aviary without fear of them mating randomly with each other. I have kept every breed of pigeon that I could purchase or otherwise obtain, and several people have kindly sent me pigeon skins from all around the world. Many works about pigeons have been published, in a variety of languages; some of these writings are especially important to us because of their great antiquity. I have also associated with several eminent pigeon breeders and have been allowed to join two London Pigeon Clubs. So I have learned much about pigeons and about breeding them.

The diversity of pigeon breeds is truly astounding (Figure 1.2). Here I will talk about just a few of them: the English carrier pigeon; the short-faced tumbler; the "runt"; the "barb"; the pouter; the turbit; and the Jacobin, tumbler, and laugher pigeons.

First compare the English carrier pigeon with the short-faced tumbler and see the wonderful difference in their beaks, which relates to corresponding differences in their skulls. The carrier pigeon, and especially the male, is also remarkable for the wonderful development of the fleshy outgrowth on the skin about its head; this is accompanied by greatly elongated eyelids, very large external openings to the nostrils, and a wide gaping mouth (Figure 1.3A).

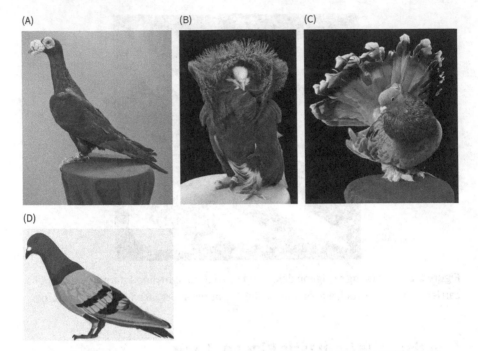

Figure 1.3 (A) English carrier pigeon. (B) Jacobin (red) pigeon. (C) Fantail pigeon. (D) Rock pigeon.

In contrast, the so-called runt is a very large bird, with a long, massive beak and large feet; some of the subspecies of runts also have very long necks, while others have very long wings. Some also have long tails, while others have singularly short tails.

The bird known as the "barb" is similar to the carrier pigeon, except that instead of a long beak it has a very short and broad one. The turbit, on the other hand, has a short and conical beak, with a line of reversed feathers down its breast, and the characteristic habit of continually expanding the upper part of its esophagus slightly, while the Jacobin pigeon has its feathers so reversed along the back of its neck that they form a hood; it also has wing and tail feathers that are unusually long compared with the size of the bird (Figure 1.3B).

The pouter has a decidedly elongated body, wings, and tail, and its enormously large crop[4] (see Figure 1.2), which it glories in inflating, may well excite astonishment and even laughter. The fantail pigeon has 30 or even 40 tail feathers, instead of the 12–14 found in all other pigeons, and it keeps those feathers expanded and erect so that in many individuals the head and tail touch (Figure 1.3C)! Also, the oil gland, common to all other pigeons, is completely absent in fantails.

---

[4] A crop is an expandable, muscular pouch near the bird's throat, used to store food.

Figure 1.4 (A) Double-crested priest pigeon. (B) Modena pigeon. (C) Frillback pigeon. (D) Scandaroon pigeon.

Behaviors also differ among the breeds: for example, the trumpeter and laugher pigeons, as their names suggest, utter a very different-sounding "coo" from members of the other breeds, and the common tumbler pigeon has the very distinctive, inherited habit of flying at a great height in a compact flock and then tumbling in the air, head over heels: No other pigeon does this! There are also several other less distinct breeds that I could talk about. But I think I have made my point about how much the behaviors of the various pigeon breeds differ from each other.

In the skeletons of the various pigeon breeds, the bones in the face vary enormously in length and width and degree of curvature; the shape of the lower jaw bone, as well as its length and width, also vary in an extremely remarkable manner. The caudal and sacral[5] vertebrae vary in number, as does the number of ribs, the width of the ribs, and

---

[5] Caudal vertebrae are those in the tail, for animals with tails; the sacrum is found in the center of the pelvis.

whether or not the ribs bear outgrowths. The size and shape of the openings in the sternum vary greatly from breed to breed, as does the amount of divergence and relative size of the two arms of the wishbone (i.e., the furcula).

Many other structures also vary a great deal among the different pigeon breeds (Figure 1.4): the proportional width of the gape of the mouth; the proportional length of the eyelids, of the opening of the nostrils, and of the tongue (and not always in exact correlation with the length of the beak); the size of the crop and of the upper part of the esophagus; the degree of development of the oil gland and even whether it is present or not; the number of primary wing and tail feathers; the length of the wing compared to the length of the tail and the length of the body; the relative lengths of the leg and foot; the number of plates on the toes; and the degree to which skin develops on the toes.

And there's more. The age at which the birds acquire their final plumage also varies among breeds, as does the state of the down with which the nestling birds are first clothed when they hatch from the egg. The shape and size of the eggs also varies considerably among pigeon breeds, as does the manner in which the birds fly. In some pigeon breeds the voice and disposition also vary remarkably, as I mentioned earlier. Finally, the males and females have come to look slightly different from each other in some breeds but not others.

The different breeds of pigeon vary so greatly in appearance that at least a dozen different sorts of pigeons might be brought to any ornithologist,[6] and if he were told that they were wild birds rather than domestically bred birds, he would certainly rank each of them as a well-defined, separate species. Moreover, I do not believe that any ornithologist would under such circumstances even place the English carrier pigeon, the short-faced tumbler, the runt, the barb, the pouter, and the fantail in the same genus! See Figure 1.5 for the distinction between species, genera, and other taxonomic categories. And yet, as great as the differences are between the breeds of pigeon, I am fully convinced that the common opinion of naturalists is correct: *all of these birds are descended from a single ancestor—the rock pigeon, Columba livia* (see Figure 1.3D). Let me explain my reasoning.

If the various pigeon breeds I have just talked about are not varieties of a single species and have not all descended from the rock pigeon, then they must have descended from at least seven or eight separate ancestral stocks; it would be impossible to make the present domestic breeds by cross-mating any of the present breeds with each other. How, for instance, could you get a pouter pigeon by crossing males and females of any other two breeds unless one of the parent stocks possessed the characteristically enormous crop of the pouter? The supposed ancestral stocks must all have been rock pigeons. But none of the two or three other known rock pigeon species have any of the characteristics found among the domestic pigeon breeds. Thus the supposed ancestral pigeon stocks must either still exist in the countries where

---

[6] An ornithologist is an expert on the biology of birds.

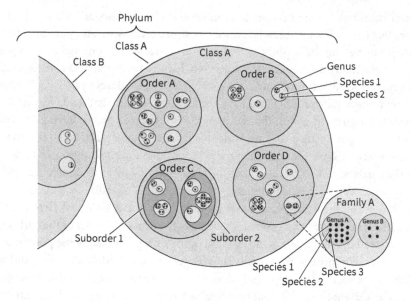

Figure 1.5 A general view of the relationship between phyla, classes, orders, families, genera, and species. Each genus may contain many species, with some genera containing more species than others. One family typically contains many genera, and some families contain more genera than others. Similarly, families are contained within distinct orders, orders are contained within distinct classes, and classes are contained within phyla. There are about 30 well-defined animal phyla. All vertebrates—including for example, dogs, cats, birds, frogs, fishes, turtles, whales, and people—are contained within the phylum Chordata.

they were originally domesticated and yet be unknown to any ornithologist—which seems rather improbable, considering the size, habits, and remarkable characteristics of these birds—or they must all have become extinct in the wild, which is also most unlikely: birds breeding on precipices and that are good fliers are unlikely to go extinct. Indeed, the common rock pigeon, which has the same habits as the domestic breeds, has not been exterminated in the wild even on several of the smaller British islands or on the shores of the Mediterranean. Thus the supposed extermination of so many bird species having similar habits with the rock pigeon seems a very rash, unlikely assumption.

Moreover, the several domesticated breeds that I have been talking about have been transported to all parts of the world, and some of them must also have been carried back again into their native country. Yet not one of them has become wild or feral,[7] even though the dovecot pigeon, which is the rock pigeon in only a very slightly

---

[7] *Feral* is a domesticated species that has reverted to living in the wild.

altered state, has indeed become feral in several places. Also, all recent experience shows that it is difficult to get wild animals to breed freely when they are first domesticated. And yet, on the hypothesis that our domesticated pigeons had many separate origins from many different ancestors, we would have to assume that at least seven or eight species were so thoroughly domesticated in ancient times by half-civilized man as to reproduce quite prolifically under confinement. Again, that is not very likely.

Another argument against the idea of our present pigeon breeds having descended from many separate ancestors, and one that carries great weight, is the fact that the various breeds I have mentioned are very much like the rock pigeon in their constitution, habits, voice, coloring, and in most parts of their structure, and yet are highly *unlike* the rock pigeon in many other aspects. For example, we may look in vain through all the wild species in the family Columbidae[8] for a beak that looks like that of the English carrier pigeon, or that of the short-faced tumbler or the barb; or for reversed feathers like those of the Jacobin; or for a crop like that of the pouter; or for tail feathers like those of the fantail. Thus it must be assumed that not only did half-civilized man succeed in thoroughly domesticating a number of very different pigeon species in ancient times, but that he intentionally or by chance picked out extraordinarily abnormal species to work with. And we must assume further that those very species have all since become either extinct or unknown. So many strange contingencies are extremely improbable.

Pigeon colors are also well worth thinking about. The rock pigeon is slaty blue in color, with white loins (see Figure 1.3D). The tail has a terminal dark bar, with the outer feathers externally edged in white at the base. The wings have two black bars. In some semi-domestic breeds, and in some truly wild breeds, the wings have not only the two black bars but are also checkered with black. These several marks do not occur together in any other species in the entire family.

Now, in every one of the domestic breeds, even in birds that have been carefully bred for many generations, all of the above marks that are seen in rock pigeons, even to the white edging of the outer tail feathers, sometimes concur perfectly developed—not very often, but every now and again. Moreover, when birds belonging to two or more distinct breeds are crossed, none of which is blue or have any of the above-specified markings, the mixed offspring very often suddenly display those characters.[9] For example (and I could give many others), some time ago I crossed some white fantail pigeons, which breed very true, with some black barb pigeons. The offspring of this cross were black, brown, and mottled. I'll talk about what I did with these "mixed fantail–black-barb" offspring shortly. I also crossed a blue barb pigeon with a spot pigeon, which is a white bird with a red tail and a red spot on its forehead, and which is notorious for breeding very true to its type; in this case the "mixed barb–spot" offspring were somewhat dark ("dusky") and mottled, but certainly not

---

[8] This family contains the doves and pigeons, about 310 species. The Columbidae is one member of the class Aves, which includes all bird species.

[9] As used here, "characters" means "traits."

blue. I then crossed a mixed fantail–black barb pigeon with a mixed barb–spot pigeon ... and they produced a bird of as beautiful a *blue* color as any wild rock pigeon! And it even had the rock pigeon's white loins, double black wing bars, and barred and white-edged tail feathers! Who would have predicted such a result? Blue varieties of barb pigeons are incredibly rare: indeed, I have never heard of a single instance of blue-colored barb pigeons in all of England.

We can understand these remarkable outcomes quite easily on the well-known principle of "reversion to ancestral characters" if all of the domestic breeds are *indeed* descended from the rock pigeon, with its blue coloration, black bars, and white edging. But if we deny this ancestral relationship, we must make one of the two following suppositions:

1. Even though no other existing pigeons are now colored and marked like the rock pigeon, we could assume that all of the eight or more imagined ancestral stocks of domestic pigeons were so-colored and so-marked, so that in each separate breed there might be a tendency to sometimes revert to the very same ancestral colors and markings, or
2. We could assume that each domestic breed, even the purest, has within the previous 12–20 generations been crossed with a rock pigeon. I say within 12–20 generations because we know of no instance in which any crossed descendants have ever reverted to an ancestor of foreign blood if they are more than 20 generations removed from that cross.

Both of these suppositions are highly unlikely.

Finally, consider the effects of crosses on future fertility (i.e., the ability of the offspring to reproduce successfully). Now it is very rare for matings between two distinct *species* to produce fertile offspring; typically the offspring from such matings are sterile. And yet the hybrids resulting from crosses between any of the existing pigeon breeds are perfectly fertile and have no difficulty in producing offspring themselves as adults; I can say this based on the results of my own studies, which were purposefully made using the most distinctly different-looking breeds available to me. Some authors believe that long-continued domestication eliminates the strong tendency toward sterility in matings between species. That conclusion may be correct if applied to species closely related to each other. But it would be rash in the extreme to extend that argument to suppose that separate species originally as distinct as carrier pigeons, tumblers, pouters, and fantails should now yield perfectly fertile offspring when crossed with each other.

In summary, we must consider the following issues:

1. The improbability of man having made seven or eight previously separate and distinct species of pigeons to breed freely under domestication;
2. These supposed seven to eight ancestral species being quite unknown in the wild, and the domesticated breeds not having become wild anywhere in nature;

3. These different breeds presenting a series of striking characteristics that are not found in any other members of the family Columbidae—which includes hundreds of species—even though they are so like the rock pigeon in most other respects;
4. The occasional reappearance of the blue color and various black marks of the rock pigeon in all of the breeds, both when kept pure and when crossed; and last,
5. The hybrid offspring from crosses between the different pigeon breeds being perfectly capable of reproducing and leaving their own offspring.

Taken together, we are led to conclude that all of our domestic pigeon breeds are indeed descended from the rock pigeon, *Columba livia*. In support of this view, I may add that *C. livia*[10] taken from nature has been successfully domesticated in Europe and in India, and that it agrees in its habits and in a great number of morphological features with all the domestic breeds. Second, although the English carrier pigeon and the short-faced tumbler differ immensely from the rock pigeon in certain characteristics, yet by comparing the several sub-breeds of these two races, particularly those brought from distant countries, we can see an almost continuous series of traits between them and the rock pigeon. We can do this in some other cases as well, although admittedly not with all breeds. Third, those characters that are mainly distinctive of each breed also vary *within* each breed; the wattle and length of beak of the carrier pigeon, for example, and the shortness of tumbler's beak, and the number of tail feathers in the fantail pigeon: all vary among individuals within each breed. The explanation of this fact—and its importance—will be made clearer later, when I talk about "selection."

Fourth, pigeons have been watched and tended with the utmost care, and loved by many people, for thousands of years in several parts of the world; indeed, the earliest known record of pigeons is in the Fifth Egyptian dynasty, about 3,000 BCE, as was pointed out to me by Professor Karl Richard Lepsius, the well-known German Egyptologist and author. But even better, Mr. Samuel Birch of the British Museum informs me that pigeons were also listed on a dinner menu from the previous dynasty! And Pliny the Elder (23–79 CE), in his *Naturalis Historia*, tells us that people paid immense prices for pigeons: "Many men are grown now to cast a special affection and love to these birds; they build turrets above the tops of their houses for dovecotes. Nay, they are come to this pass, that they can reckon up their pedigree and race, yea they can tell the very places from whence this or that pigeon first came."

Pigeons were also much valued by Akbar the Great in India, about the year 1600; at least 20,000 pigeons were taken wherever the court traveled. "The monarchs of Iran and Turan sent him some very rare birds," says the court historian, who continues: "His Majesty by crossing the breeds, which method was never practiced

---

[10] *C. livia* is the abbreviation for *Columba livia*. After using the full name once, it is standard practice to abbreviate the genus name thereafter in the same piece of writing.

before, has improved them astonishingly." About this same time, the Dutch were as eager about pigeons as were the old Romans. So people have been raising pigeons for a very, very long time—more than 2,000 years. The paramount importance of these facts in explaining the immense amount of variation that pigeons have undergone will be obvious when I talk about "selection" later.

I have discussed the probable origin of our domestic pigeons at some length because when I first kept pigeons at home and watched the several kinds, knowing well how truly they breed from one generation to the next, I found it just as difficult to believe that they had all descended from a common ancestor as any naturalist would in coming to a similar conclusion about the many finch species, or other groups of birds, in nature.

One circumstance that has struck me in particular is that nearly all the domestic animal breeders that I have talked to—not just the breeders of pigeons—along with the cultivators of plants whom I have spoken with or whose treatises I have read, are all firmly convinced that the several breeds with which each has worked are descended from just as many ancestrally distinct species (i.e., four breeds having four separate ancestors; eight breeds having eight separate ancestors). Ask, as I have asked, a celebrated raiser of Hereford cattle whether his cattle might not have descended from Longhorn cattle (Figure 1.6), or whether both may have descended from some common ancestral stock, and he will laugh you to scorn. Similarly, I have never met a pigeon breeder, or a poultry, duck, or rabbit fancier, who was not fully convinced that each main breed was descended from a distinctly separate ancestral species. The Belgian scientist Professor Jean Baptiste Van Mons, in his treatise on pears and apples, shows how utterly he disbelieves that the several sorts of apple—a Ribston pippin apple or a codlin apple, for instance—could ever have resulted from the seeds of the same tree.

I could give innumerable other examples of such misleading opinions. The explanation for these opinions, I think, is simple: from long-continued study of their organisms, these breeders have become strongly impressed with the *differences* between the several races, and though they well know that within each race the members vary slightly from each other—for breeders win their prizes by selecting such slight differences for show—yet they ignore all general arguments and refuse to sum up in their minds the effect of such slight differences accumulating over many successive generations. Now many naturalists know far less about the laws of inheritance than breeders have learned through practical experience, and know no more than breeders do about the intermediate links between varieties in the long lines of descent. Yet they freely admit than many of our domestic races have the same parents in common. May these naturalists not learn a lesson of caution when they ridicule the idea of species in the natural world being direct descendants of other species?

Now we are ready to deal with a key question: How have pigeon breeds as different from each other as the pouter, the common tumbler, and the runt all been created from a single ancestor—the rock pigeon?

(A)

(B)

Figure 1.6 (A) Polled Hereford bull. (B) Texas Longhorn.

Principles of Selection Anciently Followed, and Their Effects

Let us now briefly consider the steps by which the various distinctive domestic races have been produced, either from one or from several related species. One of the most remarkable features in our domesticated races is how easily we see that they are adapted, not to the animal's or plant's own good, but to our own use or fancy. Some variations useful to us have probably arisen suddenly, in a single step. Many botanists, for instance, believe that the common weed known as Fuller's teasel (*Dipsacus fullonum*), with its specialized hooks that really cannot be rivaled by any mechanical contrivance that I know of, is simply a variety of the wild member of the same genus and may have changed to this degree suddenly, in a single seedling. So it has probably been with the turnspit dog, a dog bred for its long body and short legs. And this is known to be the case with the ancon sheep, which also have unusually long bodies and short legs, with the forelegs not only short but also crooked.

But when we compare the dray horse with the race horse, the dromedary with the camel, the various breeds of sheep suited for cultivated land or for mountain pasture; when we see that the wool of one breed of sheep is good for one purpose and that of another breed is good for a quite different purpose; when we compare the many breeds of dogs, each good for people in different ways; when we compare the gamecock, so pertinacious in battle, with other breeds that rarely quarrel, such as the bantam, which is so small and elegant; when we compare the host of agricultural, culinary, orchard, and flower garden races of plants, most useful to us at different seasons and for different purposes, or so beautiful in our eyes, we must, I think, look further than to mere variability. We cannot suppose that all the breeds were suddenly produced as perfect and as useful as we now see them, in a single step. Indeed, in many cases we know that this has not been their history. The key to understanding these differences is in our power of accumulative selection: nature gives us successive variations, and we add them up over time in certain directions that are useful to us. In this sense we may be said to have made these many useful breeds for ourselves.

The great power of this principle of selection is not hypothetical. Several of our most distinguished breeders have modified their breeds of cattle and sheep to a large extent and within a single lifetime. To fully realize what they have done, you really need to read some of the many treatises devoted to this subject and inspect the animals personally. Breeders habitually speak of an animal's organization as something "plastic," something that they can mold almost as they please. If I had more space I could quote numerous passages to this effect from highly competent authorities. Take William Youatt, for example, who was probably better acquainted with the works of agriculturists than almost any other individual I know of, and who, being a well-respected veterinarian, was himself a very good judge of animals. He speaks of the principle of selection as "that which enables the agriculturist, not only to modify the character of his flock, but to change it altogether. It is the magician's wand, by means of which he may summon into life whatever form and mould he pleases." Similarly, Lord John Somerville, speaking of what breeders have done for sheep, says "It would seem as if they had chalked out upon a wall a form perfect in itself, and then had given it existence." In Saxony (a state in Germany) the importance of the principle of selection in regard to merino sheep is so fully recognized that men follow it as a trade: the living sheep are each placed on a table and are carefully studied, like a picture is studied by a connoisseur. This is done three times several months apart, and the sheep are each time marked and classed, so that only the very best individuals may ultimately be selected for breeding.

What English breeders have actually accomplished with their domesticated animals is proven by the enormous prices given for animals with a good pedigree, and these have been successfully exported to almost every quarter of the world. The improvements are by no means generally due to crossing different breeds with each other. And even when a cross has been made, the most careful selection among different traits is even more indispensable in that situation than it is when breeds have not been crossed. If selection consisted merely in separating some very distinct variety and breeding from it, the principle would be so obvious as to hardly be worth noticing. **Instead, its importance lies in the great effect produced by the gradual accumulation in one direction, during successive generations, of differences so small as to be absolutely undetectable by an uneducated eye—differences that I myself have attempted to discern but in vain.** Not one man in a thousand has the accuracy of eye and the judgment necessary to become a successful breeder. If someone is gifted with those qualities and studies his subject for years, and indeed devotes his lifetime to it with indomitable perseverance, he will succeed and may make great improvements. But if he lacks any of those qualities, he will assuredly fail. Few people would readily believe in the natural capacity and the years of practice needed to become even a skillful pigeon fancier.

The same principles are followed by horticulturists, who work with plants, but the variations here are often more abrupt. No one supposes that our choicest plants have been produced by a single variation from the stock of some ancestral plant. Indeed we have proof that this has not been so in several cases in which exact records have been kept. To give but one small example, consider the steadily increasing size of the common gooseberry fruit, which is now some 800% heavier than it was in ancient times. We similarly see an astonishing improvement in many florists' flowers simply by comparing the flowers of the present day with drawings made only 20 to 30 years ago. Once a race of plants is pretty well established, the seed raisers do not pick out the best plants, but merely go over their seedbeds and pull up the "rogues," as they call the plants that deviate from the proper standard. With animals this same sort of selection is, in fact, likewise followed; hardly anyone is so careless as to breed the next generation from his worst animals!

With plants, there is yet another way to observe the gradual accumulated effects of selection over many generations—namely by comparing the diversity of flowers in different varieties of the same species in a single flower garden; or the diversity of leaves, pods, or tubers, or whatever part of the plant is valued, in the kitchen garden in comparison with the flowers of the same varieties; or the diversity of the fruits produced by any one species in an orchard, in comparison with the leaves and flowers of the same varieties of that species. See how different the leaves of the cabbage are, but how extremely alike are the flowers. Notice how the flowers of the different varieties of heartsease (*Viola tricolor*) are so unlike each other, but how the leaves are so similar. See how much the fruits of the different kinds of gooseberries differ in size, color, shape, and hairiness, while the flowers themselves present only very slight differences. It's not that the varieties that differ largely in one feature do not differ at all in other features; this is hardly ever, and perhaps never the case—and I speak here after careful observation. As a general rule, it cannot be doubted that the continued selection of slight variations in particular features, either in the leaves, the flowers, or the fruit, will produce offspring differing from each other chiefly in those characteristics and not in others.

Now some readers have objected that the principle of selection has been honed to a rigorous, methodical practice for scarcely more than 75 years. Well, yes, certainly the practice has become more common in recent years; many treatises have now been published on the subject. And yes, the result has been, in a corresponding degree, rapid and important. But the principle is certainly not a modern discovery. I could give several references to works of great antiquity, in which the full importance of the principle of selection is acknowledged. Long ago, in rude and barbarous periods of English history, choice animals were often imported, and in fact laws were passed to prevent their exportation; the destruction of horses smaller than a certain size was ordered, similar to the previously described "rouging" of plants by nurserymen.

Indeed, the principles of selection are clearly given in an ancient Chinese encyclopedia, and similarly explicit rules for selection are laid down by some of the Roman classical writers. From passages in Genesis, it is clear that people paid careful attention to the color of domestic animals even in those early days. Savages now sometimes cross their dogs with wild canine animals to improve the breed, and they did so formerly as well, as attested to by ancient passages from Pliny the Elder. The primitive peoples of South Africa match their draught cattle by color, as some Eskimos do with their teams of dogs. The Scottish missionary David Livingstone states that good domestic breeds are highly valued by people living in the interior of Africa, people who have never before associated with Europeans.

Although some of these facts do not show actual selection, they certainly show that the breeding of domestic animals was carefully attended to even in ancient times and is now attended to by even the most primitive peoples. It would, indeed, have been strange to learn that attention had not been paid to breeding, for the inheritance of good and bad qualities in offspring is so obvious.

Unconscious Selection

At the present time, eminent breeders try by methodical selection, generation after generation, to make a new strain or sub-breed that is superior to anything else that exists in the country, and they do so with a distinct object in view. But for our purpose, a form of selection that results from everyone trying to possess and breed only from the best individual animals—something we may call Unconscious Selection—is more important. Thus, of course, a man who intends to keep pointers tries to get the best dogs he can to start with and afterward breeds only from his own best dogs. But he has no wish or expectation of permanently altering the breed. Nevertheless we may infer that this process, continued during centuries, would improve and modify any breed, in the same way that Robert Bakewell and other early agriculturists did methodically modify, using this very same process, the forms and qualities of their cattle even during their lifetimes. Slow and imperceptibly small changes of this kind can never be recognized unless actual measurements or careful drawings of the breeds in question were made long ago to serve for comparison. In some cases, however, unchanged or slightly changed individuals of the same breed exist in less civilized districts, where the breed has been less improved.

There is reason to believe that the King Charles's spaniel has been unconsciously modified to a large extent since King Charles II ruled England, Scotland, and Ireland in the seventeenth century. Some highly competent authorities are convinced that another dog breed, the setter, is directly derived from the spaniel and has probably been slowly altered from it. It is also widely recognized that the English pointer has been greatly changed within the last century; in this case the change has, it is believed, been brought about chiefly through deliberate matings with the foxhound. But the important point for us is that the changes have been brought about unconsciously and gradually, and yet so effectively that, although the old Spanish pointer certainly came from Spain, Mr. George Borrow tells me that in his travels he has not seen any native dog in Spain that resembles our pointer.

By a similar process of selection, and by careful training, English racehorses have come to surpass—both in fleetness and size—the parent stock of Arabian horses; indeed, the English horses, by the official regulations of the Goodwood Races in England, are now required to carry heavier weights than the Arabian racehorses to make the races fairer. Similarly, the famous cattle breeder Lord Spencer and others have shown that English cattle are now heavier and mature earlier compared with the stock formerly kept in this country. And by comparing the accounts given in various old treatises of the former and present state of carrier and tumbler pigeons in Britain, India, and Persia, we can easily trace the stages through which they have insensibly passed and have thereby come to differ so greatly from their common ancestor, the rock pigeon.

The well-respected veterinarian William Youatt gives an excellent illustration of the effects of a course of selection in his treatise on sheep (*Sheep: Their Breeds, Management, and Diseases*, 1837), one that may be considered as "unconscious" in that the breeders could never have expected (or even wished) to produce the result that ensued—namely, the production of two distinct strains of sheep. As Mr. Youatt remarks, "The two flocks of Leicester sheep kept by Mr. John Buckley and Mr. Joseph Burgess have been purely bred from the original stock of Mr. Bakewell for more than 50 years. There is not a suspicion existing in the mind of any one at all acquainted with the subject, that the owner of either of them has deviated in any one instance from the pure blood of Mr. Bakewell's flock. And yet the difference between the sheep [now] possessed by these two gentlemen is so great that they have the appearances of being quite different varieties."

Even if somewhere there are savages so barbarous as never to think of the inherited character of the offspring of their domesticated animals, any one animal particularly useful to them in some way, for any special purpose, would nevertheless be carefully protected and fed during the famines and other incidents to which savages are so liable; those chosen animals would thus generally leave more offspring than the inferior ones, so that there would be a kind of unconscious selection going on over the generations. We see the great value set on non-human animals even by the barbarians of Tierra del Fuego, by their killing and devouring their old women in times of famine; the old women were clearly valued less than the dogs at such times.

In plants, we can see the same gradual process of improvement, again achieved quite simply through the occasional preservation of the best individuals, simply by noting the increased size and beauty that we now see in the varieties of the heartsease, rose, pelargonium, dahlia, and other plants when compared with the older varieties or with their parent stocks (Figures 1.7 and 1.8).

Nobody would ever expect to get a first-rate heartsease or dahlia from the seeds of a wild plant. Similarly, nobody would expect to raise a first-rate melting pear from the seeds of the wild pear: indeed, the pear, though cultivated in classical times, appears from Pliny's description to have been a fruit of very inferior quality, and I have seen

(A)   (B)

Figure 1.7 (A) Wild rose. (B) Domestic rose.

(A)   (B)

Figure 1.8 (A) Wild dahlia (*Dahlia sorensenii*). (B) A dahlia star sunset flower.

great surprise expressed in horticultural works at the wonderful skill of gardeners in their having produced such splendid fruits from such poor starting materials. The art has been simple and, as far as the final result is concerned, has been followed almost unconsciously. It has consisted in always cultivating the best-known variety, sowing only its seeds, and, whenever a slightly better variety chanced to appear, selecting it, and so onward over many generations. But the gardeners of the classical period, who always cultivated the best pears that they obtained, never thought what splendid fruit we should be eating today; and yet we owe our excellent fruit at least partly to their having naturally chosen and preserved the best varieties they could find in every generation.

A large amount of change, thus slowly and unconsciously accumulated, explains, I believe, the well-known fact that in a number of cases we cannot recognize (and therefore do not know) the wild parent stocks of the plants that have been cultivated in our flower and kitchen gardens for the longest times. If it has taken centuries or even thousands of years to improve or modify most of our plants up to their present standards of usefulness to people, we can understand how it is that neither Australia, the Cape of Good Hope, nor any other region inhabited by uncivilized man has afforded us a single plant worth culturing. It is not that these countries, so rich in species, do not by strange chance possess the ancestral stocks of any useful plants; rather, it is simply that the native plants have not been improved by continued selection up to a standard of perfection comparable with that acquired by the plants in countries that were civilized long ago.

Domestic animals kept by uncivilized tribes almost always have to struggle for their own food, at least during some times of year. And in two countries very differently circumstanced, some individuals of the same species, differing slightly in their physiology or structure, would often succeed better in the one country than in the other; two sub-breeds might then eventually be formed by a process of "natural selection," as I will explain more fully later in this book. This, perhaps, partly explains why the varieties kept by primitive peoples are more like true species than are the varieties kept in civilized countries, something previously noted by a number of other authors.

Once we understand the important part that selection by people has played, then it becomes obvious how it is that our domestic races of animals and plants now show structural and behavioral adaptations to human wants and fancies. We can, I think, further understand the frequently abnormal characters of our domestic races and why they differ so greatly in external characters but so relatively slightly in their internal parts or organs. People cannot select, or can select only with much difficulty, for structural differences that are internal and cannot be seen; indeed, people rarely care for what is only internal. And of course we can only select variations that are first given to us in some slight degree by nature. Nobody would ever try to make a fantail pigeon until he saw a pigeon with a tail developed in some slight degree in an unusual manner. No one would try to create a pouter until he first saw a pigeon with a crop of somewhat unusual size, and the more abnormal or unusual any character was when it first appeared, the more likely it would be to catch his attention.

But to use an expression like "trying to make a fantail" is, I have no doubt, in most cases utterly incorrect. The person who first selected a pigeon with a slightly larger tail never dreamed what the descendants of that pigeon would eventually look like through many generations of selection, partly unconscious and partly methodical. Perhaps the parent bird of all fantails had only 14 tail feathers somewhat expanded, like the present Java fantail, or was more like individuals of other distinct breeds, in which as many as 17 tail feathers have been counted. Perhaps the first pouter pigeon did not inflate its crop much more than the turbit now inflates the upper part of its esophagus—a habit of turbots that is ignored by all fanciers, as it is not one of the points of interest in that breed.

Remember, it does not take a great deviation of structure to catch the fancier's eye: as I noted earlier, the fancier can perceive extremely small differences. And it is part of human nature to value any novelty in one's possession, however slight. Nor must we assume that the value which breeders formerly saw in any slight difference among individuals of the same species be judged by the value that we now set on those differences, once distinct breeds have been firmly established. With pigeons, many slight variations still occasionally appear, but these are now typically rejected as faults or undesirable deviations from the standard of "perfection" that now characterizes that breed.

These views appear to explain why we know hardly anything at all about the origin or history of any of our domestic breeds. But, in fact, a breed, just like the dialect of a language, can hardly be said to *have* a distinct origin. Someone preserves and breeds from an individual with some slight but interesting deviation of structure, or takes more care than usual in matching his best animals and thus improves them. The improved animals then mate, and slowly their offspring spread in the immediate neighborhood. But they will as yet hardly have a distinct name, and, from being so little valued, no one will have paid any attention to their history. When further improved over time, however, by the same slow and gradual process, they will spread more widely and will eventually be recognized by someone as being distinct and valuable and will then probably first receive a local name. In semi-civilized countries, with little communication over long distances, the spreading of a new sub-breed would be a slow process. As soon as the points of value are once acknowledged, however, the principle that I am calling "unconscious selection" will always tend to slowly add to the characteristic features of the breed, whatever they may be. But there will be an infinitely small chance that anyone would have kept records of such slow, varying, and insensible changes.

Circumstances Favorable to Man's Power of Selection

Let me now say a few words about the circumstances that favor—and those that stand in the way of—man's power of selection. A high degree of variability in characteristics among individuals is obviously favorable, as it freely provides the materials for selection to work on. Even small individual differences are sufficient, though, with extreme care, to allow large modifications to eventually be accumulated in almost any desired direction. But having large numbers of individuals in the population is also important: Because variations manifestly useful or pleasing to us appear only occasionally, they are much more likely to appear when keeping a large number of individuals in that population. When only a few individuals are available, all will be allowed to breed, whatever their quality may be, and this will effectively prevent selection from taking place. In keeping with this idea, William Marshall has written with respect to Yorkshire sheep, in his *The Rural Economy of Yorkshire* (1796), "As they generally belong to poor people, and are *mostly in small lots*, they never can be improved." On the other hand, nurserymen, from keeping large numbers of the same plant, are generally far more successful than amateurs in raising new and valuable varieties.

A large number of individuals of any particular animal or plant species can be reared only where the conditions favor its propagation. But probably the most important element in promoting selection is that the animal or plant should be so highly valued by man that he pays the closest attention to even the slightest deviations in its qualities or structure. Unless such attention be paid, nothing can be achieved. I have seen it gravely remarked that it was "most fortunate" that the strawberry began to vary just when gardeners began attending to this plant. But I'm sure that the strawberry has in fact varied ever since it was first cultivated; it's just that the slightest variations had for a long time been neglected. As soon, however, as gardeners picked out individual plants with slightly larger, earlier, or better fruit and raised seedlings from those particular plants and again picked out the best seedlings and bred again from them, and so on, then (with some aid by crossing distinct species) those many admirable varieties of the strawberry were raised that have appeared during the last 50 years.

With animals, the ease of preventing unwanted crosses is also an important element in forming desirable new races—at least in a country that is already stocked with other races. In this respect, being able to enclose the land is helpful. In contrast, wandering savages or the inhabitants of open plains rarely possess more than one breed of the same species because all members of the species can freely mate and breed. In this regard, pigeons are unusually easy to work with since they can be mated for life. Thus many distinct races of pigeons may be improved and kept true even with all of the pigeons housed together in the same aviary, as mentioned earlier; this must have greatly aided the formation of new breeds by pigeon fanciers over the decades. Pigeons, I might add, can be propagated in great numbers and very quickly, and inferior individuals may be freely rejected . . . and eaten!

In contrast, cats, from their nocturnal rambling habits, cannot be selectively mated very easily, and, even though they are much valued by women and children, we rarely see a distinct breed long kept up. The distinct breeds that we do sometimes see are almost always imported from some other country, often from islands. Although I do not doubt that some domestic animals vary in their characteristics less than others, yet the rarity or absence of distinct breeds of the cat, the donkey, peacock, goose, and others may be attributed mainly to selection not having been brought into play. For cats, the lack of selection is due to the difficulty of pairing them, as previously noted. For donkeys there is the problem that poor people keep only a few animals at a time and pay little attention to their breeding; in contrast, this animal has been surprisingly modified and improved by careful selection in certain parts of Spain and the United States, showing that modifications are indeed possible. For peacocks, the problem is that they are not very easily reared, and a large stock of peacocks is not kept. With geese, the problem is that they are valuable for only two purposes—for food and for their feathers—and especially that there has been no pleasure felt in displaying distinct breeds.

Some authors have claimed that the maximum amount of variation in our domesticated animals and plants is soon reached and can never be exceeded afterwards. However, it would be somewhat rash to claim that the limit of variation has been attained in any one case; for almost all of our animals and plants have been greatly improved in many ways within the recent past, which implies variation. It would be equally rash to claim that characters now increased to their utmost limit could not, even after remaining fixed for many centuries, begin to vary again under a changed environment. No doubt, as Mr. Alfred Russel Wallace has noted, a limit to at least some traits must eventually be reached. For instance, there must be a limit to the fleetness of any terrestrial animal, as this will be determined by the amount of friction to be overcome, the weight of the body to be carried, and the power with which the muscle fibers can contract. But what concerns us here is that the varieties of a given domesticated species differ more from each other in almost every character that man has attended to and selected for than do distinct species within a single genus.

Isidore Geoffroy Saint-Hilaire has proved this in regard to size, and so it is with color and probably also with the length of hair. With respect to fleetness, which depends on many bodily characters, the undefeated English racehorse Eclipse was far fleeter, and a powerful dray horse is incomparably stronger—even though all horses belong to a single species—than are the members of any two natural species belonging to the same genus. And so it is with plants: the seeds of the different varieties of bean or corn plants probably differ more in size than do those from distinct species within any genus in the same two families (see Figure 1.5). This observation also applies to the fruit of the several varieties of the plum, for example, and still more strongly to seeds of the melon.

Summary

Let me now sum up my thoughts on the origin of today's domestic races of animals and plants. Variability among individuals in traits of interest to us is an essential ingredient, and those desired traits must be inherited by offspring. Variability among individuals is governed by many unknown laws, including correlated growth, environmental effects, increased use or disuse of particular parts, and in some cases the intercrossing of what were originally distinct species. But whatever the causes of that variability, the accumulative action of *selection*—whether applied methodically and quickly or unconsciously and slowly (but efficiently) over many, many years— seems to have been the predominant power in creating the great variety of domesticated animals and plants that we have with us today.

Key Issues to Talk and Write About

1. Find out two interesting things about one of the people that Darwin mentions in this chapter. Choose from the following:
   Robert Bakewell
   Pliny the Elder
   Isidore Geoffroy Saint-Hilaire
   Alfred Russel Wallace
   Jeffries Wyman
2. Based on your reading of this chapter, what facts would you use in trying to convince someone that the different breeds of pigeon were really separate species?
3. What facts would Darwin use in trying to convince someone that the pouter, tumbler, fantail, and runt pigeons are all highly modified descendants of the rock pigeon?
4. How does Darwin explain the fact that some body parts of related animals or the fruits of some related plants are extremely different from each other, while other parts are extremely alike?
5. Suppose you wished to create a breed of pigeon that resembled a pouter, but with much larger toes than the pouters we have today. According to Darwin, how would you try to create such a modification?
6. Why is variation in traits among individuals of a species such an important part of Darwin's argument?
7. Figure 1.5 shows the relationship between phyla, classes, orders, families, genera, and species. Make a table comparing how humans and pigeons are classified into these categories. Just to get you started: both are in the animal phylum Chordata.
8. Why is Darwin so preoccupied in this chapter with convincing us that variations can be inherited?
9. Species today are generally defined by their reproductive isolation from other species (i.e., by their inability to mate with members of other species). How does Darwin seem to be defining species in this chapter?
10. Read the paragraph that begins "If the various pigeon breeds ...". Now write a one-sentence summary of that paragraph. First list what seem to be the several major points that Darwin is making. Then try to get all of those points into a single sentence. The sentence should be accurate, contain all the important

points, and be in your own words, as though you were explaining something to a friend. It should also make perfect sense to a reader who has never read the original paragraph. It can be done!

Bibliography

Marshall, W. 1796. *The Rural Economy of Yorkshire*. London.
Youatt, W. 1837. *Sheep: Their Breeds, Management, and Diseases*. London.

2

Variation in Nature

In this chapter, Darwin emphasizes that characteristics vary greatly among individuals of all species and gives many examples showing that there are often no clear boundaries between species and the varieties within a species. Deciding whether to call something a separate species or merely a variety within a species can be very subjective. He shows also that the most common and wide-ranging species tend to show the most variation among individuals. Variation is also greater for species belonging to larger genera (i.e., those containing many species) in any particular location than to those belonging to smaller genera. Also, many of the species in the larger genera resemble varieties in having restricted ranges and in being rather similar looking. These facts make sense only if varieties within species gradually become distinct new species and if species continue to generate new varieties over time.

In the previous chapter I talked about how people have gradually molded organisms in marvelous ways by selecting for small differences in particular traits over many, many generations. Before discussing how these same principles can apply to organisms in nature, I will first discuss the extent to which organisms in nature exhibit variation in traits. And indeed they do, and not only between species, but within species as well.

First, what are "species"? And what are "varieties"? And what do we mean by the term "variation"? Although no one definition of "species" has yet satisfied all naturalists, the term generally includes the unknown element of a distant act of creation. It's also hard to define the term "variety" precisely, but we generally assume that varieties are organisms that are closely related to each other but nevertheless differ in some conspicuous characteristics. Some authors have used the term "variation" to imply modifications that are directly due to changes in the physical environment and that are presumably not inherited by the offspring of those individuals. But who can say whether the unusually small shells of snails living in the brackish waters of the Baltic Sea, or the similarly "dwarfed" plants living on the tops of high mountains, or the thicker fur seen on individuals of a particular species living farther north than other individuals of the same species would not in some cases pass along those traits to their offspring, at least for several generations? In such a case I would call the form a *variety*. For me, varieties are defined by differences, in closely related organisms, that are passed to their descendants. Only such heritable differences in natural populations will concern us here.

The Readable Darwin. Second Edition. Jan A. Pechenik, Oxford University Press. © Oxford University Press 2023. DOI: 10.1093/oso/9780197575260.003.0003

The differences need not be large to merit our attention. Indeed, small differences among individuals in nature are of great importance. It seems unlikely that the sudden and considerable deviations of structure that we occasionally see in our domestic productions, particularly with plants, are ever permanently propagated in nature. Almost every part of every living being is so beautifully related to its complex conditions of life that it seems as improbable that any one part should have been suddenly produced as perfectly as we see it now as that a complex machine could be instantly invented by humans in a similarly perfected state.

Individual Differences

Now nobody supposes that all members of the same species are cast in exactly the same mold. Within every species we can see at least slight differences among individuals, even among the offspring from one set of parents. These individual differences are highly important for my argument, as they are often inherited and can thus provide the material for what I am calling "natural selection" to act on and accumulate, just as we can select for and accumulate—in any desired direction—individual differences in our domesticated animals and agricultural productions.

These individual differences generally affect what most naturalists consider to be an organism's unimportant parts. But I can provide a catalog of facts showing that an organism's "important" features, whether physiological or morphological, can also vary among individuals of the same species. Indeed, I am now convinced that even the most experienced naturalist would be surprised at how many cases of variability he could collect, even in functionally important structures, as I myself have done over many years of study. It should be remembered that taxonomists are not happy when they find variability in important characters: their focus is on finding *similarities* within groups, not differences. Remember, too, that not many people will laboriously examine and compare internal and other important organs in many individuals of the same species. If they did so, they would be surprised at what they would find. For example, I should never have expected that the branching of the main nerves close to the great central ganglion would vary as much as they do among individuals of a single insect species. Similarly, my friend and colleague, the well-respected entomologist Sir John Lubbock, has recently documented a fair degree of variability in these same main nerves among certain "scale insects," members of the insect genus *Coccus* (Figure 2.1), which may almost be compared to the irregular branching of the stem of a tree. Sir Lubbock has recently shown that the muscles also vary considerably among individuals in the larvae of certain insect species.

Some authors have argued that important organs never vary within a species, but when they make such claims they argue in a circle: these same authors, for practical purposes, rank only those characters that do *not* vary among individuals as important ones while ignoring those characters that do vary. With such an approach, no instance of an important part varying will ever be found!

Figure 2.1 Scale insects (yellow), *Coccus viridis.*

Questionable Species

Perhaps the most important organisms for my argument are those that seem in many respects to be distinct species but which naturalists do not like to rank as distinct species because they so closely resemble some other forms or are so closely linked to other forms by organisms showing intermediate gradations. The boundary between varieties and species is indeed often uncertain, even as "species" and "varieties" are themselves difficult to define precisely, as mentioned earlier. In practice, when a naturalist can unite two forms together using other individuals that have characteristics intermediate between the two, he typically treats the one as a variety of the other, ranking the most common one as the species and the other as the variety; sometimes, however, he ranks the one that was described first as the species and the other as the variety. But it is sometimes very difficult to decide whether or not to rank one form as a variety of another, even when they are closely connected by individuals with intermediate characteristics. Nor will the commonly assumed hybrid nature of the intermediate links always remove the difficulty. In very many cases, one form is ranked as a variety of another not because intermediate links are known, but rather because the observer *supposes* that such intermediate forms must exist somewhere or may formerly have existed. And this opens a rather wide door for the entry of doubt and conjecture. Indeed, I refer to organisms which may or may not be true species as "doubtful" or "questionable" species.

In determining whether a form should be ranked as a species or as a variety, the opinion of naturalists having sound judgment and wide experience seems the only guide to follow. However, most well-marked and well-known varieties have been ranked as formal varieties by some workers but as separate species by at least some other competent judges. Thus, in such cases, we must simply take the majority opinion, which is not fully convincing or satisfying.

Varieties of such a questionable nature are quite common. Compare the several floras of Great Britain, France, or the United States, drawn up by different botanists, and see what a surprising number of forms have been ranked by one botanist as good species and by another as mere varieties of a species. The esteemed botanist Mr. Hewett Cottrell Watson, to whom I am much obliged for assistance of all kinds, has marked for me 182 British plants that are generally considered to be varieties but which have all been ranked by some botanists as separate species. Even in making this list, though, he has omitted many trifling varieties that have been ranked by some botanists as separate species and has entirely omitted some genera that are unusually polymorphic (i.e., highly variable in some characters within individual populations). Under genera, including the most polymorphic forms, the well-respected botanist Mr. Charles Cardale Babington discerns 251 species, whereas the equally respected botanist Mr. George Bentham gives only 112 species, a difference of 139 questionable forms!

Among animals that physically join for reproduction and which are highly mobile, forms that are ranked by one zoologist as a species and by another as a variety are rarely found within the same region but are common in separated areas. It is incredible to see how many of those birds and insects of North America and Europe that differ only slightly from each other have been ranked by one eminent naturalist as undoubtedly distinct species and by another as varieties of one species or, as they are often called, as "geographical races"! The esteemed naturalist Mr. Alfred Russel Wallace[1] has shown, in a series of valuable papers, that a variety of animals found among the different islands of the Malay Archipelago, especially members of the Lepidoptera (butterflies), can be placed into four categories: (1) variable forms, which show much variation among individuals living on individual islands; (2) local forms, which are quite uniform and distinct within each island but differ somewhat from island to island, with the most widely separated forms being clearly distinct; (3) geographic races, which clearly differ from each other from island to island but not so radically as to define them as separate species or merely as varieties; and (4) true representative species, which differ from each other in a number of clearly marked ways. However, nobody has come up with any set of criteria to clearly and objectively distinguish among variable forms, local forms, geographical races, and true species. In practice, the distinctions just aren't that easy to make.

---

[1] Alfred Russel Wallace is the naturalist who independently came up with very similar ideas about evolution through natural selection and, in fact, sent Darwin the draft of a paper on this topic in 1858, prompting Darwin to write his book on the origin of species.

Many years ago, I compared birds found on separate islands of the Galápagos Archipelago in the equatorial Pacific Ocean, both with each other and with birds found on the American mainland. Others have now made similar comparisons. Reviewing these comparisons, I was again much struck by how entirely vague and arbitrary the distinction was between species and varieties. Similarly, on the little Madeira group of islands, about 360 miles off the coast of North Africa, many of the insects characterized as varieties in Mr. Thomas Vernon Wollaston's admirable work would clearly be ranked as distinct species by many other entomologists. Even Ireland has some animals that are regarded as varieties by most zoologists but that have been ranked as separate species by others. Several very experienced ornithologists consider our British red grouse as only a strongly marked race of a Norwegian grouse species, whereas most ornithologists rank it as an undoubtedly distinct species peculiar to Great Britain. A large distance between the homes of two similar but distinct forms leads many naturalists to rank both as separate, distinct species. But exactly how large a distance is required to justify such a distinction? If that between America and Europe is sufficient, will the distance between the European continent and the Azores, or Madeira, or the Canaries, or Ireland be sufficient as well?

We must admit that many forms considered by highly competent judges to be mere varieties have so perfectly the character of species that they are indeed ranked by other highly competent judges as good and true species. But to discuss whether they are rightly called species or varieties, before any definition of these terms has been generally accepted, is vainly to beat the air.

A good number of strongly marked varieties or questionable species are well worth thinking about more carefully; indeed, several different and very interesting lines of argument have been used in attempting to determine their true rank.[2] Here I will give but one example, in considering the relationship between two flowering plants: the primrose (*Primula vulgaris*) (Figure 2.2A) and the cowslip (*Primula veris*) (Figure 2.2B). These plants are very different in appearance, smell differently and have different flavors, flower at different times, grow in different sorts of places, are found at different heights on mountains, and have different geographical ranges. Moreover, based on the results of the many experiments conducted by that most careful observer and plant hybridization expert Karl Friedrich von Gärtner, they can be hybridized, but only with difficulty. We could hardly wish for better evidence that the two forms are specifically distinct. On the other hand, they are united by many intermediate links, and it seems unlikely that all of those links are simply hybrids. There is also what seems to me to be an overwhelming amount of experimental evidence showing that the primrose and cowslip have in fact descended from common parents and consequently should be ranked as varieties of a single species. So the distinction between species and varieties is again not at all clear. Close investigation should eventually bring naturalists to agreement about how to rank such problematic organisms.

---

[2] Darwin omitted this paragraph from the sixth edition of *The Origin of Species*, but it seems worth including here.

(A)

(B)

Figure 2.2 (A) The primrose (*Primula vulgaris*). (B) The cowslip (*Primula veris*).

Intriguingly, it is where the organisms are best known that we find the greatest number of questionable species. I have been struck by the fact that if any animal or plant in nature be highly useful to people or attract our attention for any other reason, then varieties of it will almost universally be recorded. And at least some of these varieties will often be ranked by some authors as separate species. Look at the common oak, for example, a tree that has been much studied: while one particular German author divides the various forms of this plant into more than 12 distinct species, others generally consider them to be mere varieties of a single species. Similarly, in England the highest botanical authorities and practical men cannot agree whether the sessile (also called the Welch) oak and the pedunculated (also called the English) oak are good and distinct species in the genus *Quercus* or mere varieties of a single species. There is as yet no consensus.

When a young naturalist begins to study a group of organisms quite unknown to him, he is at first much perplexed to determine which differences to consider as species-specific traits and which ones as varieties for he knows nothing of the amount or kinds of variation shown within the group. But if he confines his attention to one class of organisms within one particular geographical area, he will soon make up his mind how to rank most of the troublesome forms. Generally, he will be initially impressed with the amount of difference in the forms that he is continually studying—just like the pigeon or poultry fanciers I talked about in Chapter 1—and thus will assign the various forms to many different species. When first starting off in this way, he has little general knowledge of analogous variation in other groups and in other geographical areas by which to correct his first impressions. However, as he extends the range of his observations, he will meet with more cases of difficulty for he will come across a greater number of closely related forms. If his observations are extended widely enough, he will eventually be able to make up his own mind about which ones to call varieties and which to call species. But he will succeed in this at the expense of admitting that there is much variation among individuals—and the truth of this admission will often be disputed by other naturalists. Moreover, when he comes to study related forms brought here from countries that are well separated from each other—in which case he can hardly hope to find any intermediate links between his questionable forms—he will have to trust almost entirely to analogy, and his difficulties will then reach a climax.

Certainly no clear line of demarcation has yet been drawn between species and subspecies—that is, the forms which, in the opinion of some naturalists, are almost, but not quite, deserving the rank of separate species. Thus they are classified as subgroups within a particular species. Neither is there any clear objective distinction between the ranks of subspecies and well-marked varieties or even between lesser varieties and simple differences among individuals. These differences all blend into each other in an insensible series, which impresses one's mind with the idea of an actual progression of forms.

Indeed, I look at such individual differences as being of great importance, though they may be of small interest to the taxonomist. Such small individual differences are in fact the first steps toward such slight varieties as are barely thought worth recording in works about natural history. **But varieties that are in any degree distinct and permanent are, to me, early and important steps leading to more strongly marked and more permanent varieties; and I see these varieties as eventually becoming separate subspecies, and then separate species.** In most cases, I attribute the gradual conversion of a variety—from a state in which it differs very slightly from its parent to one in which it differs more—to the action of natural selection in accumulating differences of structure in certain definite directions over long periods of time; I will explain this idea more fully later on. Thus, I believe that a well-marked variety is essentially an incipient species—a new species in the making. Whether or not you accept this idea will depend on the degree to which you are convinced by the facts and views given throughout this book.

We need not suppose that all varieties or incipient species will eventually become acknowledged as separate species. They may become extinct before that happens, for example, or they may stay as varieties for very long periods of time, as has been shown by Mr. Wollaston with the varieties of certain fossilized land snails in Madeira. If a variety were to flourish so greatly as to eventually become more numerous than the parent species, it would then be ranked as the species, and the original species would be ranked as the variety. Or the variety might eventually become so successful as to supplant and exterminate the parent species. Or both might coexist in comparable numbers and come to be ranked as two independent species. I will return to this matter later.

From these remarks, it will be seen that I look at the term "species" as one arbitrarily given for the sake of convenience to a set of individuals that resemble each other closely and that it does not fundamentally differ from the term "variety," a term given to less distinct and more fluctuating forms.[3] The term "variety," again, in comparison with mere differences among individuals, is also applied arbitrarily and merely for the sake of convenience. Variability is present wherever we look for it, even among individuals within any particular variety; it is just a matter of degree.

Wide-Ranging, Much Diffused, and Common Species Vary the Most

Guided by theoretical considerations, I thought that some interesting results might be obtained regarding the nature and relationships of the species that vary the most by tabulating all the varieties in several well-documented plant groups. This turned out to be a much more complicated business than I had expected, but my good colleague Dr. Joseph Hooker has examined my tables and thinks that the statements I am about to make from them are fairly well established. These tables form the basis for most of the discussion that follows in this chapter.

The Swiss botanist Alphonse de Candolle and others have shown that plants which have very wide geographical ranges generally present us with many varieties; this might have been expected, as those plants are exposed to a wide range

---

[3] We now tend to define species by their reproductive isolation (i.e., their inability to mate successfully with members of other groups). In contrast, varieties of a single species can mate and successfully reproduce.

of physical conditions, and, even more importantly for my argument, they come into competition with many different sets of organisms in different parts of their range. But my tables further show that, in any particular region, the species that are the most common—that is, those that present the greatest number of individuals in the area—and the species whose members are the most widespread within their own country often give rise to varieties sufficiently distinct to have been recorded in botanical works. Thus it is the most flourishing, that is to say the most dominant, species—those that range widely over the area, occupy the most diverse habitats in that area, and are the most numerous in individuals—which most often produce well-marked varieties, which, as I have said, I consider to be incipient species. This, perhaps, might have been anticipated: in order for varieties to become permanent to any degree, they must have to struggle with the other inhabitants of the country and win. Thus, species that are already dominant in an area will be the most likely to produce offspring which, although they may be in some slight degree modified, will still inherit those advantages that enabled their parents to dominate their compatriots in the first place.

Species in Larger Genera Vary More Frequently Than Those in Smaller Genera

Now some genera contain many species—I will call these the "larger genera"—while other genera (the "smaller genera") contain many fewer species (see Figure 1.5). If the plants inhabiting any particular country and described in any botanical manual are divided into two groups, with all those in the larger genera (containing many species) being placed on one side of the page and all those in the smaller genera (containing fewer species) being placed on the other side of the page, a somewhat larger number of the most common and more widespread species will be found on the side listing the larger genera. That is, the genera containing the most species tend to also contain the most common and widespread species. Again, this might have been expected, for the mere fact that whenever we see many species of the same genus inhabiting any one country, there must be something in the conditions of that country—biological or physical, or both—that are especially favorable to members of that genus. Consequently, we would expect to find a larger proportion of dominant species in the larger genera—and indeed we do. But actually I am surprised at seeing so clear a finding as there are many factors acting to obscure such as result. Here I will mention only two of these factors. First, freshwater plants and salt-loving plants (e.g., salt-marsh plants) generally have very wide ranges and are much diffused, but this seems only to reflect the sorts of specialized habitats that they occupy rather than having anything to do with the size of the genera to which the species belong. Second, plants that are low on the scale of organization (mosses, for example) are generally much more widely diffused than plants higher on the scale, and, here again, there is no close relation to the number of species within the genera. And yet even with these sorts of complicating factors, my tables still show that at least a small majority of the most dispersed and dominant species belong to the larger genera.

In looking at species as being essentially strongly marked and well-defined varieties, I logically expected that the species of the larger genera in each country would more often present varieties than the species found in the smaller genera: wherever many closely related species (i.e., those in the same genus) have been formed, many varieties or incipient species ought, as a general rule, to be now forming. Where many large trees grow, we expect to find saplings. Where many species of a particular genus have been formed through variation, circumstances there have clearly favored variation. On the other hand, if each species was formed through a special act of creation, there is no apparent reason why more varieties should occur in a group having many species than in one having only a few.

As a way of testing my expectation, I have arranged the plants of 12 countries and the beetles (members of the order Coleoptera) of two districts into two nearly equal groups, with the species of the larger genera again on one side and those of the smaller genera on the other side. It indeed turns out that a larger proportion of the species on the side of the larger genera present varieties than do those on the side of the smaller genera.

Moreover, the species belonging to the larger genera invariably present a larger *average* number of varieties than do the species belonging to the smaller genera. Both these results follow even when all the smallest genera (each containing only between one and four species) are completely excluded from the tables.

Clearly, species are little more than strongly marked and long-lasting varieties. For wherever many species of the same genus have been formed, or where, if we may use the expression, the "species-manufacturing apparatus has been active," we ought generally to find that factory still in action, particularly as we have every reason to believe that the process of manufacturing new species is a slow one. And this certainly is the case if we think of varieties as incipient species: my tables clearly show the general rule that wherever many species of a genus have been formed, the species of that genus present an exceptional number of varieties; that is, of incipient species beyond the average. It is not that all large genera are now varying much and are thus increasing in the number of their species or that no small genera are now varying and increasing; for if this had been so it would have been fatal to my theory. Geology plainly tells us that small genera have often increased greatly in size over long periods of time and that large genera have often reached a maximum number of species, declined, and then disappeared. **My point here is simply that where many species of a genus have been formed, on average many new species are still forming.**

Many Species Included in the Larger Genera Resemble Varieties: They Are Closely Related and Have Restricted Ranges

Several other relationships between the species belonging to large genera and their recorded varieties deserve our attention. We have seen that there are no infallible criteria by which to distinguish species from distinct varieties; in those cases in which intermediate links have not been found between questionable forms, naturalists must reach a determination based on the amount of difference between them, judging by analogy whether or not that amount is sufficient to raise one or both to the formal rank of species. Thus the amount of difference is one very important criterion in settling whether two forms should be ranked as species or as varieties. Now the Swedish botanist and mycologist Elias Fries has remarked with regard to plants from genera containing many species (i.e., in large genera) that the amount of difference between the species is often exceedingly small. The English entomologist John Obadiah Westwood has made the same point with regard to insects. I have tried to test this through some calculations, and, as far as my imperfect results go, they confirm those views: in genera containing an especially large number of species, the amount of difference between the species is indeed quite small. I have also consulted some sagacious and experienced observers, and, after careful thought, they all agree with this view. In this respect, therefore, the species of the larger genera resemble varieties more than do the species of the smaller genera. Said another way, in the larger genera, in which a number of varieties or incipient species greater than the average are now in the process of being manufactured, many of the species already manufactured still to a certain extent resemble varieties for they differ from each other by a less than usual amount of difference.

In addition, the species found among the larger genera are related to each other in the same manner as the varieties of any one species are related to each other. No naturalist pretends that all the species of a genus are equally distinct from each other; they may generally be divided into sub-genera, or sections, or even lesser groups. As Mr. Fries has remarked, little groups of species are generally clustered like satellites around certain other species. And what are varieties but groups of forms, unequally related to each other and clustered around certain other forms—that is, around their parent species?

Undoubtedly there is one most important point of difference between varieties and species: the amount of difference between varieties, when compared with each other or with their parent species, is much less than that between the species of the same genus. But when we come to discuss the principle that I am calling "Divergence of Character" in Chapter 4, we shall see how this may be explained, and how the relatively smaller differences between varieties will tend to increase over time into the greater differences between species.

There is one other point that I think deserves mention. Varieties generally have much restricted ranges: indeed, if a variety were found to have a wider range than that of its supposed parent species, their designations should be reversed and the species be called the variety and the variety be called the species. But it seems that species that are very closely allied with other species, so that the different species essentially resemble varieties of each other, also often have much restricted ranges. For instance, the English botanist Mr. Watson, mentioned earlier, has marked for me 63 plants that are ranked as species in the well-sifted *The London Catalogue of British Plants* (4th edition), but which he considers as so closely related to other species that their status as separate species is questionable; these 63 reputed species range on average over 6.9 of the provinces into which Mr. Watson has divided Great Britain. Now, in this same catalog, 53 acknowledged *varieties* are recorded, and these range over 7.7 provinces, whereas the species to which these varieties belong range over 14.3 provinces. Thus, the 53 acknowledged varieties have very nearly the same restricted average range as have those very closely related forms marked for me by Mr. Watson as questionable species, but which are almost universally ranked by other British botanists as good and true species.

Summary

In summary, **varieties have the same general character as species,** for they cannot be distinguished from species except by, first, discovering intermediate linking forms and, second, by showing a certain amount of difference, which is largely undefined: when two forms differ very little from each other, they are ranked as varieties, even when intermediate forms linking the two have not been discovered. But the amount of difference considered necessary to rank two forms as species is quite unspecified. In genera having more than the average number of species in any country, the species within these large genera have more than the average number of varieties. In large genera, the species are apt to be closely, although unequally, allied together, forming little clusters around certain species. Species very closely allied to other species apparently have restricted ranges. In all of these several respects, the species of large genera present a strong analogy with varieties within species. **We can clearly understand these analogies if we accept that what are now distinct species once existed as mere varieties,** and thus originated as varieties; on the other hand, these well-documented analogies would be utterly inexplicable if each species had been independently created.

We have also seen that it is the most flourishing or dominant species of the genera containing the most species (the "larger" genera in my terminology) that, on average, vary the most, and it is these "varieties," as we shall see in more detail later, that tend to become converted into new and distinct species. The larger genera thus tend to become larger over time, and, throughout all of nature, the forms of life that are now dominant tend to become still more dominant over time by leaving many modified and dominant descendants. But through steps to be explained later in this book, the larger genera also tend to eventually break up into small genera. And thus the forms of life throughout the world slowly become divided into groups within groups (see Figure 1.5).

Key Issues to Talk and Write About

1. Find out two interesting things about one of the people that Darwin mentions in this chapter. Choose from the following:
   Elias Fries
   Joseph Hooker
   Alphonse de Candolle
   Alfred Russel Wallace
   Hewett Cottrell Watson

2. Read the paragraph that begins "Intriguingly, it is where the organisms are best known that we find ...." Now, write a one-sentence summary of that paragraph: How would Darwin make his key points if he only had one sentence in which to do so? First list the two or three major points that Darwin is making. Then try to get all of those points into a single sentence. The sentence must be accurate, contain each of the major points, and be in your own words, as though you were explaining something to a friend. It should also make sense to a reader who has never read the original paragraph.

3. How does Darwin define "large genera" (see the section "Species in Larger Genera Vary More Frequently Than Those in Smaller Genera")? How does Darwin's example about the amount of variability seen among species within large genera fit in with his argument that what we now see as distinct species were once merely varieties of some other, ancestral species?

4. Here is a sentence from the original version of Darwin's Chapter 2: "That varieties of this doubtful nature are far from uncommon cannot be disputed." Try rewriting that sentence to make it clearer.

5. We now generally define species as being reproductively isolated units, whereas varieties within a species (the various breeds of dogs, for example) are able to mate successfully with each other and produce viable offspring. How does Darwin explain the difficulty of distinguishing between varieties and species based on physical characteristics?

Bibliography

Watson, H. C., ed. 1859. *The London Catalogue of British Plants* (4th edition). London.

3

The Struggle for Existence

In this chapter, Darwin emphasizes that, for all plants and all animals in nature, far more individuals are born than can possibly survive: predation, competition, and physical stresses all take their toll, often in amazingly complex and surprising ways. Competition will be most intense among individuals that are the most similar to each other (i.e., among members of the same species). This competition will, in a sense, have individuals pushing each other, over the course of many generations—through survival, reproductive success, and the inheritance of favorable characteristics—to become better and better adapted to their way of life.

Let me set the stage: How does the "struggle for existence" relate to what I am calling "natural selection"?

I have already shown, in Chapter 2, that in the natural world, individuals within any group vary in a great many traits. Indeed, this is something that everyone seems to agree with. But the mere existence of this variation among individuals, although it provides the foundation for all my thinking, doesn't explain how new species arise in nature or how individual organisms have become so perfectly adapted to their surroundings and lifestyles. We see such beautiful adaptation very clearly in the woodpecker (Figure 3.1) and in the mistletoe, for example, and in the simplest parasite that clings to the hairs of a dog or a sheep, or to the feathers of a bird, and we see the perfect structure of beetles that dive down through the water to feed, and in the feathered plant seeds (Figure 3.2) that are wafted away by even the gentlest breeze. Basically, we see beautiful, marvelous adaptations everywhere we look, in every part of the living world.

How have all those exquisite adaptations come about? And how can small variations in various traits eventually give rise to new species? How can varieties, which I have called "incipient species" (i.e., species in the making), eventually become converted into distinct, separate species—that is, species that generally differ from each other more than do varieties of the same species?

It all follows from the *struggle for life*, a struggle most eloquently discussed in the recent writings of the eminent geologist Sir Charles Lyell and the Swiss botanist Augustin Pyramus de Candolle. Because of this struggle, even the smallest of variations can help improve the chances of an individual's survival. **Offspring that then inherit those particularly advantageous traits from their parents will have a better**

The Readable Darwin. Second Edition. Jan A. Pechenik, Oxford University Press. © Oxford University Press 2023.
DOI: 10.1093/oso/9780197575260.003.0004

Figure 3.1 The red-bellied woodpecker, male (*Melanerpes carolinus*).

Figure 3.2 Seeds of the common dandelion (*Taraxacum officinale*) being dispersed away from the parent plant.

chance of surviving than those that do not, because—and here is one of my main points—from the great many individuals of any species that are born in any one year, only a small number can survive. I call this basic principle "natural selection," to help us see its connection to our own great powers of selection in breeding animals and plants for our own use (see Chapter 1). But perhaps Mr. Herbert Spencer's term "survival of the fittest" is more accurate.[1]

In Chapter 1, I gave examples of how we have molded domesticated animals and plants for our own use, selecting for the gradual accumulation of slight but useful variations by carefully choosing which individuals to breed with other individuals in each of many generations, taking advantage of the natural variation among individuals given to us by the hand of nature. But what I have now called *natural* selection is as immensely superior to our feeble efforts at selection as the works of nature are to those of art, as you will see more clearly as you read further in this book.

Everything about the struggle for existence follows from this simple fact, as shown by the Swiss botanist Alphonse de Candolle (the son of Augustin Pyramus de Candolle who was mentioned earlier) and Sir Lyell: **All organisms are exposed to severe competition, predation, and physical stress. We must keep this constantly in mind: If we don't integrate this idea into our brains—if we don't keep this continuous struggle in mind at all times—the whole point of our observations on animal and plant distributions, rarity, abundance, extinction, and variation will only be dimly seen, or will be completely misunderstood.**

When we look on the face of nature, seemingly bright with gladness, we often see what appears to be a superabundance of food. What we don't think about is that the birds that are merrily singing all around us must in fact eat insects and seeds to survive and thus are constantly destroying other forms of life. And we forget how these songsters, or their eggs and babies, are themselves destroyed by predators, including other birds. And we often forget that although food supplies may indeed be plentiful at some times of the year, that is certainly not true in all seasons and in every year, and. at those times, individuals must indeed compete and struggle with each other for their continued existence.

The Term "Struggle for Existence" Used in a Larger Sense

When I talk about the "struggle for existence," I'm not just talking about how different organisms depend on each other, and I'm not just talking about the struggle to stay alive; I'm also talking about success—or not—in leaving offspring that grow up to reproduce successfully themselves. Two wild dogs, for example, may fight over food when food is in short supply, and that fight may determine which dog eats and which does not, and in fact may determine which one lives to reproduce.

But a plant on the edge of a desert also struggles for life, this time against dryness—for it depends on moisture for its existence. A plant may produce 1,000 seeds every year, but if only 1 of those 1,000 seeds survives to reproductive maturity, then that

---

[1] This term is commonly misunderstood: yes, natural selection is about survival, but it is especially about fitness for reproductive success, which includes survival.

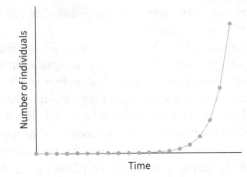

Figure 3.3 An example of exponential growth. The population increases over time in proportion to the number of individuals already present: the more individuals present at any given time, the faster the population grows.

one surviving plant has truly struggled with other plants, both of its own species and of other species, for its survival. Similarly, mistletoe seedlings growing on a single branch compete with each other for space on that branch. But even more interesting is that mistletoe is moved about and transported to other areas by birds, so that the continued existence of mistletoe depends on its being spread by those feathered chariots. Thus we can say that the mistletoe, in this sense, also "struggles" with other fruit-bearing plants in trying to get birds to eat only them, thus distributing its own seeds instead of those produced by other plants.

The term "struggle for existence" applies in all of these instances.

Exponential Rates of Increase

A struggle for existence in the natural world is inevitable because of the high rate at which most organisms reproduce themselves. Every animal species, whether its members produce several eggs or many eggs each season, must suffer great destruction during some phase of its life, and during some seasons, at least in some years. The same must be true for plants, whether they produce several seeds or many seeds. Otherwise, based on the principle of exponential increase, their numbers would quickly become so enormously great that no one country could support them all (Figure 3.3)! By "the principle of exponential increase" I mean that population growth rates increase proportionally as the size of the population increases, as first explained by Thomas Malthus in 1798. For example, if we have 10 females in a population and each female produces two eggs, the population will grow by another 20 individuals in the next generation if all of the eggs hatch and no individuals die. But if we have 1,000 females in the population, that population will grow by 2,000 individuals in the next generation under the same rules. When we have 10,000 females in the population, that population will grow by 20,000 individuals in the same amount of time, again assuming no mortality: the increase in numbers is proportional to the size of the population.

A struggle for existence is inevitable whenever more individuals are produced than can possibly survive. Individuals struggle with others of the same species, or with the individuals of some other species, or with the physical stresses of life, such as temperature changes or dehydration—but struggle they must. Here, the rule of Mr. Malthus applies with tremendous force to all living organisms in the natural world, both animals and plants; in this situation there can be no artificial increase in food supplies and no means of keeping individuals from mating and reproducing. Although some species may well be increasing in numbers now, more or less rapidly, all cannot do so, and none can do so forever for the world would not hold them.

There is no exception to the rule that every species naturally increases at so high a rate that, without high rates of mortality, the Earth would soon be covered by the off-spring of a single pair of parents. Even humans—which breed fairly slowly compared with most other animals—have doubled their population numbers in just the past 25 years; at this rate, in the absence of mortality, within a thousand years there would be no standing room for our descendants! Indeed, the Swedish botanist and zoologist Carl Linnaeus calculated (in the 1730s) that if an annual plant produced only two seeds in its lifetime—something that no plant actually does—and if their seedlings produced two seeds themselves the following year, and so on each year, without any mortality there would be 1 million plants in only 20 years!

Here is another example. The elephant is believed to be the slowest breeder of all known animals. Let's assume that it doesn't start reproducing until it is 30 years old and that it then continues to reproduce every year for the next 60 years, leaving six offspring each year. Suppose, too, that the animals die when they become 100 years old. According to my calculations, then, after about 740 or 750 years, there would be almost 19 million elephants alive on this planet, all descended from that first pair.

But this is not just theorizing on paper. In fact, we have many actual examples of astonishingly rapid increases in the numbers of some animals in the real world when living under favorable conditions for even just two or three seasons. Even more striking is the evidence from many of our domesticated animals that have escaped from domestication and lived subsequently in the wild in some parts of the world; indeed, if the reports about the remarkably high rates of increase of slow-breeding cattle and horses in South America, and more recently in Australia, had not been very well documented, they would have been too incredible to be believed!

The same is true with plants: I know of many cases in which introduced plant species have become common throughout entire islands in less than 10 years. Indeed, several such plants, including the cardoon (*Cynara cardunculus*) (Figure 3.4) and a species of tall thistle, were introduced from Europe to the plains of La Plata in Argentina only a few years ago and are now the commonest plants in that new habitat; they now dominate many, many square miles of surface and in fact pretty much exclude all other plants. Similarly, some plants introduced from America sometime after its discovery now range in India over enormous distances, from Cape Comorin

Figure 3.4 The cardoon (*Cynara cardunculus*).

to the Himalaya Mountains. How can we explain this? Certainly in these and in many other such examples that I could give, it's not that the egg production of the introduced animals or plants has suddenly and temporarily increased to any great extent. No, the obvious explanation is that the conditions of life in the new lands have been extremely favorable to the invaders: so favorable, in fact, that both old and young have survived very well, and most of the young have managed to reproduce successfully. Their exponential rate of increase in their new homes, the result of which is always remarkable and surprising, completely explains their extraordinarily rapid increase in numbers and the tremendous spread of their populations in their new homes.

In the natural world pretty much every full-grown plant produces seed every year. Among animals, too, there are very few species that don't mate and reproduce each year. Thus we can state with confidence that all plants and animals tend to increase at an exponential rate—and that all species would rapidly dominate every place that supported their existence—unless the exponential tendency to increase was somehow suppressed by destruction at some point in the life cycle.

Cows, sheep, and the other large domestic animals that we are all so familiar with are not good examples of life in the wild at all because we see no great destruction falling on them. But in fact we kill thousands of those animals every year for food; an equal number would be killed in one way or another in the natural world.

Note, too, that it generally matters little whether an organism produces thousands of eggs or seeds each year or only a very few. In both cases the offspring would eventually dominate the landscape; one would take a few more years to do so than the other, but eventually both would reach huge population sizes. The condor lays only a few eggs each year, while the ostrich lays a dozen, and yet in some countries the condor population is larger than that of the ostrich. Why, the fulmar petrel (*Fulmarus*

Figure 3.5 The fulmar petrel (*Fulmarus glacialis*).

glacialis) (Figure 3.5) lays only one egg each year and yet may be the most numerous bird in the world!

Producing many eggs *is* of some importance, of course, to species whose food supply fluctuates from year to year for it allows them to rapidly expand their population size when conditions are good. But the main benefit of producing a great many eggs or seeds is to make up for the great destruction that occurs at some point in the life cycle, usually at some early point. If many eggs or young are routinely killed by natural forces, then many must also be produced, or the species will become extinct. You could have no change at all in the number of trees in an area for a species that lived an average of 1,000 years, even if it produced only one seed in its lifetime, as long as all those seeds germinated successfully and in suitable locations. Thus, in every case we can think of, the average number of animals or plants now living in an area depends very little and only indirectly on the number of eggs or seeds it produces.

Never forget that every living thing may be trying its utmost to increase in numbers, that each lives by struggle at some point in its life, and that in every generation, sometimes at repeated intervals, all populations experience heavy destruction. Reduce that level of natural destruction even slightly, and the number of individuals living in an area will increase greatly in a very short time.

Nature of the Checks to Population Growth Increase

The factors acting against the natural tendency of each species to increase in numbers are most obscure. Look at the most vigorous species: by as much as it swarms in numbers, so much will its numbers tend to increase still further in the next generation. What, then, limits its growth? Unfortunately, we can't identify all the various checks to population growth, even in a single instance.

I hope to discuss the topic more fully in the future, but here I will just bring up some of the chief points. First, it seems that, for animals, the eggs or the very young usually suffer the most. Similarly, with plants we see a vast destruction of seeds. My observations suggest that seedlings have the most difficulty germinating in areas that are already thickly stocked with other plants. Seedlings are also destroyed in vast numbers by various enemies. For example, I cleared an area of ground 3 feet by 2 feet in my garden, so that there could be no choking from other plants, and I then separately marked each of the seedlings of our native weeds as they germinated. Out of 357 such seedlings that sprouted, 295 (nearly 83%) were destroyed, chiefly by slugs and insects.

When any area of lawn that has long been mowed or closely grazed by cattle is let to grow, the more vigorous plants gradually kill the less vigorous plants, even if those other plants are full grown. Thus, out of 20 species of plants growing on a little 3-foot by 4-foot plot of turf at my home, 9 species perished simply from other species being allowed to grow up freely around them.

The amount of food available must of course set the upper limit to population size in any area, for any animal species. But often it is not the availability of food that determines population size for a species, but rather the extent of *predation* by other animals. Thus, there seems little doubt that the population of partridges, grouse, and hares on any large estate depends chiefly on their destruction by predatory foxes, weasels, and other vermin. Indeed, if not a single game animal were shot during the next 20 years in England, and at the same time none of these predators was destroyed, there would, in all probability, be fewer game animals than at present even though hundreds of thousands of such game animals are now killed every year. On the other hand, as in the case of the elephant and rhinoceros, for example, sometimes predation is not so important in regulating population growth; even the tiger in India very rarely dares to attack a young elephant protected by its mother. The importance of predation in regulating population size clearly varies with species and situation.

Climate also plays an important role in determining species numbers; indeed, I believe that long periods of extreme cold or drought are probably the most effective of all checks on population growth. I estimated that the winter of 1854–1855 destroyed 80% of the birds on my own property; this is a tremendous amount of destruction when we remember that epidemics in humans rarely kill more than 10% of the individuals in a population.

At first glance, the action of climate might seem largely unrelated to the struggle for existence that we have been discussing. However, to the extent that climate change acts chiefly to reduce food availability, it can bring on the most severe struggle between individuals, both of the same species and of different species, that subsist on the same kinds of food. Even when climate—periods of extreme cold, for example—acts on organisms directly, it will be the least healthy individuals, or those that have gotten the least food through the advancing winter, who will suffer the most.[2]

---

[2] Darwin's thoughts here are certainly relevant to the changes in climate that are now being seen around the world. The impacts of these changes on food availability and competition for food will surely be substantial.

When we travel from north to south, or from a damp region of the country to a dry one, we invariably see some species gradually getting rarer and rarer and finally disappearing altogether; with the change of climate being so conspicuous, we are tempted to attribute the shift in species numbers entirely to its effect. But this is a false view: we forget that each species, even where it is most abundant, is constantly suffering enormous destruction at some period in its life, either from enemies or from competitors for its space or food. If those enemies or competitors are in the least degree favored by any slight change in climate, they will increase in numbers and, as each area is already fully stocked with inhabitants, the other species will decrease in abundance. When we see a species decreasing in numbers as we travel southward, we may feel sure that the cause lies quite as much in other species being favored as in this species being hurt by the changing climate.

We see the same thing when traveling from south to north, although to a somewhat lesser degree, for the numbers of species of all kinds—and therefore the numbers of competitors—decreases as we travel northward. Thus, in going northward, or in ascending a mountain, we meet with stunted forms, due to the directly injurious action of climate, more often than we do in proceeding southward or in descending from a mountain. When we reach the Arctic regions, or snow-capped summits, or absolute deserts, the struggle for life is almost exclusively against the physical environment rather than with other species.

However, climate often acts largely indirectly, by favoring other species: We may clearly see this in the prodigious numbers of plants in our gardens that, although they have no difficulty coping with our climate, nevertheless never become "naturalized," (i.e., part of the natural landscape in their new homes); they just cannot compete with our native plants, nor can they avoid being destroyed by our native animals.

When a species, owing to highly favorable circumstances, shows an inordinate increase in numbers in a small area, epidemics often ensue; certainly we see this often with our game animals. And here we have a limiting check on population growth that seems to be independent of the struggle for life. But even some of these so-called epidemics appear to be caused by parasitic worms, which have for some reason, possibly in part though the ease of spread among the crowded animals, been disproportionately favored. And thus we see here another sort of struggle, this one between the parasite and its host.

For many species, population size must be very large relative to the number of its enemies if the species is to be maintained. Thus, for example, we can easily raise plenty of corn[3] and rapeseed in our field because there are so many seeds compared with the number of birds that feed on them. Nor can the birds, though having a superabundance of food at this one season, increase in number proportionally to the supply of seed as their numbers are held back in the winter, when food is scarce. But anyone who has tried knows how troublesome it is to get seed from only a few wheat plants or other such plants in a garden; I myself have lost every seed in such cases.

---

[3] Here, Darwin is actually referring to a cereal grain, not what we typically think of as corn.

This idea of a large population size being essential to the preservation of a given species explains, I believe, some singular facts in nature—for example, that some very rare plants can sometimes be extremely abundant in the few spots where they do occur, and that some plants can be extremely abundant even on the extreme edges of their ranges. In such cases it seems that a particular plant can exist only where the conditions of its life were so favorable that many can exist together and thus save each other from utter destruction. I should add that the good effects of frequent intercrossing and the ill effects of mating with close relatives probably come into play in some of these cases; but I will say no more about this intricate subject here.

Complex Relations of All Animals and Plants to Each Other in the Struggle for Existence

We have many records showing how complex and often unexpected the checks and relationships are between organisms that have to struggle with each other in the same area. I will give only one example, one which has interested me a great deal despite its simplicity. One of my relatives in Staffordshire, England, allowed me to explore his estate in detail. On that estate there was a large and extremely barren heath (Figure 3.6), one that had never been touched by the hand of man. But on the same estate there were several hundred acres of exactly the same sort that had been enclosed by fencing 25 years earlier and planted with Scotch fir trees. The change in the native vegetation of the planted part of the heath was most remarkable, and more than is generally seen in passing from one quite different sort of soil to another. Not only were the proportional numbers of the various heath plants wholly different in the two areas, but 12 plant species (not even counting the grasses and grass-like sedges) that flourished in the plantations could not be found at all on the heath. The effect on the insect populations must have been even greater, for six species of insect-eating birds that were very

Figure 3.6 Barren heathland in Dartmoor National Park, United Kingdom.

common on the plantations were not seen at all on the heath. And the heath itself was visited by two or three distinct insectivorous bird species that were not seen on the plantations. Thus we see how potent the effect can be of simply introducing a single new tree species, nothing else having been done other than to enclose the land so that cattle could not enter.

But in an area near Farnham, in Surrey, England, I plainly saw just how important enclosure itself can be. Here there are extensive heaths, with a few clumps of old Scotch firs on the distant hilltops. Within the last 10 years, large spaces on those heaths have been enclosed, and self-sown fir trees are now springing up in multitudes, growing so close together in fact that all cannot live. After I learned that these young trees had not been deliberately sown or planted within the enclosures, I was so surprised at their great numbers that I looked at them from several different vantage points, taking in hundreds of acres of unenclosed heath in each view; literally, I couldn't see a single Scotch fir in any of the unenclosed areas except for the old planted clumps. But on looking closely between the stems of the heath in these unenclosed areas, I found a multitude of seedlings and little trees that had been perpetually grazed down by cattle. In one square yard, at a point some hundred yards distant from one of the old Scotch fir clumps, I counted 32 little trees. One of them, with 26 growth rings on its stem, had during many years tried to raise its head above the other stems of the heath and had failed to do so. No wonder then, that as soon as some portion of the land was enclosed, keeping out all cattle, it became thickly clothed with vigorously growing young firs. Yet the unenclosed heath was so extremely barren and so extensive that no one would ever have imagined that cattle could have so closely and effectively searched it for food.

Here then we see that cattle absolutely determine the existence of the Scotch fir in this part of England. But in some parts of the world, insects determine the existence of the cattle. Perhaps Paraguay, near the center of South America, offers the most curious example. Neither cattle nor horses nor dogs have ever run wild in Paraguay, although they certainly swarm southward and northward in a feral, undomesticated state. Two naturalists, Félix de Azara of Spain and Dr. Johann Rudolf Rengger of Switzerland, have shown that this is caused by large populations of a certain fly in Paraguay: the fly lays its eggs in the navels of cattle and other mammals. Population growth of these flies, numerous as they are, must be habitually checked by some agent, probably birds. Thus, if insect-eating birds (whose numbers are, in turn, probably regulated by hawks or other beasts of prey) were to increase in Paraguay, the fly population would decline. Cattle and horses could then become wild, and that would greatly alter the vegetation of the region, as indeed I have observed elsewhere in South America. This again would largely affect the insects and thus affect the insectivorous birds, just as we have seen in Staffordshire, and so onward in ever-increasing circles of complexity.

We began this series with insectivorous birds, and we have ended with them. Of course, in nature the relationships can never be this simple. Battle within battle must ever be recurring with varying success, and yet, in the long run, the forces are so nicely balanced that the general abundances and distributions of organisms often remain uniform for long periods of time, though assuredly the merest trifle would often give the victory to one species over another. Even so, so profound is our ignorance, and so great our presumption, that we marvel when we hear about the extinction of any organism. And as we do not see or understand the cause of that extinction, we invoke biblical floods and other cataclysms that desolate the world, or invent laws about predetermined, "fixed life spans" of species.[4] But no: the causes in fact all relate to the struggle for existence.[5]

Allow me to give just one more example showing how plants and animals, so different from each other in the scale of nature, are bound together by a web of complex relations. Later in this book I will show that the exotic plant *Lobelia fulgens*, the "cardinal flower" (Figure 3.7A), is never visited by insects in this part of England and, consequently, because of its peculiar structure, it never reproduces successfully here. Similarly, many of our orchids absolutely require moths to visit and remove their pollen masses, and thus to fertilize them. I also have reason to believe that bumblebees are indispensable to the fertilization of the heartsease (*Viola tricolor*) (Figure 3.7B), for I have seen no other types of bee visit this flower. From experiments that I have recently conducted at my home, I have found that some kinds of clover are fertilized only by bees. But only bumblebees visit the red clover (*Trifolium pratense*) (Figure 3.7C), as other bees cannot reach the nectar within the long, tubular flowers of this species. Therefore I have little doubt that if the entire genus of bumblebees became extinct, or even just very rare, in England, the heartsease and red clover would also become very rare or perhaps disappear entirely.

Now the number of bumblebees in any district depends to a great degree on the number of field mice living there because the mice destroy the bee's honeycombs and nests. Indeed, Mr. Henry Wenman Newman, who has studied the habits of bumblebees for many years, believes that "more than two-thirds of them are thus destroyed all over England." As everyone knows, the number of mice largely depends on the number of cats: as Mr. Newman says, "Near villages and small towns I have found the nests of bumble-bees more numerous than elsewhere, which I attribute to the number of cats that destroy the mice." Thus it is quite likely that the presence of large numbers of cats in a district might determine, through the intervention first of mice and then of bees, the commonness of certain flowers in that district—remarkable!

In the case of every species, many different checks probably come into play, each acting at different times of life and during different seasons or years; perhaps one check or some few are especially potent, but all ultimately combine in determining the average number of individuals in an area, or even the existence of the species. In some cases it can be shown that very different checks act on the same species in different areas.

[4] In trying to explain extinctions, some authors had suggested that species are created with fixed life spans: i.e., that they are essentially programmed for automatic extinction after a certain number of years.

[5] While that has indeed generally been the case, five major extinctions are now known to have been caused by large-scale cataclysms, such as that occurring about 66 million years ago, which killed off all of the non-feathered dinosaurs and about 75% of all other animal and plant species on Earth.

Figure 3.7 (A) Cardinal flower (*Lobelia fulgens = L. cardinalis*). (B) Heartsease (*Viola tricolor*). (C) Red clover (*Trifolium pratense*).

When we look at the diversity of plants and bushes clothing an entangled bank along a river, we are tempted to attribute their proportional numbers and kinds to mere "chance." But how false a view this is! Everyone has heard that when an American forest is cut down, a very different vegetation springs up. But it has also been observed that ancient Indian ruins in the Southern United States, which must first have been cleared of trees many years ago, now display the same beautiful diversity and kinds of trees in the same proportion as in the surrounding virgin forests; the forest seems to have returned now to its former state. What a struggle between the several kinds of trees must have gone on here during long centuries, each tree annually scattering its seeds by the thousands; what war between insect and insect—and between insects, snails, and other animals with birds and beasts of prey—all striving to live and reproduce, and all feeding on each other or on the trees, or on the seeds and seedlings of those trees, or on the other plants that first clothed the ground and thus checked the growth of the trees! Throw up a handful of feathers, and all must fall to the ground according to definite laws; but how simple is that problem compared with trying to understand the actions and reactions of the innumerable plants and animals that have determined, over the course of centuries, the proportional numbers and kinds of trees now growing on the old Indian ruins!

The dependency of one organism on another, as with a parasite dependent on its host, lies generally between beings very different from each other in the scale of nature. This is also often the case with those that may strictly be said to struggle with each other for existence, as in the case of locusts and grass-feeding quadrupeds, such as cattle. *But almost invariably the struggle will be most severe between individuals of the same species, for they frequent the same districts, require the same foods, and are exposed to the same dangers.* In the case of varieties of the same species, the struggle will generally be almost equally severe, and we sometimes see the contest quickly decided. For instance, if several varieties of wheat are sown together, and the mixed seed is resown for the next generation, those varieties that best suit that particular soil or climate, or are naturally the most fertile, will beat the others and so yield more seed; consequently, in a few years, these varieties will quite supplant the other varieties. It takes a good deal of work to keep up a mixed stock of even such extremely close varieties as the variously colored sweet peas: each variety must be harvested separately each year, and the seed then mixed in due proportion; otherwise the weaker kinds will steadily decrease in numbers and eventually disappear.

And so it is again with the varieties of sheep: certain mountain varieties will apparently starve out other mountain varieties so that they cannot be kept together. The same result has followed from keeping together different varieties of the medicinal leech: one variety eventually wins out over the others. It may even be doubted whether the varieties of any one of our domesticated plants or animals have so exactly the same strength, habits, and constitution that the original proportions of a mixed stock could be maintained for even a half-dozen generations if they were allowed to struggle together like beings in a state of nature and if the seed or the young were not deliberately sorted by us every year.

(A)   (B)

Figure 3.8 (A) Mistle thrush (*Turdus viscivorus*). (B) Song thrush (*Turdus philomelos*).

The Struggle for Life Is Generally Most Severe Between Individuals and Varieties of the Same Species

This idea, that the struggle for life is more severe the more closely the competing individuals resemble each other, leads in many interesting directions. **Because species within the same genus are usually similar in habits and constitution, and always have similarities in structure, the struggle will generally be more severe between species of the same genus when they come into competition with each other, than between species of different genera.** For example, the recent extension over parts of the United States of one particular swallow species has caused the population of another swallow species to decrease. Similarly, the recent increase of the mistle thrush (Figure 3.8A) in parts of Scotland has caused populations of the song thrush (Figure 3.8B) to decrease. And how frequently we hear of one species of rat taking the place of another species under the most different climates! In Russia, the small Asiatic cockroach has driven out everywhere before it a larger species in the same genus, while in Australia the imported honeybee is rapidly exterminating the small, stingless native bee. One species of the charlock weed (*Sinapis arvensis*) (Figure 3.9) will commonly supplant another, and so on in other cases. We can dimly see why the competition should be most severe between allied forms which fill nearly the same place in the economy of nature,[6] but probably in no one case could we precisely say why one particular species has been victorious over another in the great battle for life.

A corollary of the highest importance may be deduced from the previous remarks: namely, the structure of every organism is related, in the most essential yet often hidden manner, to that of all the other organisms with which it competes for food or residence, or from which it has to escape, or on which it preys. This is obvious in the structure of the teeth and talons of tigers and in the morphology of the legs and

---

[6] If Darwin were writing today he would talk instead of species occupying the same niches.

Figure 3.9 Charlock weed (*Sinapis arvensis*).

(A)   (B)

Figure 3.10 (A) The tiger's teeth are adapted to the types of prey that it catches for food. (B) The louse's legs are specialized for gripping the hairs on the tiger's body.

claws of the parasite that clings to the hair on the tiger's body (Figure 3.10). But in the beautifully feathered seed of the dandelion (members of the genus *Taraxacum*) and in the flattened and fringed legs of the water beetle (Figure 3.11), the morphological features might at first seem merely related to the art of living in air or water. Yet the true advantage of feathered seeds no doubt relates closely to the land being already thickly clothed by other plants; the feathering of the seeds allows them to be more widely distributed and to sometimes fall on unoccupied ground. In the water beetle, the structure of its legs, so well adapted for diving, allows it to compete successfully with other aquatic insects, to hunt for its own prey, and to escape being eaten by other animals.

Figure 3.11 (A) Water beetle (*Dystiscus latissimus*). (B) Feathered seed of a dandelion.

The storage of large amounts of nutrients within the seeds of peas, beans, and many other plants seems at first sight to have no obvious relationship to interactions with other plants. But from seeing the strong growth of young plants produced from such seeds when sown in the midst of long grass, I suspect that the seed's nutrients are chiefly used to promote rapid growth of the young seedling as it struggles with the other plants growing vigorously all around it.

Competition and predation are powerful forces. Look at any plant in the middle of its range: Why does it not double or quadruple its numbers? We know that it can perfectly well withstand a little more heat or cold, or a little more dampness or dryness, for elsewhere it ranges into slightly hotter or colder areas and slightly damper or drier areas. So what is holding it back? In this case we can clearly see that if we wished in our imagination to give the plant the power of increasing its numbers in any particular area, we should have to give it some advantage over its competitors or over the animals that prey on it. On the edges of its geographical range, a change of constitution with respect to climate would clearly be an advantage to our plant; but we have reason to believe that only a few plants or animals range so far that they are destroyed by the rigor of the climate alone. Not until we reach the extreme confines of life, in the Arctic regions or on the borders of an utter desert, will competition with other individuals and other species cease. In most cases, even if the land is extremely cold or dry, yet there will still be competition between some few species, or between the individuals of the same species, for the warmest or the dampest spots.

Thus, we can also see that when a plant or animal is placed in a new area among new competitors, the conditions of its life will generally be changed in some essential manner even though the climate may be exactly the same as it was in its former home. If we wished to increase its average numbers in its new home, we would have to modify it differently here than if it was still in its native country; we would now need to do something to give it an advantage over a different set of competitors or help it to better deal with its enemies.

It is a good exercise to try in our imagination to give any particular organism some advantage over another. Probably in no single instance will we know what to do to ensure success. It will convince us of our ignorance of the mutual interactions of organisms in the wild, a conviction as necessary as it seems to be difficult to acquire. All that we can do is to keep steadily in mind that each living being is striving to increase at an exponential rate and that each individual, at some time in its life, during some season of the year, during each generation or at intervals, has to struggle for life, and that populations must periodically suffer great destruction. This seems terribly grim. But when we think about this great struggle, perhaps we can console ourselves with the full belief that the war of nature is not incessant, that no fear is felt, that death is generally prompt, and that the vigorous, the healthy, and the happy survive and multiply.

Key Issues to Talk and Write About

1. What does Darwin mean by "the struggle for existence"? Why does he claim that such struggles are inevitable?
2. According to Darwin, what factors control the number of individuals of any particular species that live in any particular area at any particular time?
3. Darwin talks about investigating the plants on one of his relative's estates (see page 52). According to Darwin, what accounts for the difference in species composition of the plants and insects on different parts of that estate? What is his evidence? How convincing do you find his conclusion?
4. Darwin argues that the presence of cats in a neighborhood might determine the abundance of flowers such as red clover (see page 54). Summarize his argument in a single sentence or two, as though you were explaining this to a friend or parent.

5. Why is Darwin convinced that struggles among individuals are typically the most intense between individuals of the same species and between individuals of different species that are found in the same genus? Summarize his argument in your own words.

6. Find out two interesting things about one of the people that Darwin mentions in this chapter. Choose from the following:

Félix de Azara

Augustin Pyramus de Candolle

Carl Linnaeus

Sir Charles Lyell

Thomas Malthus

Herbert Spencer

7. Take a look at one of the following two paragraphs:

Page 45, the paragraph that begins, "When we look on the face of nature, seemingly bright with gladness...."

Page 52, the paragraph that begins, "We have many records showing how complex and often unexpected the checks...."

For the chosen paragraph, what are the two or three main points that Darwin wishes to get across?

Now summarize those points in a single sentence: your sentence should include all the major points, be accurate, make sense to someone who has not read the original paragraph, and be in your own words.

8. Here are three sentences from the original version of *The Origin of Species*. Rewrite each sentence to make it shorter and clearer.

   a. "The causes which check the natural tendency of each species to increase are most obscure."

   b. "Now the number of mice is largely dependent, as everyone knows, on the number of cats."

   c. "I have also found that the visits of bees are necessary for the fertilization of some kinds of clover."

4

Natural Selection, or the Survival of the Fittest

In this chapter, Darwin emphasizes that natural selection works slowly and over extremely long periods of time and that it works by favoring variations that provide advantages to the organisms that have them in the struggle for nutrients and space and against predators and physical stresses and even in the struggle for mates. Such "sexual selection" explains why individuals of one sex (usually the males) sometimes look so different from individuals of the other sex of the same species. Natural selection also works not just on adults but also on all other stages of development—eggs, seeds, embryos, larvae, and juveniles—and can act to either increase or decrease the complexity of various organs. In all cases, the best-adapted individuals will typically survive and propagate their offspring more successfully to the next generation, and those offspring will inherit the characteristics that made their parents so successful. Darwin also discusses the importance of outbreeding (matings among unrelated individuals) in promoting the vigor of offspring and the negative impacts of self-fertilization and inbreeding (matings between close relatives). He also describes the evolutionary advantages of "divergence in character" over time: individuals that differ in certain ways from their fellows can exploit new niches and reduce competition with others of their own species, eventually driving intermediate forms to extinction. Finally, he provides a detailed example of natural selection in action, creating new species, new genera, and even new families and orders from a few ancestral species in a single genus, over many thousands of generations. A great many species go extinct along the way because of their inability to compete with those newer creations that are better adapted to their conditions of life.

How will the struggle for existence, briefly discussed in the previous chapter, act on the variation among individuals? Can the principle of selection—which, as we have seen in Chapter 1, is so very potent in human hands—apply equally well in nature? I think we shall see that it can indeed act in a very similar way and most effectively.

In reading this chapter, keep in mind the seemingly endless number of slight variations and individual differences that occur in our domesticated animals and plants, and also in those in the wild, as well as the tendency for individual traits to be passed along to the owner's offspring. Remember, too, that the variability we see to so great an extent in our domesticated animals and plants is not directly produced by us: we can neither cause such varieties to occur nor prevent their occurrence; all we can do

The Readable Darwin. Second Edition. Jan A. Pechenik, Oxford University Press. © Oxford University Press 2023.
DOI: 10.1093/oso/9780197575260.003.0005

is preserve particular variations that do occur and cause them to accumulate in off-spring, generation after generation, as discussed in Chapter 1. And keep in mind how infinitely complex and closely linked are the mutual relations of all living things to each other in nature and to the physical demands and opportunities of their lives (see Chapter 3), and think how the infinitely varied diversities of structure within a population might be of use to individual organisms as the conditions around them change over time.

Considering how variations in the traits of individual plants and animals have been so useful to us—in farming, for example—how improbable can it be that other variations can also be useful in some way to each organism in the great and complex battle for life in nature? Some variations will surely be helpful to those individuals that have them. Remember that many more individuals are born in nature than can survive (see Chapter 3): Can we doubt, then, that individuals having even slight advantages over others would have the best chance of surviving and successfully leaving offspring and of passing those traits on to those offspring? On the other hand, it must also be true that any variation that is in the least harmful to its owner would be ruthlessly destroyed through the owner's death or unsuccessful reproduction. This **gradual preservation of favorable individual differences and variations, and the destruction of those that are harmful, I have called natural selection, or "survival of the fittest."**

Some authors have misunderstood or objected to the term "natural selection," and some have even imagined that natural selection *induces* variation. **But no: natural selection only implies the preservation of variations that come about naturally, if those variations benefit the individual that possesses them.** No one objects to agriculturists talking about the potent effects of *man's* selection for particular traits; but remember that man can select for such particular traits only after such individual differences first occur in a population naturally.

Others have objected that the word "selection" implies conscious *choice* by the animals that become modified, as though the animals were doing the selecting. It has even been argued that as plants have no ability to make decisions on their own, natural selection cannot apply to them! But of course there is no deliberate choice here: selection is essentially imposed on the animals and on the plants. That is, individuals with certain characteristics are more likely to survive and leave offspring, many of which will inherit those characteristics, while individuals with certain other characteristics are more likely to die without reproducing.

Some have said that I speak of natural selection as an active power or deity; but who ever objects when an author says that the "attractions of gravity" rule the movements of the planets? Everyone knows what is meant and implied by such metaphors and knows that they are almost always used for their brevity. Indeed, it's difficult to avoid personifying the word "nature." But when I say "nature" I mean only the aggregate action and product of many natural laws, and by "laws" I mean the sequence of events as ascertained by us. Such superficial objections to "natural selection" will eventually be forgotten once people get used to the term.

We can best understand the probable course of natural selection over long periods of time by considering the case of a country undergoing some slight physical change—in climate, for example. Populations of most species will quickly change in size, and some species will probably go extinct. From what we have already learned about the intimate and complicated interactions that bind the organisms of each region together, it is clear that any change in the population size of some species will seriously affect all others in the area, independently of the direct effect of the change in climate itself. If the region were accessible at its borders, new forms would certainly immigrate into the area, and this would seriously disturb the interactions among some of the original inhabitants; I have already shown how powerful the influence of a single introduced tree or mammal can be on indigenous populations (see Chapter 3). On the other hand, in the case of an island or of some other parcel of land partially surrounded by other physical barriers that prevent new and better-adapted organisms from entering the area, climate changes would create new niches that would assuredly be filled up eventually if some of the original inhabitants, or their offspring, were in some manner modified; if those areas had been open to immigrants from elsewhere, those same new niches would have been seized upon by the intruders. In such cases, even slight modifications that better adapted the individuals of any particular species to the new climatic conditions would tend to be preserved and become more and more common over the generations. And natural selection would have free scope for the work of such improvement. Again, we see the importance of natural variation: without natural variation in traits, natural selection can do nothing. And please don't forget that the term "variation" includes *all* individual differences, not just those that are favorable to an organism. As we can produce great results with our domesticated animals and plants by adding up individual differences in any given direction over many generations, so can natural selection—but natural selection can do this even more easily, from having incomparably longer periods of time to bring such changes about.

Although climate change can certainly be a driving force for natural selection, I do not believe that it requires any great physical change—such as substantial shifts in climate or any unusually large degree of isolation (to prevent immigration)—for selection to work. For as all the inhabitants of any particular region are struggling together with nicely balanced forces, even extremely small modifications in the structure or habits of an individual over time could be enough to give it an advantage over others and increase the likelihood of its survival and reproduction; some of those slight modifications would then be passed on to at least some of its offspring. And still further modifications along the same lines would often still further increase the advantages for the offspring of that individual as long as the species continued under the same conditions of life and profited by similar means of subsistence and defense.

There is no place on Earth in which all native inhabitants are now so perfectly adapted to each other and to the physical conditions under which they live that none of them could become still better adapted or "improved" over the generations. Indeed, in all places that have been studied, at least some native populations have been overwhelmed and conquered by immigrant species; the immigrants have taken firm possession of the land. As such invasive or introduced species have thus in every country successfully beaten some of the native species in the struggle for life, we may safely conclude that the native species could have been better adapted to their environment if they were to have better resisted the intruders. But they weren't, and they didn't.

As people *can* produce and certainly *have* produced many great results by our methodical and unconscious means of artificial selection among domesticated animals and plants (see Chapter 1), what changes might natural selection not bring about? People can only act on external characters that we can see and select for. But nature, if I may be allowed to personify the natural preservation or "survival of the fittest," cares nothing for mere appearances unless those appearances are useful to the individual in some way. And nature can act on every internal organ and on every physiological difference: on the whole machinery of life.

We select only for our own good. But nature always selects for the good of the individual that she acts on. And every selected character is fully acted on by her, as is implied by the fact of its selection. In our agricultural and animal domestication programs, we keep native species originating from many different climates together in the same place, and we seldom exercise each selected character in any specific and fitting manner for it under domestication: we feed the same food to both long- and short-beaked pigeons; we do not exercise an unusually long-backed or long-legged quadruped[1] in any unusual manner; we expose sheep with both long and short wool coats to the same physical environments. Similarly, we don't allow the most vigorous males to struggle with each other for the females in mating; we select who will mate with whom. And we do not rigidly destroy all inferior animals each generation but instead protect, as far as we can, each one during each varying season. We typically begin the selection process by selecting some modification prominent enough to catch the eye or be plainly useful to us. Under nature, though, the slightest differences of structure or constitution may well turn the nicely balanced scale in the struggle for life,, and so be preserved.

How fleeting are the wishes and efforts of man! How short his time! And consequently how poor will his results be, compared with those accumulated by nature during the whole of geological time! Can we wonder, then, that nature's productions should be far "truer" in character than our productions and that they should be infinitely better adapted to the most complex conditions of life,, and should plainly bear the stamp of far higher workmanship?

It may be said metaphorically that natural selection is scrutinizing—daily and even hourly—every variation throughout the entire world, rejecting those that are bad and preserving and adding up all that are good, silently and insensibly working whenever and wherever the opportunity allows for the improvement of every living organism

---

[1] Any four-legged animal, such as cats, dogs, horses, and cows.

(A)   (B)

Figure 4.1 (A) Leaf insect (*Phlogophora meticulosa*) camouflaged against a leaf. (B) Peppered moth (*Biston betularia*) camouflaged on an oak tree.

in relation to the physical and biological conditions of its life. We can see nothing of these terribly slow changes in progress until huge amounts of time have passed, and then so imperfect is our view into long-past geological ages that we see only that the forms of life are now quite different from what they had been.

In order for any species to become greatly modified over time, a new variety, once formed, must continue to show variation in individual characteristics among its members as before, so that helpful variations can be preserved in each generation, and so on over time. Seeing that individual differences in the same traits do indeed recur in each subsequent generation, this assumption of continued variation seems fully warranted. Whether or not it is in fact true we can judge only by determining the degree to which the hypothesis agrees with and explains what we see in nature.

Natural selection can act only for the good and benefit of individuals, but the characteristics it acts on need not be of obvious importance to us; characteristics and structures that seem to us to be of trifling importance can be acted on in the same way. When we see that leaf-eating insects are green, while insects feeding on tree bark are mottled gray, how can we not believe that these color patterns help protect these insects from their predators by helping to camouflage them against their normal backgrounds (Figure 4.1)? Similarly, we see that the rock ptarmigan (Figure 4.2A) is white in winter, while the red grouse (Figure 4.2B) is the color of the heather in which it lives. Such coloration must be protective. Grouse, if not destroyed at some point in their lives, would eventually form massive populations. But they are in fact known to be eaten, especially by birds of prey, such as the hawk. And hawks are guided to their prey by eyesight—so much so that in Europe, people are warned not to keep white pigeons as these are most likely to be seen and eaten by hawks. Thus, natural selection should be extremely effective in eventually giving an appropriate color to every kind of grouse and in keeping that color, when once acquired, true and constant, if in fact the color is beneficial to its possessor.

Nor should we think that the occasional destruction of an animal of any particular color would produce little effect on a population. People who raise sheep know

(A)

(B)

Figure 4.2   (A) Rock ptarmigan (*Lagopus muta*) in winter plumage. (B) Red grouse (*Lagopus lagopus*).

how essential it is to destroy a lamb with the faintest trace of black in a flock of white sheep: one dark animal is all it takes to attract predators. We have also seen (Chapter 1) how the color of hogs in Virginia determines whether they shall live or die when feeding on the "paint-root" plant (*Lachnanthes tinctoria*). In plants, botanists typically think that the downy fuzz on the surface of fruit and the color of the flesh are characters of the most trifling importance; and yet we hear from that excellent horticulturist Andrew Jackson Downing (*The Fruits and Fruit Trees of America*, 1845) that smooth-skinned fruits in the United States suffer far more damage from curculio beetles (Figure 4.3) than those with external fuzz. Similarly, purple plums suffer far more from a certain disease than do yellow plums, while another disease

Figure 4.3 Plum curculio beetle (*Conotrachelus nenuphar*).

attacks yellow-fleshed peaches far more than those with flesh of other colors. If such seemingly slight differences make such a great difference in cultivating the several varieties of fruit for us, then surely, in a state of nature, where trees would have to struggle with other trees and with a host of enemies, such differences would effectively determine which variety—whether a smooth or downy fruit, or a yellow or purple-fleshed fruit—would succeed.

In looking at many small points of difference between the structures of various species, some of which may seem quite unimportant and even trivial to us, we also need to bear in mind that when one part varies—and when those variations are accumulated through natural selection—then other modifications will also occur, sometimes of a very unexpected nature.

What else can we say about variation? Variations that appear at any particular period of life under domestication tend to reappear in the offspring at the same point in development—for example, in the shape, size, and flavor of the seeds of the many varieties of our plants used for cooking and agriculture; in the caterpillar and cocoon stages of the various varieties of silkworm; in the eggs of poultry, and in the color of the down on the skin of their chickens; and in the horns of our sheep and cattle when nearly adult. Similarly, in the wild, natural selection will be able to act on and modify organisms at any stage of development and at any age, through the accumulation of variations profitable at that age and by their future inheritance by the next generation at a corresponding age. If it benefits a plant to have its seed more and more widely disseminated by the wind, I can see no reason why this could not be effected through *natural* selection, in the same way that a cotton planter can increase and improve the down on the pods of his cotton trees by *artificial* selection. Natural selection may modify and adapt the larva of an insect to a variety of contingencies that are wholly different from those that concern the mature insect and these modification may bring about, through correlation, changes in the structure of adults. Conversely, modifications in the adult may—over the generations—affect the structure of the larvae. But in all cases, natural selection will ensure that those modifications shall not be injurious: for if they were injurious, the species would become extinct.

Natural selection will modify structures in the young in relation to the parent and of the parent in relation to the young. In social animals, like bees and ants, it will adapt the structure of each individual in ways that benefit the whole community, if the community benefits from those selected changes. Contrary to what I have read in various natural history writings recently, what natural selection cannot do is modify the structure of one species for the good of another species: it must first and foremost benefit the owner of the modification.

A structure used only once in an animal's life, but which plays an important role at that time, might still be modified to any extent by natural selection. For example, consider the great jaws possessed by certain insects, used exclusively for opening the cocoon after metamorphosis, or the hard tip at the beak of baby birds before they hatch, used for breaking out of the egg. I have seen reports that more of the best short-beaked tumbler pigeons perish in the egg than are able to hatch out of it successfully; pigeon fanciers must in fact help the birds escape from the egg. Now if for some reason it came to be advantageous for a full-grown pigeon to have a much shorter beak, the process of modification would be very slow, and there would simultaneously be the most rigorous selection for young, unhatched birds that had the most powerful and the hardest beaks for breaking out of the eggs; baby birds with weak beaks would be unable to hatch and would inevitably die. Or perhaps more delicate and more easily broken eggshells would be selected for instead; shell thickness is in fact known to vary, just as every other structure does.

Note that with all organisms there must also be a great deal of destruction by random chance. Such fortuitous destruction can't possibly have much of an influence on the course of natural selection because by definition no selection is taking place when all organisms are equally likely to die. For instance, various animals eat a vast number of eggs or seeds every year at random; these could become modified by natural selection over time only if they varied in some manner that protected them from their enemies. Yet many of these eggs or seeds would perhaps, if they hadn't been eaten, have yielded individuals better adapted to their conditions of life than any of those that happened to survive. So again, a vast number of mature animals and plants, whether or not they be the best adapted to their environments, must be destroyed every year by purely accidental causes, which would not be in the least degree mitigated by certain changes of structure or constitution that would in other ways benefit the species. But let the destruction of the *adults* be ever so heavy, or again let the destruction of eggs or seeds be so great that only a hundredth or a thousandth of them survive to develop, yet of those that *do* survive, it will still be the best-adapted individuals that will tend to propagate their kind more successfully to the next generation. If the numbers are kept down dramatically by the causes just mentioned, as must often be the case, then nearly all individuals will die and natural selection will be powerless to bring about any changes in beneficial directions. But this in no way means that natural selection cannot be very efficient at other times and in other ways. We have no reason to suppose that all species always undergo modification and improvement at the same time in the same area.

Sexual Selection

Physical peculiarities often appear in males *or* females when organisms are raised under domestication, and those characteristics soon become hereditarily associated with that one sex. Thus it is entirely feasible that the two sexes of a species can become modified in different ways through natural selection. Indeed we know that this occurs, which leads me to say a few words about what I have called "sexual selection." This form of selection depends not on a struggle for existence, but rather on a struggle between the individuals of one sex—generally the males—for mating with the opposite sex. The loser of the struggle in this case does not die; it simply fails to leave offspring, or at most leaves very few offspring.

Sexual selection is therefore less vigorous than natural selection. Generally, the most vigorous males—those that are best adapted for their places in nature—will most likely succeed in mating and leave the most offspring. But in many cases victory will depend not so much on general vigor, but rather on having special weapons that are found only on males. A hornless stag or a spurless rooster[2] would have a poor chance of leaving many offspring because females select males based on the size and quality of those characteristics. Sexual selection, by always allowing the victor to mate and produce offspring, must surely lead to males with indomitable courage, spurs of great length, and wings of particularly great strength to strike opponents, in nearly the same manner as the brutal cockfighter produces especially strong and fierce roosters (gamecocks) by carefully selecting only the strongest and most aggressive males for mating in each generation.

I don't know how far down the scale of nature this law of battle extends, but certainly male alligators have been described as fighting, bellowing, and whirling around like Indians in a war dance for the possession of females. Similarly, male salmon have been observed fighting all day long, and male stag beetles (Figure 4.4) sometimes bear wounds from the huge mandibles of other males. That inimitable observer of insect

---

[2] Male roosters, turkeys, and other members of the avian order Galliformes have a sharp, bony, rearward projection from the lower leg, called a spur. It is used for defense.

Figure 4.4 Male stag beetle (*Lucanus cervus*).

Figure 4.5 Male sockeye salmon (*Oncorhynchus nerka*) from the north Pacific Ocean.

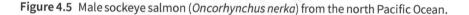

behavior, Monsieur Jean-Henri Fabre, has often seen the males of certain hymenopterous insects[3] fighting to possess a particular female while the female sits by watching, an apparently unconcerned beholder of the struggle who later shyly leaves with the winner for mating.

The battles are perhaps most severe between the males of polygamous animals,[4] which accordingly often possess special weapons for fighting. The males of carnivorous animals are already well armed, but even there sexual selection may lead to special means of defense, as with the mane of male lions and the hooked jaws of the male salmon (Figure 4.5)—for the shield may be as important for victory as the sword or the spear.

---

[3] Hymenoptera is an order of insects containing wasps, bees, and other species that have two pairs of membranous wings and an egg-laying organ in females (the ovipositor) that is specialized for stinging or piercing.

[4] Polygamous refers to a single male routinely mating with more than one female.

The contest for mates among birds tends to be more peaceful. All those who have studied the subject believe that the males of many species use their singing to compete intensely with each other in attracting females. Some birds, including the rock thrush of Guiana and the birds-of-paradise of New Guinea (Figure 4.6) gather together in large groups while successive males show off for the females—with the most elaborate care—their gorgeous plumage. They also perform strange antics before the females who, after standing by as spectators, eventually choose the most attractive partner. People who have paid careful attention to birds in confinement well know that they often show clear individual preferences and dislikes. Sir Robert Heron, for example, has described how a particular peacock with feathers of an unusually mottled coloration was eminently attractive to all of the hens in his menagerie. I can't go into all of the details here, but if a human breeder can in a short time select for beauty

Figure 4.6 Male bird-of-paradise (*Paradisaea minor*). Most members of this family (the Paradisaeidae) are found in New Guinea and its neighboring islands.

and an elegant posture in his miniature chickens (called bantams), according to his standard of beauty, then I can see no good reason to doubt that female birds, by selecting the most melodious or beautiful males according to their standard of beauty, might produce—over hundreds or thousands of generations—a marked change in those characteristics. Some well-known laws regarding the plumage of male and female birds of particular species in comparison with the plumage of their young can partly be explained through the action of sexual selection on variations occurring at different ages, variations that are then transmitted to the males alone or to both sexes at corresponding ages. Unfortunately I don't have space here to discuss this topic in greater detail.

I believe, then, that when the males and females of any animal species have the same general habits of life but differ in structure, size, color, or ornamentation, such differences have been mainly caused by sexual selection; that is, by individual males with certain specific characteristics having had, in many successive generations, some slight advantage over other males in gaining mates—either in their weapons, means of defense, or in their general charms—characteristics that they have then transmitted only to their male offspring. That doesn't mean that I would attribute *all* differences between the sexes to sexual selection. For example, we see in our domestic animals some peculiarities arising and becoming associated with the male sex alone that seem not to have been deliberately encouraged through selection by man. Similarly, it's hard to believe that the tuft of hair on the breast of the wild male turkey can be of any use, and I doubt that the female birds view it as attractive. Indeed, had the tufts appeared under domestication, they would have undoubtedly been called a monstrosity and been actively selected against.

Examples of Natural Selection, or the Survival of the Fittest, in Action

In order to clarify how I believe natural selection acts, I would like to give one or two hypothetical illustrations. Consider a wolf, which preys on a variety of other animals, capturing some by craft, some by strength, and some by fleetness. And let us also suppose that the swiftest prey—deer, for instance—had from some change in the surrounding countryside increased in numbers, or that other, slower prey had for some reason decreased in numbers, during that season of the year when the wolf had the greatest difficulty in finding food. Under such circumstances, wolves would have to focus all their attention on deer, and the swiftest and slimmest wolves would have the best chance of surviving and so be preserved (i.e., be selected for), as long as they remained strong enough to master their prey at this or some other time of the year, when they might be compelled to prey on animals other than deer. I can see no more reason to doubt that this would be the result than that human breeders should be able to improve the fleetness of their greyhounds by careful and methodical selection from generation to generation, or by that kind of unconscious selection that follows from each person trying to keep the best dogs without any thought of eventually modifying the breed, as discussed in Chapter 1. I may add that, according to Mr. James Pierce, author of "A Memoir on the Catskill Mountains" (1823), there are, in fact, two varieties of wolf inhabiting the Catskill Mountains in the United States—one with a light greyhound-like body shape, which pursues deer as prey, and the other more bulky, with shorter legs, which more often attacks the shepherd's sheep.

Note that in the previous illustration I spoke of the slimmest individual wolves having an advantage in that particular situation and not of any single strongly marked variation having been preserved. Although I have long appreciated the great importance of individual differences, it wasn't until I read a very interesting article in the *North British Review*, in 1867, that I appreciated how rarely such single variations could be perpetuated to future generations. The author of this review gives an example using a pair of animals that produce 200 offspring during their lifetime, of which, from various causes of destruction, only two on average survive to reproduce in turn. This is a rather extreme estimate for most of the higher animals but must be quite common for most of the lower forms. He then shows that if a single individual were born into the population that varied in some way that made it twice as likely to survive in comparison with individuals without that characteristic, the chances would still be very much against its survival: twice a very small probability is still a very small probability. Suppose that it did survive and reproduce, and that half of its offspring inherited the favorable variation. Even so, as the article goes on to show, the young would still have only a very slightly better chance of surviving and breeding, and this chance would continue to decrease in the succeeding generations.

I can't dispute the logic of those arguments. If, for instance, a bird of some kind could obtain its food more easily by having its beak curved, and if one were born with its beak strongly curved and flourished as a consequence, even so there would be a very poor chance that this one individual would survive long enough to mate and thus to pass this characteristic on to its offspring. On the other hand, there can be little doubt, judging from what we see taking place when humans selectively breed our domesticated animals and plants, that such beneficial characteristics would indeed follow from the preservation during many generations of *a large number* of individuals with heritable, strongly curved beaks and from the more common destruction of a still larger number of birds with the straightest—and therefore most disadvantageous—beaks.

Some have also suggested that mating between individuals with different characteristics will eliminate variations of all kinds in the next generation. I will have more to say about this later, but here let me just note that most animals and plants stay within a small area and do not needlessly wander about. We see this even with migratory birds, which almost always return to the same spot every year. Consequently, each newly formed variety would generally be at first local, as seems to be the common rule with varieties living in nature. Thus, similarly modified individuals would stay together in a small group and would often breed together. If the new variety was successful in the battle for life and for mates, it would then slowly spread from the central areas, competing with and conquering the unchanged individuals on the edges of an ever-increasing circle.

Let me give another, more complex illustration of how natural selection works, this time concerning plants. Certain plants excrete a sweet juice from specialized glands, a juice that apparently eliminates something injurious from the sap. Some legumes, for example, secrete such juices from specialized glands at the base of the stipules, while the common laurel has similar glands on the back of its leaves. Although the glands produce only small amounts of this juice, it is greedily sought by insects, whose visits have no apparent benefit for the plants.

Let's suppose that this same juice eventually came to be secreted from inside the flowers of a certain number of plants of some particular species. Insects would now get dusted with pollen while seeking the nectar and would then often end up transporting that pollen from one flower to another so that the flowers of two individuals of the same species would get mated, or "crossed." We know that such crossing gives rise to especially vigorous seedlings, which consequently would have the best chance of flourishing and surviving. The plants that produced flowers with the largest juice glands (often called "nectaries") would secrete the most nectar and thus be most often visited by insects. They would in consequence most often be crossed with the flowers on other plants and so, in the long run, would gain the upper hand against other local members of the species. Eventually they would form a local variety. Any individual flower that had its stamens and pistils placed so as to favor in any degree the transport of pollen by the particular insects that visited them would likewise be favored.

I could instead have used an example of insects visiting flowers to feed on pollen rather than nectar. As pollen is produced solely for the fertilization of eggs, its destruction by visiting insects would seem to be a clear and simple loss to the plant. And yet, if even just a little pollen were carried from flower to flower by the pollen-devouring insects, at first only occasionally and later more regularly, thus resulting in a cross between plants, then it might still be a great gain to the plant to be thus robbed of its pollen, even if 90% of the pollen were eaten by the insects beforehand. Those individuals that produced more and more pollen, and had larger anthers for its production,, would be selected for and come to be increasingly common in the population over many generations. If this process continued for long periods of time, then once our plant had become highly attractive to insects, those insects would routinely carry pollen from flower to flower, and without at all intending to do so. Indeed that is exactly what they do. I will give just one example, one which also illustrates a step in the separation of the sexes in plants.

Some holly trees bear only male flowers, each of which has four stamens (Figure 4.7) producing only a little pollen, and a rudimentary female portion, the pistil. Other holly trees bear only females flowers, with a full-sized pistil, but they also have rudimentary male parts: four stamens with shriveled anthers, in which not a grain of pollen can be detected. Having found a female holly tree exactly 60 yards from a male

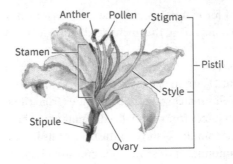

Figure 4.7 Parts of a flower.

tree, I looked at the stigmas of 20 flowers, taken from a number of different branches, under the microscope; on every single one of these flowers I saw at least a few pollen grains, and on some a profusion of grains. As the wind had been blowing for several days in the opposite direction, from the female tree to the male tree, the pollen could not have been carried to the female by the wind. Even though the weather had been cold and boisterous and therefore not especially favorable to bees, every female flower that I examined had been fully fertilized by the bees, which had flown from tree to tree in search of nectar.

But returning now to our imaginary case, as soon as the plant had been rendered so highly attractive to insects that pollen was regularly carried from flower to flower, another process might begin. No naturalist doubts the advantage of what has been called the "physiological division of labor." Thus we may believe that it would be advantageous for a plant to become more specialized and produce male parts alone—stamens—in one flower or on one whole plant, and female parts alone—pistils—in another flower or on another plant. When domesticated plants are placed under different environmental conditions, sometimes the male organs and sometimes the female organs do indeed become more or less impotent. Now if we suppose this to occur in ever so slight a degree in nature, then as pollen is already carried regularly from flower to flower, and as a more complete separation of the sexes of our plant would be very advantageous on the principle of specialization and division of labor, then those individuals showing this tendency more and more would be continually favored or selected for in each generation until at last a complete separation of the sexes might be achieved. I don't have the space here to show the various steps through which plants of various types are now in the process of separating the sexes, but I may add that, according to Dr. Asa Gray, the great American botanist at Harvard University, some species of holly in North America are in fact presently in exactly such an intermediate condition.

Let us now consider the nectar-feeding insects. Let's suppose that the plant for which we have slowly been increasing the nectar production through continued selection is a common one and that certain insects depend in large part on that plant's nectar for food. I could give many details showing how anxious bees are to save time when obtaining nectar. For example, they have a habit of cutting holes and sucking the nectar at the bases of certain flowers, which, with a very little more trouble, they could have entered by the "mouth" of the flower. Bearing such facts in mind, it is easy to believe that, under certain circumstances, individual differences in the curvature or length of the bee's proboscis, too slight to be even noticed by us, might benefit a bee or other insect so that certain individuals would now be able to obtain their food more quickly than others. Thus the communities to which they belonged would flourish and throw off many swarms inheriting the same peculiarities. Similarly, the tubes of the corolla of the common red and crimson clovers (*Trifolium pratense* and *T. incarnatum*, respectively) (Figure 4.8) do not obviously differ in length; and yet the honeybee (members of the genus *Apis*) can easily suck the nectar out of the crimson clover, but not out of the common red clover, which is instead visited only by bumblebees (genus *Bombus*). Thus entire fields of red clover offer in vain an abundant supply of precious nectar to the honeybee. And it's not that the honeybee dislikes the nectar of these flowers, for I have repeatedly seen, in the autumn, many honeybees sucking nectar from them through holes previously bitten in the base of the tube by bumblebees. The difference in the length of the corolla in the two kinds of clover—which determines whether or not honeybees can visit—must be very trifling, for I have been assured (although I have not myself made such an observation) that when red clover has been mown, the flowers of the second crop are somewhat smaller, and honeybees can then successfully visit these flowers for their nectar. I have also been told that the Ligurian bee, which is usually considered to be a mere variety of the common honeybee and which freely breeds with it, is able to reach and suck out the nectar of the red clover. Thus, in a region where this kind of clover was abundant, it might be a great advantage for honeybees to have a slightly longer or differently built proboscis. On the other hand, as the fertility of this red clover absolutely depends on bees visiting their flowers, if bumblebees were to become rare in any particular region it might be a great advantage to the plant to have a shorter or more deeply divided corolla, so that the honeybees should now also be able to suck nectar from the flowers. Thus I can understand how a flower and a bee might slowly become, either simultaneously or one after the other, modified and adapted to each other in the most perfect manner simply by the continued preservation of all those individual plants and bees in each generation that presented slight deviations of structure that were mutually favorable to each other.

I am well aware that my doctrine of "natural selection," as illustrated in the above imaginary examples, is open to the same objections that were initially leveled at the geologist Sir Charles Lyell's noble views on "the modern changes of the Earth, as illustrative of geology." But we now seldom hear people talking about erosion and uplifting—agencies that we still see at work on this planet—spoken of as being trifling or insignificant when used to explain the excavation of the deepest valleys on

(A)

(B)

Figure 4.8   (A) A bumblebee feeding on a red clover (*Trifolium pratense*). (B) Honeybee on crimson clover (*Trifolium incarnatum*).

our planet or the formation of long lines of inland cliffs. Natural selection acts only through the preservation and accumulation of small inherited modifications, each in some way beneficial to the preserved individual. Just as modern geology has almost banished such ideas as the excavation of a great valley by a single, ancient, huge wave of water, so will the principle of natural selection eventually banish anyone's belief in the continued creation of entirely new organisms, or of any great and sudden modification in their structure.

On the Outbreeding of Individuals

Here I must briefly change the subject a bit and say something about the importance of matings between individuals that are not closely related to each other.[5]

In the case of animal and plant species with separate sexes, it is easy to see why two individuals must usually come together to fertilize the eggs. The only exception is the curious and little-understood case of parthenogenesis, in which eggs can develop normally without being fertilized by sperm. But many organisms are hermaphrodites, with each individual having both male and female reproductive organs. For such hermaphroditic species, it is not at all obvious why two individuals must be involved in getting the eggs fertilized: Why shouldn't each individual simply fertilize its own eggs? And yet it seems that even among hermaphrodites, two individuals do at least occasionally come together for reproduction, as first suggested more than 50 years ago by the botanists Christian Konrad Sprengel, Thomas Andrew Knight, and Joseph Gottlieb Kölreuter. We will presently see in some detail why this is so important, but for now I will be brief. Among insects, vertebrates, and some other large animal groups, two individuals generally come together to produce offspring, even in hermaphroditic species. Yet many other hermaphroditic animals do not routinely pair, and the vast majority of plants are in fact hermaphrodites. What evidence do we have that two hermaphroditic individuals ever get together to reproduce?

Based on my extensive reading and many of my own studies, it is clear that crosses between different varieties of plants and animals greatly increase the vigor and reproductive capacity (i.e., the fertility) of their offspring, something that is also well-known by most professional breeders. The same result holds for matings between individuals of the same variety but belonging to different strains. It is also well understood that matings among closely related individuals ("inbreeding") reduce both vigor and fertility in the offspring. These facts alone make me think that it is a general law of nature that no living organism routinely fertilizes its own eggs generation after generation; crosses with another individual seem indispensable, even if they occur only infrequently.

Assuming that this is indeed a law of nature, we can, I think, understand several large classes of facts that would otherwise be inexplicable. For example, every plant hybridizer knows that exposing flowers to moisture reduces the chances of successful pollination (probably through the loss of pollen).

Why, then, do such a multitude of flowers have their anthers and stigmas fully exposed to the weather? And why do the plant's own anthers and pistil stand so near each other as almost to ensure self-fertilization? Such placement of these flower parts can really make sense only if it is essential for flowers to at least occasionally cross with other flowers: having the anthers and stamens fully exposed to the air increases

---

[5] Outbreeding, called "intercrossing" by Darwin and sometimes referred to as "outcrossing," refers to matings between individuals that are not closely related to each other, thus increasing genetic diversity within the lineage. Inbreeding, in contrast, refers to matings between closely related individuals

the likelihood that one flower's pollen will travel elsewhere and that the eggs of one flower will be fertilized by pollen from another.

On the other hand, many flowers have their reproductive organs well enclosed, as in the great "pea family" (the Fabaceae, formerly known as the Leguminosae) and so cannot be pollinated by wind. Such flowers, however, almost inevitably present beautiful and curious adaptations that promote visits by insects. Indeed, so necessary, for example, are the visits of bees to peas and many other members of this great family that preventing such visits greatly diminishes the plants' reproductive output. It is essentially impossible for insects to fly from flower to flower without carrying pollen from one flower to the other, to the great indirect benefit of the plant. I can ensure fertilization of such flowers just by touching a painter's camel's-hair pencil first to the anthers of one flower and then to the stigma of another; bees and other insects must act in the same way as they fly from flower to flower. This does not in any way suggest that bees produce a multitude of hybrids between distinct plant species, for if a plant's own pollen is placed on a stigma with pollen from another plant species, the plant's own pollen invariably destroys the influence of the foreign plant's pollen, as shown by the German botanist Karl Friedrich von Gärtner.

Intriguingly, flowers of the common barberry plant show a curious behavior that would seem to ensure self-fertilization: the stamens of the barberry flower either suddenly spring toward the pistil of the same flower, or they move toward it one after another. However, insects are often required to bring this springing behavior about, as shown by Mr. Kölreuter, and insects, as we have already noted, are excellent vehicles for transferring pollen from flower to flower. Indeed, although the members of this plant genus seem well-designed for self-fertilization through this clever contrivance, it is well known that if closely related forms are planted near each other it is almost impossible to raise pure seedlings, so often do they naturally cross.

In many other cases, not only is self-fertilization not promoted, but there are special contrivances that effectively prevent the stigma from receiving pollen from the same flower. This has been documented in the work of Mr. Sprengel and other botanists, and I have seen it myself. For example, in the cardinal flower (*Lobelia cardinalis*; see Figure 3.7A), there is a beautifully elaborate contrivance through which all of the incredibly numerous pollen grains are swept out of the fused anthers of each flower before the stigma of that particular flower is ready to receive them, thus nicely preventing self-pollination; as this flower is never visited by insects, at least in my garden, it never sets seed naturally. And yet, I can raise plenty of seedlings by simply transferring pollen from one flower to the stigma of another. A different species of *Lobelia* is visited by bees in my garden and seeds regularly without any help from me.

In many other cases, though, there is no special mechanical contrivance to prevent the stigma of one flower from receiving pollen from the same flower. And yet, as a number of botanists, including Mr. Sprengel and, more recently, Friedrich Hildebrand, have shown, and as I can confirm from my own studies, either the anthers burst and release their pollen before the stigma of that same flower is ready to be fertilized, or the stigma is ready to receive pollen before the pollen grains of that same flower are ready to be released; thus these plants in fact have functionally separated sexes and must be habitually crossed, even though they are, strictly speaking, hermaphroditic. It is the same with the reciprocally dimorphic and trimorphic plants previously alluded to, in which each plant bears two or three distinctly different sorts of flowers with the styles and stamens at different heights: one flower will, for example, have a short style and long stamens, while another flower will have a long style and short stamens. Such flowers must be crossed if they are to produce seeds. How strange are these facts! And how strange that the pollen and stigmatic surfaces of the same flower, though found so close together as if to ensure self-fertilization, should be in so many cases mutually useless to each other! But how simple it is to explain these facts if we understand the advantages, or even the indispensability, of at least occasional crosses with other individuals.

If several varieties of some plants, including the cabbage, radish, and onion, are allowed to seed near each other, most of the seedlings thus raised turn out to be combinations of the varieties—mongrels—as I have found through experiments in my own garden. For example, I raised 233 seedling cabbages from a number of plants of different varieties growing near each other; only 78 of those seedlings were true to their variety, and some of those were not even perfectly true. Yet the ovary-containing pistil of each cabbage flower is surrounded not only by its own six pollen-bearing stamens, but also by those of the many other flowers on the same plant. Moreover, the pollen of each flower readily gets on the stigma of the same flower without any help from insects; indeed, I have found that plants carefully protected from visits by insects produce the normal number of pods. How, then, is it possible that such a vast number of the seedlings are mongrels? It must arise from the pollen of a distinct variety having a greater potency than the flower's own pollen.[6] This is part of the general law that I discussed earlier, that substantial benefits must derive from outbreeding among distinct individuals of the same species. When distinct *species* are crossed,, the case is reversed: a plant's own pollen is almost always more potent than the pollen coming from flowers of a different species. I will return to this subject in a later chapter (Chapter 9).

In the case of a large tree covered with innumerable flowers, it may be objected that pollen could seldom be carried from tree to tree, and at most only from flower to flower on the same tree, thus promoting self-fertilization since flowers on the same tree can be considered as distinct individuals only in a limited sense. I believe that this objection is valid but that nature has largely provided against such self-pollination by giving to trees a strong tendency to bear flowers with separated sexes. When the sexes are thus separated, although male and female flowers may be produced on the same tree, pollen must be regularly carried from flower to flower for fertilization to occur; this will increase the likelihood of pollen being at least occasionally carried from tree to tree. I recently determined that, in England, trees belonging to all orders have their sexes more often separated than other plants do.

---

[6] Today we call this phenomenon "self-incompatibility."

Similarly, at my request Dr. Joseph Hooker, of the Royal Botanical Gardens, tabulated the trees of New Zealand, and Dr. Gray those of the United States, and the results were similar. On the other hand, Dr. Hooker informs me that the rule does not hold good in Australia. But if most of the Australian trees are dichogamous (i.e., having pistils and stamens that mature at different times), the same result would follow as if they bore flowers with separated sexes. I have made these few remarks on trees simply to call attention to the subject.

Let us turn briefly now to animals. Various terrestrial species, such as land snails and earthworms, are hermaphroditic. However, all of these animals pair for reproduction; I have yet to find a single terrestrial animal that can fertilize its own eggs. This remarkable fact, which offers so strong a contrast with terrestrial plants, makes perfectly good sense if an occasional cross with other individuals is indeed indispensable; there is nothing analogous to the action of insects or wind on plant reproduction by which an occasional cross could be achieved for terrestrial animals; it always requires the meeting of two individuals.

Self-fertilizing hermaphrodites are common among aquatic animals, but since they release their sperm into the water, water currents offer an obvious means for least an occasional cross between individuals. As in the case of flowers, I have so far failed—even after consulting Professor Thomas Henry Huxley, one of the highest authorities in biology—to discover a single hermaphroditic animal with its organs of reproduction so perfectly enclosed that it would be impossible for another individual to access them on occasion. For a long time, barnacles seemed to be a possible exception to this general belief, but I have now been able to prove, through a bit of good luck, that even here, two barnacles do sometimes cross with each other, even though both are normally self-fertilizing hermaphrodites.

It must have struck most naturalists as a strange anomaly that, among both animals and plants, some species are hermaphrodites while others within the same family (and some even within the same genus!) have separate sexes (i.e., they are dioecious), even though they are extremely similar in all other aspects of their biology. But if, in fact, all hermaphrodites do at least occasionally intercross, the difference between hermaphroditic and dioecious species is, as far as function is concerned, very small.

From these several considerations and from the many other facts that I have collected but don't have sufficient space to discuss here, it appears that with both animals and plants at least an occasional intercross between distinct individuals is a very general, if not universal, law of nature.

Circumstances Favoring the Production of New Forms Through Natural Selection

This is an extremely intricate subject. A great amount of variability among individuals will obviously favor the process of natural selection by offering traits to be selected for or against. Having a large number of individuals in a population is, I believe, also highly important for success by making it more likely that useful variations will appear within any given period and compensating for lesser amounts of variability within each individual. Though nature grants long periods of time for natural selection to work, she does not grant an indefinite period: as all living beings are competing with each other for space, shelter, and food, if any one species does not become modified and improved over time as much as its competitors are so modified and improved, it will go extinct. And of course unless favorable variations are inherited by at least some of each individual's offspring, natural selection can accomplish nothing.

In the case of a human breeder methodically selecting for particular traits, his work will completely fail if the individuals are allowed to freely intercross with each other. But when many men, without intending to alter the breed at all, have a very similar standard of perfection in mind that they all work toward, and all try to procure and breed from only the best of their animals, improvement surely but slowly follows from this unconscious process of selection even when no special effort is made to prevent interbreeding with inferiors.

It will be the same in nature, for within any confined area if there is some niche not already perfectly occupied by some animal or plant species, all individuals varying in the proper direction, regardless of degree, will more likely be preserved and reproduce. But if the area is large, its various districts will almost certainly present different environmental conditions; if so, any newly formed varieties will interbreed with others on the edges of each such district. But we shall see in Chapter 6 that intermediate varieties, living in intermediate districts, will in the long run generally be supplanted by one of the neighboring varieties.

Outbreeding will chiefly affect those animals that unite for each birth and wander about a good deal, and which do not reproduce very quickly. Thus with birds and other animals of this nature, particular varieties will generally be confined to rather large areas that are particularly far apart and at least somewhat isolated, located, for example, in separated countries. Indeed, I find this to be the case. With hermaphrodites and organisms that unite for each birth but move about less and leave many offspring, a new and improved variety might be quickly formed in any small area and might first maintain itself there within a small group and later spread further, so that members of the new variety would mate mostly with each other. Thus we are likely to see more varieties occurring over much smaller distances. Following this principle, nurserymen always prefer to save the seed produced by a large number of plants of any particular variety, thus reducing the chance of outcrossing with other varieties.

Even with animals that come together for each birth and that do not increase their numbers quickly, we must not assume that free outbreeding among individuals would always take place and eliminate the effects of natural selection. Indeed, I now have a considerable body of facts showing that, within the same area, two varieties of the same species may long remain distinct from each other just by living in slightly different places within that area or from breeding at slightly different seasons or from the individuals of each variety preferring to mate with others within that variety.

Outbreeding is important in nature, in keeping individuals of the same species, or the same variety, true and reasonably uniform in character over time. Obviously it will act far more efficiently with those animals that unite for each birth, but, as I stated earlier, there is good reason to believe that all animals and all plants exhibit outbreeding at least on occasion. Even if this happens only infrequently, the young thus produced will gain so much in vigor and fecundity compared with the offspring from long-continued self-fertilization that they would have a better chance of surviving and then propagating their kind. Thus the influence of crosses between individuals will always be great, even if they happen rarely.

Isolation also plays an important role in allowing species to be modified through natural selection. In any small isolated area, the conditions of life will generally be almost uniform; thus natural selection will tend to modify all of the varying individuals of any one species in the same direction. This will deter crossbreeding with inhabitants of surrounding areas. The German explorer and natural historian Moritz Friedrich Wagner has recently shown that the role of isolation in preventing crossbreeding between newly formed varieties is probably even greater than I had imagined, although for reasons already noted, I disagree with his suggestion that physical isolation is essential for the formation of new species.

Isolation is also important in preventing potentially better adapted organisms from immigrating into an area after environmental or physical conditions have changed— following a change in climate, for example, or a change in elevation of the land. Thus any newly created niches will remain open, to be filled eventually by the gradual selection of suitable variants from descendants of the original populations. Finally, isolation from other areas will allow plenty of time for a new variety to be improved gradually, at a very slow rate. On the other hand, to be effective at promoting evolution by natural selection, the isolated area cannot be too small for then the total number of inhabitants living in that area will be small as well; this will greatly slow the production of new species through the process of natural selection, by decreasing the chances of favorable variations arising in the population in the first place: if there is no variation, there is nothing to be selected for or against.

Note that the mere passage of time alone does not guarantee that species will be modified. Organisms do not change their traits automatically over time, through any innate law. The passage of time simply increases the likelihood that beneficial variations will eventually arise within a population and be selected for, accumulate, and eventually become fixed.

Now let us see how all of these remarks about isolation apply in nature, by considering what can happen on any particular small, isolated area such as an oceanic island. Although only a small number of species will be found living on that island, as we shall see in a later chapter on Geographical Distribution (Chapter 13), a very large proportion of those species will be found only there and nowhere else. Thus oceanic islands might at first seem especially likely to produce new species. We shouldn't jump to that conclusion though: we really can't know whether small isolated areas like islands or large open areas like continents have been most favorable for producing new varieties and species unless we know the amount of time taken to do so in each case—and this is something that we simply do not know.

Although isolation is very important in producing new species, I think that the largeness of the area involved is even more important, especially for producing species that can endure for long periods of time and spread widely. Not only will favorable variations be more likely to appear in a large and open area simply because of the greater number of individuals living in such areas, but the diversity of habitats and niches will also be more complex in such an area because of the large number of species that already live there. If some of those many species become modified and improved over time, that will favor modification and improvement in the species that they interact with as well; those that fail to improve under such pressure will eventually go extinct.

Each new form, once it has become substantially improved, will be able to spread over the large open and continuous area and thus will come into competition with many other forms. Moreover, areas that are now large and continuous will often have been substantially fragmented in the past, owing to previous rises and declines in sea level and in the elevation or subsidence of land, and thus have benefitted from the helpful effects of isolation long ago.

Finally, my reasoning suggests that although small isolated areas have in some respects favored the production of new species, yet such modification will generally have been quicker on larger areas like continents. Even more importantly, the new forms produced on larger areas of land will already have been victorious over many competitors, and will thus spread the most widely and give rise to the greatest number of new varieties and species; accordingly, they will play a more important role in the changing history of the organic world.

This superior competitive ability of forms that have developed on large continents is clearly seen in a number of situations. The species of the smaller continent of Australia, for example, are now being outcompeted by those introduced from the larger European and Asiatic areas, as discussed more fully later in my chapter on Geographical Distribution (Chapter 12). The superior competitive ability of species produced on large continents also explains why such species have so often become dominant after being introduced onto islands. On a small island, competition will have been less severe, and, consequently, there will have been less subsequent modification and extinction. We can understand, then, why the flora presently found on the island of Madeira resembles, according to the Swiss paleobotanist Oswald von Heer, the now extinct European flora from the Tertiary period (about 65 million to 2 million years ago): weaker forms that have persisted on the islands have been outcompeted to extinction on the larger continents.

Similarly, competition among freshwater organisms should also have been relatively less severe than elsewhere, because all freshwater basins taken together make a far smaller area than that of the ocean or the land. Consequently, without the impetus of severe competition, new forms should have been produced more slowly in freshwater environments and gone extinct more slowly as well. Not surprisingly then, in freshwater habitats we find seven genera of primitive ganoid fishes, with their odd, thick bony scales, the last remnants of a once far-more-common order. And it is in freshwater, too, that we find some of the strangest forms known in the world: the platypus (Figure 4.9A), which belongs to the mostly extinct family Ornithorhynchidae, and the lungfish (Figure 4.9B), a member of the order Lepidosireniformes. In marine habitats, most such forms have been driven into extinction. These anomalous freshwater forms can be considered as "living fossils"; they have survived to the present day by inhabiting a confined area and from having been exposed to a smaller variety of—and therefore less severe—competitive interactions.

Let me now summarize the circumstances that favor, and those that impede, the production of new species through the process of natural selection. For terrestrial species it seems that a large continental area, and one that has undergone many oscillations of level, will have been most likely to have produced many new life forms, fitted to endure for a long time and to spread widely. While the area was part of a continuous large continent, there will have been many kinds of individuals and many individuals of each kind, creating severe competition among the various forms. When these large areas were converted by the sinking of land or the rising of sea level into separate large islands, each of those islands will have supported many individuals of the same species. Matings between individuals on the edges of each species range will have been impossible, and, after physical changes of any kind, immigration from the outside would not have been possible either. Thus any new niches opening up on each island would have to have been filled through the gradual modification of the original inhabitants, and there will have been sufficient time for the varieties in each to have become well modified and perfected. When the islands were later reconverted into a

(A)   (B)

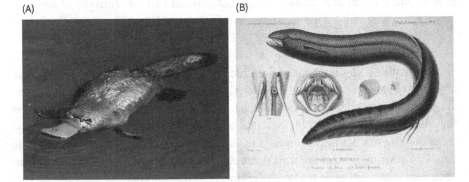

Figure 4.9 (A) Australian duck-billed platypus swimming. (B) South American lungfish (*Lepidosiren paradoxa*). In this and in most other lungfish species, the gills are incapable of substantial gas exchange; instead, each fish has two lungs and breathes air. The Australian lungfish has only a single lung but is still able to use its gills in respiration.

large continuous continental area, by elevation of the land or a lowering of sea level, competition will again have become intense: the best adapted and most improved varieties will have been able to spread, while the less improved forms will have been outcompeted to extinction. The relative proportions of the various forms on the reconstituted continent will again have been changed. And once again there will have been a fair field for natural selection to act in improving the inhabitants still further and thus in producing new species.

I fully admit that natural selection generally acts extremely slowly and can act only when some niches in an area can be better occupied by the modification of some of the existing inhabitants of that area. The creation of such areas will often depend on physical changes, which generally take place very slowly, and on the immigration of better adapted forms into the area being in some way prevented. As a small number of individuals become modified, the interactions with other inhabitants will often be disturbed. This will in turn create new niches, ready to be filled up by better adapted forms. But all of this will take place very slowly. Although every individual differs in some slight degree from other members of its species, it would often be a long time before differences of the right sort in particular traits appeared in a population. And such progress would often be greatly slowed by free crossbreeding.

Many will exclaim that these several limitations are sufficient to neutralize the power of natural selection. I, however, do not believe so. But I do believe that natural selection will generally act very slowly and only on a few of the inhabitants of any given region. I further believe that these slow, intermittent results agree well with what geology tells us about the rate and manner at which the inhabitants of the world have in fact changed over long periods of time.

Slow though the process of natural selection may be, if feeble humans can accomplish as much as we have by artificial selection, I can see no limit to the amount of change, or to the beauty and complexity of the co-adaptations between all organic beings, one with another and with their physical conditions of life, which may have been brought about in the long course of time through nature's power of selection—that is, through survival of the fittest.

Extinction Caused by Natural Selection

I will discuss this subject more fully in a later chapter on geology (Chapter 11), but as extinction is intimately connected with natural selection I must say something about it here as well.

Natural selection acts solely by preserving variations that are in some way advantageous to the organisms that possess them, and which consequently endure. Owing to the exponential rate of increase of all living organisms, each habitat is already fully stocked with inhabitants. It follows, then, that as the favored forms increase in number, so, generally, will the less favored forms decrease and eventually become rare. *Rarity, as geology tells us, is the precursor to extinction.* We can see that when any form is represented by only a few individuals it will run a good chance of complete extinction, for example during a time of great climate fluctuation or from a temporary increase in the number of its enemies. But we may go further than this: as new forms are produced, many old forms *must* become extinct, unless we assume that specific forms can increase in numbers indefinitely. But geology plainly tells us that the total number of species has not increased indefinitely over time. I shall presently attempt to show why the number of species throughout the world has not become immeasurably great.

We have seen that species with the greatest number of individuals also have the best chance of producing favorable variations within any given period; if only a very small percentage of variations are favorable, then such variations are not likely to appear in populations with only a small number of individuals. And I showed in Chapter 2 that it is the most common and widespread or dominant species that presently show the greatest number of recorded varieties. Thus species with relatively few individuals (the so-called rare species) will be less quickly modified or improved within any given period and, in consequence, will probably be beaten in the race for life by the modified and improved descendants of the more common species.

From these several considerations, I think it inevitably follows that as new species are formed over time through natural selection, others will become rarer and rarer, and finally go extinct. The closest competitors with those undergoing modification and improvement will naturally suffer the most. And as seen in the chapter on "The Struggle for Existence" (see Chapter 3), it is the most closely allied forms—varieties of the same species and species of the same genus or of related genera—that, from having nearly the same structure, physiologies, and habits, generally come into the severest competition with each other. In consequence, each new variety and each new species, during the progress of its formation, will generally press hardest on its closest relatives and tend to exterminate them.

We see the same process of extermination among our own domesticated animals and plants, through the selection of improved forms by humans. I could give many curious instances showing how quickly new breeds of cattle, sheep, and other animals—and new varieties of flowers as well—completely take the place of older and inferior kinds that simply have appealed less to their human breeders. In Yorkshire, England, we know with certainty that the ancient black cattle were displaced by long-horn cattle,, and that these were, as noted so nicely by William Youatt, "swept away by the short-horns as if by some murderous pestilence." Extinction is an inevitable part of the process of selection.

Divergence of Character

The principle of "divergence of character" is extremely important to my argument. I believe it explains several important facts. In the first place, although

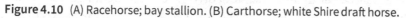

Figure 4.10 (A) Racehorse; bay stallion. (B) Carthorse; white Shire draft horse.

varieties—even strongly marked ones—may have somewhat the character of species, they differ far less from each other than do good and distinct species. But if
varieties are incipient species, as I have suggested—new species in the making—
then how does the smaller difference between varieties eventually become enlarged
into the greater difference between species? That this does routinely happen we
must infer from this fact: most of the innumerable species throughout nature are
very clearly different from each other, whereas varieties—the supposed prototypes
and parents of future well-marked species—present only slight and ill-defined differences. Mere chance might cause one variety to differ in some character from its
parents and, by chance, cause the offspring of this variety again to differ from its
parents in the very same character and to a greater degree; but this alone would
never account for so habitual and large a degree of difference as that seen between
species of the same genus. So how can we explain this transformation of a variety
into a new species?

  As usual, I have turned to our domestic animals and plants for some help in answering this question. The situation is fully analogous. Clearly, the production of
races so different from each other as, for example, shorthorn and Hereford cattle,
or racehorses and cart horses (Figure 4.10), or the various breeds of pigeons (see
Chapter 1), could never have been brought about through the mere chance accumulation of similar variations in each of many successive generations. In practice, a pigeon fancier is perhaps struck by one particular pigeon having a slightly shorter beak
than others, while another fancier is struck by a pigeon having a rather longer beak;
on the acknowledged principle that "fanciers do not and will not admire an average
standard, but like extremes," they both go on choosing and breeding from birds with
longer and longer beaks, or from birds with shorter and shorter beaks, generation
after generation after generation. This has certainly happened with the sub-breeds of
the tumbler pigeon.

Similarly, we may suppose that, at an early period of history, the men of one nation or district required swifter horses, while those of some other place required stronger and bulkier horses. The horses would not look very different from each other early in the process. But over time, from the continued selection of the fastest horses in the one case and of stronger ones in the other case, the differences would be greater, and greater, and greater, and eventually we would end up with two sub-breeds. Ultimately, after the lapse of centuries, these sub-breeds would become converted into two well-established and distinct breeds. As the differences became greater, the inferior animals with intermediate traits, being neither especially swift nor very strong, would not have been used for breeding and thus will have tended to disappear from the population. Here, then, we see in man's productions the action of what may be called "the principle of divergence" causing differences—at first barely detectable—to steadily increase with each generation and the breeds to then diverge more and more in character over time, both from each other and from their common parent.

But how, it may be asked, can such a principle apply in nature, without the involvement of selection by humans? I believe that it does indeed apply and does so most efficiently, although it was a long time before I saw how. It follows from this simple circumstance: the more the descendants of any one species become diversified in structure, physiology, and habits, the more will they also be better enabled to seize on niches and habitats not previously occupied by other members of that species, and so be enabled to increase in number.

We can clearly see this in the case of animals with simple habits. Take the case of some carnivorous quadruped whose population size has long ago reached the maximum that can be supported in some particular place. Unless there is some change in the climate or some other conditions in the area, then if its natural power of increase is allowed to act, the animals can increase in numbers only if their varying descendants are able to seize on places presently occupied by other animals: some of these descendants being enabled to feed on new kinds of prey, either dead or alive, for example; some inhabiting different lifestyles than their parents—climbing trees or frequenting water, or even becoming less carnivorous and feeding more on plant material. The more diversified in habits and structure the descendants of our carnivorous animals become, the more new niches they will be enabled to occupy.

What applies to one animal will apply throughout all time to all animals, as long as they vary in individual characteristics—for otherwise natural selection cannot act, and can accomplish nothing.

The same arguments apply to plants. Experiments show that if a plot of ground is sown with just one species of grass, and a similar plot is sown with seeds of several grass species belonging to a number of distinctly different genera, a greater number of plants and a greater total weight of dry plant material can be raised in the diversified plot than in the single-species plot. Imagine, then, that we have a species of grass that continues to vary and that there is a gradual selection for those varieties that come to differ from each other—even to only a slight degree—in the same manner as do the members of distinct species and genera. Over time, then, a greater number of individual plants of this species, including its modified descendants, would succeed in living on the same piece of ground. And we know that each species and each variety of grass is every year releasing almost countless numbers of seeds and is thus striving to the utmost, in a sense, to increase in number. Consequently, in the course of many thousands of generations, the most distinct varieties of any one species of grass would have the best chance of surviving, reproducing, and increasing in numbers, and thus of overwhelming and replacing the less distinct varieties of that same species. And once varieties become very distinct from each other, they eventually earn recognition as separate species.

The truth of this basic principle—that the greatest amount of life in any area can be supported by great diversification of structure and function—is seen under many natural circumstances. In an extremely small area, especially if it is freely open to immigration, the contest between individuals must be very severe; in such situations we always find great diversity in its inhabitants. For instance, I found that a small piece of turf, only 3 feet wide and 4 feet long, that had been exposed to the same conditions year after year for many years, supported 20 species of plants belonging to 18 different genera and eight different orders, which shows just how much these plants differed from each other. And so it is with the plants and insects on small and uniform islands or in small freshwater ponds. Similarly, farmers find that they can raise the most food on any given area of land by periodically sowing the seeds of plants belonging to the most different orders; nature similarly follows what may be called a simultaneous crop rotation. Most of the animals and plants that live together on any small piece of suitable ground could live on it, and may be said to be striving to the utmost to live there; but where they come into the closest competition, the advantages of diversification of structure, with the accompanying differences of habit and constitution, determine that the inhabitants that jostle each other the most closely shall, as a general rule, show great variation in how they live and thus belong to what we call different genera and different orders.

The same principle is seen in the naturalization of plants through human intervention in foreign lands. You might think that the invasive plants that have become successfully established in any new land would generally be species that are closely related to the indigenous, native species, particularly if the native species had been specially created for life in its own country. Wouldn't successful invaders then be expected to share similar traits? We might also have expected that successfully introduced non-native plants would belong to a few groups more especially adapted to certain conditions in their new homes. But the actual case is quite different; as the Swiss botanist Alphonse de Candolle has well remarked in his great and admirable work, successful invasions increase diversity in an area in proportion to the number of the native genera and species, and far more *in new genera* than in new species; thus plant diversity in the area is increased greatly through these invasions. Differing characteristics permit long-term coexistence.

To give but one example, the most recent edition of Dr. Gray's *A Manual of the Botany of the Northern United States* (1856), lists 260 naturalized plant species[7] belonging to 162 different genera. These non-native plants differ considerably from the indigenous species: at least 100 of the 162 successfully invading genera are not represented at all among the native species in that area, thus greatly increasing the number of plant genera now living in the United States and increasing plant diversity enormously.

Considering the nature of the introduced or invasive plants and animals that have struggled successfully with the native species in any area and become "naturalized" (i.e., become permanent members of those communities) may help us to understand how some of the native species would now have to become modified if they were to gain an advantage over their new compatriots. We may at least infer that substantial diversification of structure, at the level of generic differences, would benefit them.

Diversification of structure among the inhabitants of the same region is advantageous in the same way that physiological division of labor among the various organs of the body is advantageous—a subject so well elucidated by that expert on the biology of lower animals Henri Milne-Edwards. No physiologist doubts that if an animal's stomach is specifically adapted for digesting vegetable matter, the organism possessing that stomach will draw most of its nutrition from that material. On the other hand, a stomach that is specifically adapted for obtaining nutrients from flesh alone will draw most of its nutrients from those materials. Thus in the general economy of any land, the more widely and perfectly the animals and plants of that area are specialized for different habits of life, the greater the total number of individuals that will be able to support themselves there by avoiding competition. A set of organisms with their organization only slightly diversified could hardly compete successfully for long with a set of organisms that were more perfectly diversified in structure. For example, the Australian marsupials are divided into groups differing fairly little from each other and represent our carnivorous, ruminant, and rodent mammals only feebly; we can surely doubt whether those marsupials would be able to compete successfully with members of these well-developed orders if they were introduced to the United Kingdom, or vice versa. In the Australian mammals we can see that the process of diversification is still an early and incomplete stage of development.

Effects of Natural Selection on the Descendants of a Common Ancestor, Through Divergence of Character and Extinction

After the previous, unfortunately brief discussion, we may assume that the modified descendants of any one species will succeed so much better as they become more

---

[7] Naturalized species are non-native plants: exotic plants that escaped from cultivation after being deliberately introduced to an area and that are now living on their own in their new habitat.

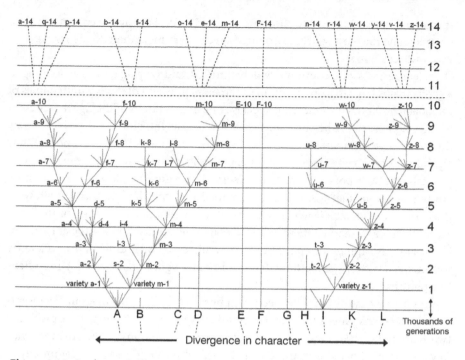

Figure 4.11 A schematic illustration of how morphological (and physiological) diversity increases over time. Organisms close to each other along the x-axis resemble each other more than do organisms further apart. The y-axis represents time, with time increasing from the bottom of the graph upward, representing thousands of generations or even more.

diversified in structure (and function) and thus become enabled to encroach upon the habitats and lifestyles presently occupied by other beings. Now let us see how this basic principle of benefitting from a divergence in character tends to act in combination with the principles of natural selection and extinction.

Figure 4.11 helps us to visualize this rather perplexing subject. The capital letters A through L[8] along the x-axis represent the 11 species of some single genus that is widely represented in its own country. These 11 species differ from each other to different extents, as is so often the case in nature, and so I have placed them at different distances away from each other on the diagram. The distance between species A and B, for example, is much smaller than that between species D and E, indicating that species A and B are more similar to each other than are species D and E. Species A and D are obviously even more different from each other. I have said that all of these species belong to a large genus because, as we saw in Chapter 2, more species, on average, vary in large genera than in genera with relatively few species. Moreover, the varying

---

[8] Darwin skipped the letter J in his diagram of 11 species, so we have skipped it as well.

species of large genera tend to show more varieties. We have also seen that the most common and the most widely distributed species vary more than do the rare species with restricted ranges.

So let species A be a common, widely diffused, and varying species belonging to a genus that is large in its own country. The branching and diverging dotted lines of unequal lengths that extend from species A near the bottom of the figure represent its varying offspring. The variations are extremely slight, but very diversified; they do not all appear simultaneously, but often after long intervals of time, nor do they all endure for the same amounts of time. Only those variations that are in some way profitable to their possessor will be preserved or naturally selected because those are the variations that will minimize competition with the "parents" by allowing those offspring to exploit new resources. If a dotted line reaches one of the horizontal lines and is marked there by a lowercase numbered letter, a sufficient amount of variation is supposed to have been accumulated by then to form it into a fairly well-marked variety—something that would be thought worth recording in a scientific publication.

Suppose that the distance between each horizontal line in the diagram represents 1,000 generations. After 1,000 generations, then, species A is shown to have produced two fairly well-marked varieties, namely varieties a-1 and m-1. Since the tendency to vary is in itself hereditary, these varieties will also tend to vary, and commonly in nearly the same manner as did their parents. Moreover, these two varieties, being only slightly modified forms, will tend to inherit those advantages that made their parent (species A) more successful and numerous than most of the other inhabitants in the same region and will also partake of those more general advantages that made the genus to which the parent species belonged a large genus in its own region. All these circumstances favor the production of new varieties.

If, then, these two varieties continue to vary, the most divergent of their variations will generally be preserved during the next thousand generations. And, after this interval, you see that variety a-1 in our diagram has now produced variety a-2, which will, owing to the principle of divergence, differ more from ancestor A than did variety a-1. Variety m-1 in the diagram is seen to have produced two new varieties, namely s-2 and m-2, which will differ not only from each other but even more so from their common parent, ancestor A. We may continue this process by similar steps for any length of time, with some of the varieties after each 1,000 generations producing only a single variety (e.g., a-6) but in a more and more modified condition; with some producing two or three varieties (e.g., the decendants of m-1); and some failing to produce any new varieties at all (e.g., species D and G). Thus the varieties or modified descendants of the common parent, ancestor A, will generally go on increasing in number and diverging in character. Figure 4.11 summarizes this process to the ten-thousandth generation and, under a condensed and simplified form, beyond that up to the fourteen-thousandth generation at the top of the figure.

I do not suppose that the actual process ever goes on as regularly as is shown in the diagram, nor that it goes on continuously. It is far more likely that each form remains unaltered for long periods of time and then again undergoes modification for a time. Nor do I suppose that it is always the most divergent varieties that are preserved: a medium form may well endure for a long time and may or may not produce more than one modified descendant. Natural selection will always act according to the nature of the places that are either unoccupied or not perfectly occupied by other beings, and this will depend on infinitely complex relations and interactions. But, as a general rule, the more that the descendants of any one species can become increasingly diversified in structure, the more places they will be enabled to succeed in and the more their modified progeny will increase in numbers. In our diagram, the line of succession is broken at regular intervals by small numbered letters marking the successive forms that have become sufficiently distinct to be recorded as varieties. But these breaks are imaginary: I might have inserted them anywhere, after intervals long enough to allow a considerable amount of divergent variation to accumulate.

All the modified descendants from a common and widely distributed species belonging to a large genus will tend to benefit from the same advantages that made their parents successful in life; thus they will generally go on multiplying in number as well as diverging in character. This is shown in our figure by the several diverging branches proceeding from species A. The modified offspring from the later and more highly improved branches in the lines of descent will often take the place of the earlier and less improved branches and so destroy them: this is represented in the diagram by some of the lower branches simply ending before they reach the upper horizontal lines. In some cases the process of modification will probably be confined to a single line of descent and the number of modified descendants will not be increased, even though the amount of divergent modification may have been great. Such a case would be represented in the diagram if all the lines proceeding from ancestor species A were removed except for the one extending from a-1 to a-10. In the same way, the English racehorse and the English pointer dog (Figure 4.12) have both apparently gone on slowly diverging in character from their ancestral stocks without either one having given off any fresh branches or races.

After 10,000 generations in our diagram, species A has produced three forms (a-10, f-10, and m-10) which, from having diverged in character during the succeeding hundreds of generations, will have come to differ largely, although perhaps unequally, both from each other and from their common parent. If we suppose the amount of change between each horizontal line in our diagram to be extremely small, these three forms may still be only well-marked varieties, so that we would now have, for example, variety a-10 and variety m-10. But suppose that the steps in the process of modification had been more numerous or greater in degree; in that case, these three forms will likely have been converted into either doubtful/questionable species or perhaps even into well-defined species (i.e., a-10 and m-10 might now be distinct species). **Thus the diagram illustrates the steps by which the small differences distinguishing varieties can slowly increase to become the larger differences distinguishing species.** By continuing the same process for an even greater number of

Figure 4.12 English pointer.

generations (as shown at the top of the diagram in a simplified manner), we get eight species, marked by the letters between a-14 and m-14, all descended—over many thousands of generations—from species A. This, I believe, is how species are multiplied and genera are formed over time.

In a large genus, more than one species would probably vary. I have assumed in Figure 4.11 that after 10,000 generations a second species (species I) has produced, by analogous steps, either two well-marked varieties (w-10 and z-10) or two species, according to the amount of change that is represented between the horizontal lines. After 14,000 generations, six new species, marked by the letters n-14 to z-14, have been produced. In any genus, the species that already differ greatly in character from each other will tend to produce the greatest number of modified descendants for these will have the best chance of seizing on new and widely different niches. In the diagram I have therefore chosen the extreme species (A) and the nearly extreme species (I) as those that have varied the most and have given rise to new varieties and species. The other nine species in our genus, marked by the other capital letters, may continue to transmit unaltered descendants for long but unequal periods of time; this is shown in the diagram by the dotted lines unequally prolonged upward (species D, E, F, and G, for example).

But, during the process of modification that is represented in the diagram, another of our principles, namely that of extinction, will also have played an important part. In any stable area that is fully stocked with species, natural selection acts by the selected form having some advantage in the struggle for life over other forms. Thus there will be a constant tendency for the improved descendants of any one species to supplant and exterminate their predecessors in each stage of descent, including the form that first gave rise to them. Recall that competition will generally be most severe between those forms that are most nearly related to each other in habits, physiology, and structure. Thus all the intermediate forms between the earlier and later stages—that is, between the less improved and more improved states of the same species—as well as the original parent species itself, will generally tend to eventually become extinct. So it probably will be with many whole collateral lines of descent, which will eventually be conquered by later and improved lines. On the other hand, the modified offspring of a species might end up living in a different area or become quickly adapted to some quite new niche in the original area, so that offspring and parents do not come into competition; if so, then both may continue to coexist.

If we now assume that our diagram represents a considerable amount of modification, the original species A and all the earlier varieties will have become extinct, having been replaced by eight new species: a-14 to m-14. Similarly, species I is now extinct, but it has been replaced by six new species, n-14 through z-14.

But we may go still further. The various original species in our genus were supposed to resemble each other to different degrees, as is so generally the case in nature: species A is more closely related to species B, C, and D than to the other species in our diagram; and species I is more closely related to species G, H, K, and L than to the others. Species A and species I were also supposed to be very common and widely distributed species, so that they must originally have had some advantage over most of the other species of the genus. Their modified descendants, 14 by the fourteen-thousandth generation in the hypothetical scenario of Figure 4.11, will probably have inherited some of these same advantages: they have also been modified and improved in a diversified manner at each stage of descent, so as to have become adapted to many related places in the natural economy of their country. It seems extremely probable, then, that they will have taken the places of—and exterminated—not only their parents A and I, but also some of the original species that were most nearly related to and most like their parents. Thus, very few of the original species will have been successful in transmitting offspring to the fourteen-thousandth generation. In our diagram, only one (species F) of the two species (species E and F) that were least closely related to the other nine original species has transmitted descendants to this late stage of descent.

In our diagram, 15 species have now descended from the original 11 species. Owing to the divergent tendency of natural selection, the extreme amount of difference in character between species a-14 and z-14 will be much greater than that between even the most distinct of the original 11 species. The new species, moreover, will be allied to each other in a widely different manner. Of the eight descendants from species A, the three marked a-14, q-14, and p-14 will be closely related since they have all recently branched off from their common ancestor, a-10; b-14 and F-14 will also be closely related, from their having diverged longer ago from a-5, but will be in some degree distinct from the three first-named species; and finally, although species o-14, e-14, and m-14 will also be closely related to each other, they will differ greatly from the other five species, having diverged at the very beginning of the process of modification. Indeed, species o-14, e-14, and m-14 may now constitute a separate subgenus or even a distinct genus.

The six descendants from species I will form two subgenera or genera. But as the original species I differed so largely from species A, standing nearly at the extreme end of the original genus on our chart, the six descendants from ancestor I will, owing to inheritance alone, differ considerably from the eight descendants of ancestor A. Moreover, the two groups are shown in the figure to have gone on diverging in different directions. The species that originally had characteristics that were intermediate between those of species A and species I have all gone extinct, except for species F, and left no descendants. This is a very important point. Thus the six new species descended from ancestor I, and the eight descendants from ancestor A, will now have to be ranked as being in very distinct genera, or even in distinct subfamilies.

This is, I believe, how two or more genera are eventually produced by descent with modification over many, many generations from two or more species of the same genus. And the two or more parent species are presumably descended from some one species of an earlier genus. In our diagram, this is indicated by the broken lines beneath the capital letters, converging in sub-branches downward toward a single point; this point (e.g., the point formed by following the lines down from species A, B, C, and D) represents what is presumably the ancestral species that eventually gave rise to our several new subgenera and genera.

It is worth reflecting for a moment on the character of the new species F-14, which has not diverged much in character and has retained the form of the original species, species F, either unaltered or altered only slightly. In this case, its affinities to the other 14 new species will be of a curious and circuitous nature. Being descended from a form that stood between the parent species A and I, now supposed to be extinct and unknown, it will now in some degree have characteristics that are intermediate between those of the two groups descended from these two species. But as these two groups have gone on diverging in character from the prototype of their parents, the new species (F14) will not be directly intermediate between the two, but rather between *types* of the two groups; every naturalist will be able to call such actual cases to mind.

In Figure 4.11, I suggested that each horizontal line represented 1,000 generations. But each line may instead represent a million generations, or even more. Each line may also represent a section of the successive layers of the Earth's crust including remains of organisms that are now extinct. We will refer again to this subject when we come to our chapter on geology (Chapter 11), and I think that we will then see that this diagram also throws considerable light on the relationships among extinct organisms, which though generally belonging to the same orders, families, or genera represented by those now living, yet are often at least somewhat intermediate in character between the members of existing groups. This makes good sense when we consider that the extinct species lived at various remote epochs when the branching lines of descent had diverged less from each other than they do now.

I see no reason to limit the process of modification that I have just explained to the formation of genera alone. If, referring again to Figure 4.11, we suppose the amount of change represented by each successive group of diverging dotted lines to be very great, then those forms marked a-14 to p-14, those marked b-14 and f-14, and those marked o-14 to m-14, will form three very distinct genera. We shall also have two very distinct genera descended from species I that differ greatly from the descendants of species A. These two groups of genera will thus form two distinct families—or even orders—of organisms, depending on the amount of divergent modification that is represented in the diagram. And the two new families (or orders) are descended from two species (species A and species I) of the original genus, and these are, in turn, supposed to be descended from some still more ancient and unknown form.

We have seen that—in each country—it is the species belonging to the larger genera, containing many species, which most often present varieties (i.e., incipient species). This indeed might have been expected: as natural selection acts through one form having some advantage over other forms in the struggle for existence, it will chiefly act on those that already have some advantage over others; moreover, the largeness of any group shows that the species within that group have inherited from a common ancestor some advantages in common. Thus the struggle for the production of new and modified descendants will mainly lie between the larger groups that are all trying to increase in size. One large group will slowly conquer another large group and reduce its numbers, thus lessening the chance of further variation and improvement among members of the conquered group. Within the same large group, the later and more highly perfected subgroups (e.g., subfamilies, or subspecies) will, from branching out and seizing on many new niches, constantly tend to supplant and destroy the earlier and less improved subgroups; small and broken groups and subgroups will then finally disappear. Looking to the future, we can predict that the groups of organisms that are now large and triumphant, and which are least broken up—that is, those that have as yet suffered the least extinction among members—will continue to increase for a long time.

But which groups will ultimately prevail, no one can predict: we know with certainty that many groups that were formerly most extensively developed have now become extinct. Looking still further out into the future, we may predict that, owing to the continued and steady increase in size of the larger groups, a multitude of smaller groups will eventually become utterly extinct and leave no modified descendants. Consequently, of the species living at any one period, extremely few will successfully transmit descendants to a remote and distant future.

According to this view, very few of the more ancient species have transmitted descendants to the present day, and, as all the descendants of the same species form a taxonomic class, we can understand how it is that so few classes now exist in each main division of the animal and plant kingdoms. Although few of the most ancient species may have left modified descendants to the present day, yet at even the most remote geological period, the Earth may have been almost as well populated with species of as many genera, families, orders, and classes as we have now. I shall have more to say about this subject later, in the chapter on classification (Chapter 14).

On the Degree to Which Organisms Tend to Advance in Complexity

Natural selection acts exclusively by preserving and accumulating variations that are beneficial under the ecological and physical conditions to which each creature is exposed at all periods of its life. The ultimate result is that each creature tends to become better and better adapted to the conditions that surround it. This improvement inevitably leads to a gradual advancement in organization for most organisms throughout the world. But here we enter on a very intricate subject, for naturalists have not yet defined to each other's satisfaction what they mean by an "advance in organization." Among the vertebrates, both the degree of intelligence and the degree to which the body structure approaches that of humans are clearly relevant to the argument. It might be thought that the amount of change that the various parts and organs pass through in their development from the embryo to maturity could serve as a valid standard of comparison, but there are cases, as with certain parasitic crustaceans,[9] in which several parts of the structure have become less well-developed, so that the mature animal cannot be considered—in terms of morphology—more advanced than its larval stage.

Development in such parasitic organisms thus involves the loss of complexity and a move toward simplification rather than what we normally think of as an "advance." The embryologist Karl Ernst Baer's standard seems the most widely applicable and the best. In his view, "advanced" refers to the amount of differentiation of the parts of the same organism—in the adult state, I should be inclined to add—and their specialization for different functions. Alternatively, we might adopt Monsieur Milne-Edwards's view of advancement as the completeness of the division of physiological labor.

But we shall see how obscure this subject is if we look, for instance, to fishes. Some naturalists rank those which, like the sharks, are most like amphibians as the most advanced, while other naturalists rank the common bony fishes as the most advanced, inasmuch as they are more strictly "fish-like" and differ most from members of the other vertebrate classes. We see the obscurity of the subject still more plainly by turning to plants, among which a standard of intelligence is of course quite excluded. Some botanists consider the highest-level plants to be those having every organ—such as sepals, petals, stamens, and pistils—fully developed in each flower, whereas other botanists, probably with more truth, consider those plants having their various organs much modified and reduced in number to be the most advanced.

If we consider the amount of differentiation and specialization of the several organs in each adult (including the advancement of the brain for intellectual purposes) in evaluating the standard of organization, then natural selection clearly leads toward that standard. All physiologists admit that it is an advantage for organisms to

[9] The Crustacea is a group of arthropods containing such animals as crabs, shrimp, barnacles, and copepods.

have more specialized organs, inasmuch as, in this state, those organs will perform their functions better. Thus the accumulation of variations tending toward increasing specialization is within the power of natural selection. On the other hand, bearing in mind that all organisms are striving to increase their numbers at a high rate and to seize on every unoccupied or less well-occupied place in the economy of nature, we can see that it is quite possible for natural selection to gradually fit an organism to a situation in which several organs would be superfluous or useless: in such cases there would be movement toward reduced complexity. Whether organization in general has actually advanced from the remotest geological periods to the present day will be more conveniently discussed later, in the chapter on geological succession (Chapter 11).

But one may object that if all organic beings thus tend to become more complex as their species develops, then how is it that throughout the world a multitude of the lowest forms still exist? And how is it that within each great class of organisms, some forms are far more highly developed than others? Why haven't the more highly developed forms everywhere supplanted and exterminated the less complex? The French naturalist Jean-Baptiste de Monet Lamarck, who believed in an innate and inevitable tendency toward perfection in all organisms, seems to have felt this difficulty so strongly that he was led to suppose that new and simple forms are continually being produced by spontaneous generation, something that science has yet to confirm. However, the continued existence of lowly organisms poses no difficulty for us: natural selection, or the survival of the fittest, does not necessarily require progressive development. It only takes advantage of such variations as arise and are beneficial to each creature under its complex relations of life. And, it may be asked, what advantage would it be to a paramecium (Figure 4.13) or to some other protozoan, or to an intestinal worm, or even to an earthworm, to be more highly organized? If it were not advantageous, then those forms would be left unimproved or but little improved by natural selection and might remain for indefinite ages in their present lowly condition. And geology tells us that some of the lowest forms, such as the ciliated protozoans and amoebae, have remained in nearly their present state for an enormous period of time. But to suppose that most of the many now-existing lowly forms have not advanced in the least since the first dawn of life would be extremely rash; indeed, every naturalist who has dissected some of these so-called lowly beings must have been struck with how wondrous and beautifully organized they were.

Nearly the same remarks apply if we look to the different grades of organization within any one great group. Within the Vertebrata, for example, how can we explain the co-existence of mammals and fish? Among mammals, how can we explain the coexistence of humans and the platypus? Among fishes, how can we explain the coexistence of the shark (Figure 4.14) and the lancelet (*Amphioxus*) (Figure 4.15), whose extremely simple structure approaches that of the invertebrates? But mammals and fishes hardly ever come into competition with each other; thus the advancement of the entire class of mammals—or of certain members in this class—to the highest grade would not lead to their taking the place of fishes. Physiologists believe that the

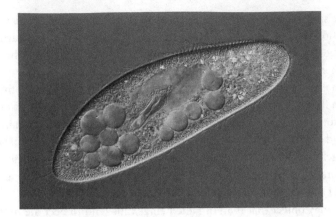

Figure 4.13 *Paramecium caudatum.*

Figure 4.14 A large grey reef shark (*Carcharhinus amblyrhynchos*).

brain must be bathed by warm blood to be highly active, and this requires aerial respiration. Thus, warm-blooded aquatic mammals are disadvantaged by having to come continually to the surface to breathe. With fishes, members of the shark family would not tend to supplant the lancelet; according to the German biologist Fritz Müller, the lancelet has an anomalous annelid (the phylum of segmented worms that includes marine worms, earthworms, and leeches) as its only companion and competitor on the barren sandy shores of South Brazil where it lives. The three most primitive orders of mammals— the marsupials, the edentata (which includes the armadillos, anteaters, and aardvarks),[10] and the rodents—coexist in South America in the same region with numerous monkeys, and they also probably don't interact with each other very

---

[10] These animals have subsequently been distributed among three new orders, including the Xenarthra.

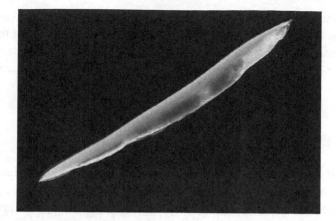

Figure 4.15 A lancelet (*Amphioxus lanceolatus*). Lancelets of various species are found all over the world in shallow temperate and tropical seas, usually half-buried in sand. They are used as a food source in Asia and are commonly studied for what they may tell us about the evolution of vertebrates.

much. Although organization, on the whole, may have advanced and may still be advancing throughout the world, yet the scale will always present many degrees of perfection; the high advancement of certain whole classes, or of certain members of each class, does not at all necessarily lead to the extinction of groups that they don't directly compete against. In some cases, as we shall see later, lowly organized forms appear to have survived or even flourished to the present day simply from inhabiting confined or peculiar habitats where they have been subjected to less severe competition and where the relatively small numbers of individuals have reduced the chances of favorable variations arising.

Finally, I believe that many lowly organized forms now exist throughout the world from a variety of causes. In some cases, appropriate variations or individual differences of a favorable sort may never have arisen for natural selection to have acted on and accumulated. In no case, probably, has there been enough time to permit the utmost possible amount of development. In some few cases there has been what we must call simplification of organization. But the main cause lies in the fact that under very simple conditions of life, a high organization would be of no value—indeed, it might even be a disadvantage, as being of a more delicate nature and more liable to be put out of order and injured.

Looking to the dawn of life, when all organisms, we may assume, presented the simplest of structures, how, it has been asked, could the first steps in the advancement or differentiation of parts have arisen? As we have no facts to guide us, speculation on the subject is almost useless. It is, however, an error to suppose that there would be no struggle for existence and, consequently, no natural selection, until many forms had been produced: variations in a single species inhabiting an isolated region might be beneficial, and thus the whole mass of individuals might be modified, or two distinct forms might arise. But, as I remarked toward the end of the introduction, no one ought to be surprised that there is much that is still unexplained about the origin of species, particularly if we allow for our profound ignorance of the mutual interrelationships and interactions among the inhabitants of the world at the present time and still more so during past ages.

Convergence of Character

Mr. Hewett Cottrell Watson, the English botanist I mentioned in Chapter 2, thinks that I have overrated the importance of divergence of character (although he does, apparently, believe in it) and that convergence, as it may be called, has likewise played a part. If two species belonging to two distinct though related genera had both produced a large number of new and divergent forms, it is conceivable that these forms might resemble each other so closely that they would all end up being classified under the same genus; in this case, then, the descendants of two distinct genera would converge into a single genus. But it would in most cases be extremely rash to explain such a close and general similarity of structure in the modified descendants of widely distinct forms through convergence. The shape of a crystal is determined solely by the molecular forces acting within it, and it is not surprising that dissimilar substances should sometimes assume the same form. But with living organisms, the form of each depends on an infinitude of complex relations, namely (1) on the variations that have arisen; (2) on the nature of the variations that have been preserved or selected, which depends of course in part on the surrounding physical conditions and even more so on the surrounding organisms with which each being has come into competition; and (3) on inheritance from innumerable ancestors, all of which have had their forms determined through equally complex relations. It is incredible to think that the descendants of two organisms that were originally very different from each other should ever afterward converge so closely as to make those descendants now nearly identical. If this had occurred in the past, then we should now encounter the same form, independently of genetic connection, recurring in widely separated geological formations; the balance of evidence argues again any such admission.

Mr. Watson has also objected that the combination of natural selection and divergence of character would tend to create an indefinite number of species. Considering only physical conditions, it seems likely that a sufficient number of species would soon become adapted to all considerable diversities of heat, moisture, and so forth, and that no new species would be formed. But I fully believe that the ecological interactions among organisms are more important; and, as the number of species in any land goes on increasing, the ecological interactions among them must become more and more complex over time. Consequently, at first sight, it does seem that there must be no limit to the amount of profitable diversification of structure that might be produced and thus no limit to the number of species that might be created. We don't know whether or not even the most prolific areas are now fully stocked with species; at the Cape of Good Hope (located at the southernmost tip of Africa) and in Australia, places that now support such an astonishing number of species, many European plants have recently become part of the natural flora, showing quite clearly that not all available niches were previously filled. But geology shows us that over the past 60 million years or so the number of shelled animal species (e.g., snails, clams, and brachiopods) (Figure 4.16) has not increased greatly or at all, and the same is true of the number of mammal species over the past 30 million years or so. What then prevents an indefinite increase in the number of species living at any given time?

There must be some limit to the amount of life that can be supported in any area, depending as it does on physical conditions. Thus, if an area is inhabited by very many species, each or nearly each species will be represented by only a few individuals. Such species will then be liable to extermination from occasionally severe fluctuations in climate or in the numbers of their enemies. The process of extermination in such cases would be rapid, whereas the production of new species must always be slow. Imagine the extreme case of having only one individual representing each of many species in England; the first unusually severe winter or the first unusually dry summer would exterminate thousands upon thousands of species. Rare species—and every species will become rare if the total number of species in any country becomes indefinitely increased—will, as I have explained earlier, present within any given period very few beneficial variations. Consequently, the process of giving birth to new species would be slowed. Moreover, when any species becomes very rare, close inbreeding will help to exterminate it; a variety of authors have suggested that such inbreeding accounts for the deterioration, for example, of the aurochs (the large ancestors of our domestic cattle) in Lithuania, of red deer in Scotland, and of bears in Norway. Finally, and perhaps most importantly, a dominant species that has already beaten many competitors in its own home will tend to spread and supplant many others. Alphonse de Candolle

Figure 4.16 A dried specimen of a brachiopod (*Terebratulina septentrionalis*). The animal's lophophore, which generates feeding currents and collects food particles, is visible inside the shell.

has shown that those species that spread widely tend to spread *very* widely indeed. Thus they will tend to supplant and exterminate several species in several areas, thereby preventing an endless increase in new species throughout the world. How much weight to attribute to these several considerations, I will not pretend to know. But, taken together, they must limit, in any given region, the tendency to an infinite and endless increase in the numbers of species alive at any given time.

Summary

The following points cannot be disputed: (1) under changing conditions of life, living organisms present individual differences in almost every part of their structure; (2) owing to the exponential rate at which individuals in a population increase, there will inevitably be a severe struggle for life at some age, or in some season, or in some year. It follows then, that a great diversity in structure, physiology, and habits is highly advantageous within a species, considering the infinite complexity of the relations of all organisms to each other. Thus it would be most extraordinary if no variations had ever been useful to each being's own welfare, in the same way that so many variations have proved useful to us.

But if variations useful to any organism do occur, then individuals possessing those variations will have the best chance of surviving in the struggle for life and reproducing. From the strong principle of inheritance, these individuals will tend to produce offspring with similar characteristics. I have called this principle "natural selection." It leads inevitably to the improvement of each creature in relationship to the ecological and physical conditions of its life and, consequently, in most cases, to what must be regarded as an advance in organization. Nevertheless, low and simple forms will long endure if they are well suited to their simple conditions of life.

Because traits tend to be inherited at the same ages that they were first developed in the parental generation, natural selection can select for advantageous traits in eggs, seeds, larvae, or juveniles as easily as in the adult. Among many animals, sexual selection will have aided ordinary selection by giving the greatest number of offspring to the most vigorous and best-adapted males. Sexual selection will also encourage the development of characteristics (i.e., traits) useful to the males alone in their struggles or rivalry with other males. These traits will in turn be transmitted to one sex or to both sexes,, according to the form of inheritance that prevails in that species.

Whether or not natural selection has really acted in this way in adapting the various forms of life to their particular environments and lifestyles must be judged by the weight of evidence provided in the following chapters. But we have already seen how it involves extinction; indeed, the large role played by extinction is very clearly seen in looking at the geological record. Natural selection also leads to divergence of character: the more that organisms differ in structure, behavior, and physiological traits, the larger the total number of organisms that can be supported in any area, through reduced competition. We see clear proof of this in looking at the inhabitants of any small piece of land and in the species that have become naturalized in foreign lands. Thus, as the descendants of any one species are modified over time, and during the incessant struggle of all species to increase their numbers, the more diversified the descendants become, the more likely they are to succeed in the constant battle for life. Thus the small differences distinguishing varieties of the same species tend to steadily increase until they eventually become large enough to form distinct species or even a new genus.

We have seen that it is the common, widely diffused, and wide-ranging species belonging to the larger genera within each class that vary the most; these tend to transmit to their modified offspring that superiority that now makes them dominant in their own habitats. Because natural selection leads to divergence of character and to much extinction of the less improved and intermediate forms of life, the nature of the affinities as well as the generally sharply defined distinctions between the innumerable organisms in each class throughout the world may be explained. It is a truly wonderful fact—the wonder of which we are apt to overlook from familiarity—that all animals and all plants throughout all time and space should be related to each other in groups, and in a very particular way: varieties of the same species are always most closely related; species of the same genus are still related, but less closely and to unequal degrees, forming sections and subgenera; species belonging to different genera are much less closely related; and genera are related to each other to different degrees, forming subfamilies, families, orders, subclasses, and classes (see Figure 1.5). The several subordinate groups within any class of organisms cannot be ranked in a single file, but seem instead to be clustered around points, and these are clustered around other points, and so on in almost endless cycles. If species had been independently created, we would not be able to explain this kind of arrangement, but it is explained very nicely through inheritance and the complex actions of natural selection, involving extinction and divergence of characteristics, as we have seen illustrated in Figure 4.11.

The relationships of all members of the same class have sometimes been represented by a great tree. I believe that this simile largely speaks the truth, with the green and budding twigs representing the existing species and those produced during former years representing the long succession of extinct species. At each period of growth, all of the growing twigs have tried to branch out on all sides and tried to smother and kill the surrounding twigs and branches, in the same manner as species and groups of species have at all times overwhelmed other species in the great battle for life. The limbs that we now see divided into great branches, and these into lesser and lesser branches, were themselves once budding twigs when the tree was young, and this connection of the former and present buds by ramifying branches may well represent the classification of extinct and living species in groups within groups. Of the many twigs that flourished when the tree was a mere bush, only two or three, now grown into great branches, still survive and bear the other branches. So it is with the species that lived during ancient geological periods: very few have left living and modified descendants. From the first growth of the tree, many a limb and branch has decayed and dropped off; those fallen branches of various sizes now represent whole orders, families, and genera that no longer have living representatives and which are known to us only as fossils. As we here and there see a thin straggling branch springing from a fork low down in a tree, and which by some chance has been favored and is still alive on its summit, so we occasionally see an animal like the platypus or the lungfish, which in some small degree connects two large branches of life and which has apparently been saved from fatal competition with other organisms by having inhabited an unusual niche with limited competition and assault from enemies. As buds give rise by growth to fresh buds, and as these, if vigorous, branch out and overgrow on all sides many a feebler branch, so I believe it has been with the great Tree of Life, which fills with its dead and broken branches the crust of the Earth and covers the surface of our living planet with its ever-branching and beautiful ramifications.

Key Issues to Talk and Write About

1. This chapter is about natural selection. As Darwin describes it, how is natural selection similar to the kind of selection (artificial selection) that people have used in creating our domesticated animals and agricultural crops? How do the two forms of selection differ?

2. Find out two interesting things about one of the people that Darwin mentions in this chapter. Choose from the following:
   Karl Ernst Baer
   Alphonse de Candolle
   Jean-Henri Fabre
   Asa Gray
   Joseph Hooker
   Thomas Henry Huxley
   Joseph Gottlieb Kölreuter
   Jean-Baptiste de Monet Lamarck
   Henri Milne-Edwards
   Fritz Müller
   Christian Konrad Sprengel
   Moritz Friedrich Wagner
   Hewett Cottrell Watson

3. Carefully read the paragraph that begins "Physical peculiarities often appear either in males or females...." (see page 70). For the chosen paragraph, what are the two or three main points that Darwin wishes to get across?
   Now summarize those points in a single sentence: your sentence should include all the major points, be accurate, make sense to someone who has not read the original paragraph, and be in your own words.

4. Following the instructions given near the end of Chapter 1 (see page 28), write a one-sentence summary of that paragraph, being careful to include all the key

points that you think Darwin is trying to get across. Try the same exercise with the paragraph that begins, "A structure used only once...." (see page 69).

5. Rewrite the following sentence from Darwin's original, to make it more concise and clear: "Natural selection acts exclusively by the preservation and accumulation of variations, which are beneficial under the organic and inorganic conditions to which each creature is exposed at all periods of life."

6. How does Darwin explain the continued existence of the platypus and other "living fossils"?

7. According to Darwin, how does isolation favor the creation of new varieties and species?

8. How does Darwin explain the fact that island populations tend to diminish or go extinct when the islands are invaded by species from the continents?

9. On July 1, 1858, two papers were presented at a meeting of the Linnean Society in London, England—one written by Charles Darwin—and the other by Alfred Russel Wallace under the title "On the Tendency of Species to Form Varieties, and on the Perpetuation of Varieties and Species by Natural Means of Selection." Now that you've read the crux of Darwin's argument in Chapters 1–4 of this book, read Alfred Russel Wallace's paper on the same topic. How were his ideas similar to those of Darwin? How were they different?

Bibliography

Downing, A. J. 1845. *The Fruits and Fruit Trees of America*. New York.

Gray, A. 1856. *A Manual of the Botany of the Northern United States: Second Edition; including Virginia, Kentucky, and all east of the Mississippi; arranged according to the Natural System. (The Mosses and Liverworts by Wm. S. Sullivant.) With fourteen plates, illustrating the genera of the Cryptogamia*. New York.

Pierce, J. 1823. A memoir on the Catskill mountains with notices of their topography, scenery, mineralogy, zoology, and economic resources. *The American Journal of Science and Arts* 6(1): 86–97.

5

Laws of Variation

Darwin wrote *The Origin of Species* before the principles of genetics were understood. Gregor Mendel published his pea plant paper in 1866, about seven years after *The Origin* was first published, but Darwin apparently never found that paper, and nobody ever sent him the information. At the time it was published, most people seemed to miss the general applicability of Mendel's findings anyway, so even those who read the paper didn't understand that his findings also applied to species with "blended" inheritance rather than just to the discrete traits that Mendel had been working with. Indeed, Mendel seems never to have tried to contact Darwin even after having read a translation of his book, so that even Mendel must have missed seeing the connection between his work and what was described in *The Origin*. Thus, much of what is contained in Chapter 5 is Darwin trying his best to make sense of the "laws of variation" without having any understanding of what those laws actually were; he had no idea of how variation was generated, controlled, or passed on to offspring. Writing this chapter must have been an extremely frustrating experience for him. Still, it is fascinating to see Darwin wrestling with the issue and to see how he categorizes the sorts of variability that need to be explained. In the first part of the chapter, he talks about possible causes of variability, and later he talks about patterns in the distribution of variability. But clearly he is always thinking about how natural selection deals with variation, and he believes that although other factors—such as the prolonged disuse of parts or gradual acclimatization to particular climatic regimes—may also be at work, natural selection has probably played the major role in shaping the patterns of variation that we see in the current populations of all organisms, both plants and animals.

I have previously sometimes spoken as if the variations—so common and affecting so many different features in animals and plants under domestication, and to a lesser degree also in nature—were due to chance. This is not the case, of course, but it serves to acknowledge quite plainly our ignorance of how each particular variation originates. Variability seems somehow related to the conditions of life to which each species has been exposed over several successive generations. But it is very difficult to decide how much changed environmental conditions, such as those of climate and food availability, have acted in shaping the patterns we see today. Certainly, the innumerable complete coadaptations of structure that we see between various organisms throughout nature cannot be attributed simply to changes in climate and food

The Readable Darwin. Second Edition. Jan A. Pechenik, Oxford University Press. © Oxford University Press 2023.
DOI: 10.1093/oso/9780197575260.003.0006

supplies. But when a variation is of the slightest use to any organism, we cannot tell how much to attribute to the accumulative action of natural selection and how much to the direct effects of environmental conditions. For example, furriers are well aware that their target animals have thicker and better fur the further north they live; but who can say how much of this difference results from selection for the warmest-clad individuals over many generations and how much to the direct action of cold weather on the growth of fur? Innumerable instances are known to every naturalist of species not varying at all even when living under the most opposite climates, which leads me to think that the direct action of the surrounding conditions plays a far smaller role in producing variation than does a natural tendency to vary, caused by factors of which we are now quite ignorant.

In one sense, the conditions of life may be said not only to cause variability, either directly or indirectly, but also to include natural selection, for environmental conditions determine whether this or that variety shall survive and leave offspring for future generations. When people do the selecting in breeding domesticated animals or plants and agricultural crops, we can clearly see that the two elements of change are distinct; variability is in some manner excited, but it is our deliberate selection of who will survive and who will mate that causes variations to accumulate in particular directions. Similarly it is selection that leads to the survival of the fittest in nature.

In this chapter, I will consider the possible roles of changing environmental conditions in producing variation as well as the use and disuse of parts, and will discuss variations that are themselves correlated with other traits. I will also review the levels of variability encountered at different levels of organization and talk again about how the steady accumulation of such variations in certain directions leads to the patterns that we currently see in so many species of both animals and plants.

Effects of Increased Use and Disuse of Parts as Controlled by Natural Selection

From the facts alluded to in Chapter 1, I think there can be no doubt that increased use of certain parts in our domestic animals has strengthened and enlarged them, and disuse diminished them, and that such modifications have been inherited.[1] For organisms living in nature we have no standard of comparison by which to judge the effects of long-continued use or disuse for we don't know what the organisms' ancestors looked like; but many animals possess structures that can best be explained by the effects of disuse. As Sir Richard Owen of the British Museum has remarked, there is no greater anomaly in nature than a bird that cannot fly, and yet there are several such bird species in this condition. The logger-headed duck of South America can

---

[1] Although the evidence seemed compelling to him at the time, Darwin was wrong on this point. We now know that physical changes caused by use or disuse are not passed along to offspring.

(A)   (B)

Figure 5.1 (A) Logger-headed duck (*Tachyeres brachypterus*) now better known as the steamer duck. These ducks are native to the Falkland Islands, slightly east of the southern tip of South America. (B) Aylesbury duck (rear) and drake (front).

only flap along the surface of the water, for example, and yet has its wings in nearly the same condition as the domestic Aylesbury duck (Figure 5.1B); remarkably, according to the Scottish naturalist Mr. Robert Oliver Cunningham, while the adult birds cannot fly, the young birds can. As the larger ground-feeding birds seldom take flight except to escape danger, it seems likely that the nearly wingless condition of some birds that now inhabit or which recently inhabited several oceanic islands has been caused by disuse,, since there are no predators on those islands. The ostrich, on the other hand, inhabits continents, not islands, and is certainly exposed to dangers from which escape by flight would be helpful; however, although it can defend itself quite well by kicking its enemies as efficiently as many four-legged animals, it cannot fly. It seems likely that the ancestor of the ostrich genus had habits like those of the large terrestrial birds known as bustards (Figure 5.2), and that, as the size and weight of its body increased during successive generations, its legs were used more often and its wings less often until the birds became incapable of flight. I believe that natural selection has also played an important role here, and I will have more to say about this shortly.

William Kirby, coauthor of *An Introduction to Entomology*, has remarked (and indeed I have also observed this myself) that parts of the terminal section of the anterior legs (the "tarsi" or feet) of many male dung beetles (Figure 5.3) are often broken off; he examined 17 specimens in his own collection, and not one had even a relic of a foot remaining on the anterior legs. In the dung beetle species *Onites apelles*, the tarsi are so routinely lost that the insect has often been described as not having them. In some other genera they are present, but only in a rudimentary condition. In the sacred beetle of the Egyptians, a member of the genus *Ateuchus*, they are totally absent. How do we explain these facts? The evidence that accidental mutilations can be

Figure 5.2 Kori bustard (*Ardeotis kori*) breeding display. Despite their size, bustard birds do fly.

Figure 5.3 Male dung Tarsus beetle (*Onthophagus coenobita*).

inherited is at present not decisive, but the remarkable cases observed by the French neurobiologist and physiologist Charles Édouard Brown-Séquard in guinea pigs, on the inherited effects of certain operations,[2] should make us cautious in denying this

---

[2] In a series of experiments, Brown-Séquard found that severing the spinal cord or certain nerves in adult guinea pigs caused what seemed to be inherited problems for the offspring.

possibility. Perhaps it will be safest to look at the rudimentary condition of the anterior tarsi in some dung beetle genera and their entire absence in *Ateuchus* not as cases of inherited mutilations but as due simply to the effects of disuse over a long period of time. Since many dung-feeding beetles are commonly found with their tarsi lost, this must happen early in life; obviously then, the tarsi cannot be of much importance to these insects.

In some cases we might incorrectly attribute modifications of structures to disuse when their reduction or absence is in fact wholly, or at least primarily, due to natural selection. The entomologist Mr. Thomas Vernon Wollaston has discovered the remarkable fact that of the more than 550 beetle species found on the island of Madeira, 200 species have such poorly developed wings that they cannot fly, and that of the 29 genera endemic to Madeira, 23 have all their species in this condition! How can we explain these remarkable facts? Well, we know the following: (1) that beetles in many parts of the world are frequently blown out to sea, where they perish; (2) that the beetles of Madeira, as observed by Mr. Wollaston, spend their time in hiding until the wind dies down and the sun shines; (3) that the proportion of wingless beetles is even larger on the fully exposed neighboring rock island of Desertas than on Madeira itself; and—a most extraordinary fact—(4) that certain large groups of beetles that are very numerous elsewhere, and which absolutely require their wings to function, are almost entirely absent on Madeira. These facts convince me that the wingless condition of so many Madeiran beetles is mainly due to the action of natural selection, probably combined with the direct effects of disuse: during many successive generations, any individual beetles that flew the least, either from their wings having been ever so slightly less perfectly developed or from indolent habit, will have had the best chance of surviving simply from not having been blown out to sea. In contrast, those beetles that most readily took to flight would have been blown out to sea the most often and been destroyed.

(A)   (B)

Figure 5.4 (A) European mole (*Talpa Europaea*); note how poorly developed the eyes are. (B) Brazilian tuco-tuco (*Ctenomys brasiliensis*); Darwin talks here about seeing this rodent when he was in Uruguay, during his voyage on *HMS Beagle*.

The insects of Madeira that are not ground feeders and which, as with certain flower-feeding coleopterans (beetles) and lepidopterans (moths and butterflies), must use their wings to find food have, as Mr. Wollaston suspects, fully developed—even enlarged—wings. This is quite compatible with the action of natural selection. For when a new insect first arrived on the island, the tendency of natural selection to enlarge or to reduce the wings would depend on whether more individuals were saved by successfully battling with the winds or by giving up the attempt and rarely, if ever, flying. Similarly, as with sailors shipwrecked near a coast, it would have been better for the good swimmers if they had been able to swim further, whereas it would have been better for the bad swimmers if they had not been able to swim at all and had instead stayed with the wreck.

A similar series of gradations can be seen among rodents and moles. The eyes of moles (Figure 5.4A) and of some burrowing rodents are rudimentary in size, and in some cases are covered by skin and fur. This state of the eyes is probably due to gradual reduction from disuse, but aided perhaps by natural selection. In South America, the tuco-tuco (Figure 5.4B), a burrowing rodent in the genus *Ctenomys* that is even more subterranean in its habits than the mole, is frequently blind. One that I kept alive myself was certainly blind; on dissection I found the cause to have been inflammation of the nictitating membrane, which is basically a transparent or translucent, supplementary eyelid. As frequent inflammation of the eyes must be injurious to any animal, and as eyes are certainly not needed by animals living below the ground, a reduction in their size, with the adhesion of the eyelids and growth of fur over them, might in such cases be an advantage; if so, then natural selection would augment the effects of disuse.

It is well known that several animals inhabiting the caves of both Carniola[3] in Europe and the American state of Kentucky in North America—and belonging to a wide range of different taxonomic classes—are blind. Some of the blind crabs in these caves retain their eye stalks, although the eye itself is gone: the *stand* for the telescope remains, but the telescope itself, with its lenses, has been lost. As it is difficult to imagine that eyes, even though useless, could in any way be injurious to animals living in darkness, their loss may be attributed to simple disuse. Professor Benjamin Silliman of Yale University captured two blind cave rats (genus *Neotoma*) more than a half mile from the mouth of a cave, thus in darkness but not in the profoundest depths. Their eyes were lustrous and large; after exposing the animals to light of gradually increasing intensity for about one month, Professor Silliman tells me that the animals acquired a dim perception of objects. Clearly the eyes remained at least partly functional.

It is hard to imagine conditions of life more similar than that in deep limestone caverns under nearly similar temperature and humidity. In accordance with the traditional view that each of the various blind animals has been separately created for

---

[3] The area known then as Carniola is essentially what we know today as Slovenia.

the American and European caverns, we would thus expect to find many similarities in their organization and affinities. But if we look at the two whole faunas, we
see that they are *not* similar in organization or affinity. With respect to insects alone,
the Danish entomologist Jørgen Matthias Christian Schiødte has remarked that they
seem "purely local," and that "the similarity that is exhibited in a few forms between
the Mammoth Cave in Kentucky and the caves in Carniola" is simply "a very plain
expression of that analogy which subsists generally between the fauna of Europe and
of North America." In my view, we must suppose that American animals with ordinary powers of vision slowly migrated over successive generations from the outer
world into the deeper and deeper recesses of the Kentucky caves,, and that European
animals independently migrated into the caves of Europe. We do, in fact, have some
evidence of this gradation of habit. As Mr. Schiødte remarks, "We accordingly look
upon the subterranean faunas as small ramifications which have penetrated into the
Earth from the geographically limited faunas of the adjacent tracts, and which, as
they extended themselves into darkness, have been accommodated to surrounding
circumstances. Animals not far remote from ordinary forms prepare the transition
from light to darkness. Next follow those that are constructed for twilight; and last of
all, those destined for total darkness, and whose formation is quite peculiar." It is important to realize that Mr. Schiødte's remarks apply not to the individuals of any one
species, but rather to different species. By the time an animal has reached, after numberless generations, the deepest recesses, disuse will, on this view, have more or less
perfectly obliterated its eyes, and natural selection will often then have brought about
other changes, such as an increase in the length of the antennae or sensory palps,[4] as
compensation for blindness.

Notwithstanding such modifications, if they have indeed been brought about by
natural selection, we might still expect to see some affinities between the cave animals of North America and other non–cave-inhabiting animals of that continent,
and, likewise, between the cave animals of Europe and other inhabitants of the
European continent. This is indeed the case with at least some of the American cave
animals, as I hear from Dr. James Dwight Dana at Yale University. Similarly, some of
the European cave insects are very closely allied to the non-cave insects of the surrounding areas in Europe. It would be difficult to explain the affinities of the blind
cave animals to the other inhabitants of the two continents on the traditional view of
the cave animals having been specially and independently created. We might instead
have expected to see a close relationship between those inhabiting the caves of the Old
and New Worlds, from the well-known close relationships among most of their other
fauna. A blind genus of ground beetles (*Anophthalmus*) offers us a particularly good,
although rare, example of such close relationships: as the entomologist Mr. Andrew
Murray observes, although the species has so far been found only in caves, those
that inhabit the several caves of Europe and America resemble each other. Perhaps

---

[4] These are sensory appendages found on the heads of crustaceans and insects, formed from certain of
the mouthparts.

the ancestors of these several species, while they were still furnished with eyes, had formerly ranged over both continents and then become extinct, except in their present secluded abodes. Far from being surprised that some cave animals should lack any closely related terrestrial counterparts, as Louis Agassiz of Harvard University has found with regard to the blind American cave fish in the genus *Ambylopsis*, and as is also the case with the blind, cave-dwelling, European amphibian *Proteus*, I am only surprised that more such wrecks of ancient life have not been preserved in these caves, owing to the less severe competition to which the very few inhabitants of these dark abodes will have been exposed.

Acclimatization

The period of flowering in plants, the season in which the plants become dormant, the amount of rain needed for seeds to germinate—these and similar habits are all hereditary with plants. This leads me to say a few words about acclimatization. As it is extremely common for distinct species belonging to the same genus to inhabit both hot and cold countries, then if it be true that all members of the same genus are descended from a single ancestor, acclimatization to different climates must happen easily over many generations.

It is notorious that each species is adapted to the climate of its own home: species from an arctic or even from a temperate region cannot endure a tropical climate, and conversely. Similarly, many succulent desert plants cannot endure a damp climate. But the degree to which species are adapted to the climates under which they live is often overrated.

We may infer this from our frequent inability to predict whether or not any particular imported plant will endure our climate here in England and from the number of plants and animals brought from different countries that do perfectly well here. We have reason to believe that species in nature are tightly restricted in their ranges at least as much by competition with other organisms as by their being adapted to particular climates. Indeed, we have clear evidence that some plants have sometimes become, to at least a certain extent, well habituated to new and different conditions. Thus the excellent botanist (and my good friend) Dr. Joseph Hooker has collected seeds from certain species of pines and rhododendrons (Figure 5.5) growing at different heights in the Himalayas of Asia and found that when the seeds were planted in this country, plants from the same species collected at different heights on the mountain differed in their ability to resist cold. Another botanist, Mr. George Henry Kendrick Thwaites, has observed similar facts in Ceylon, and Mr. Hewett Watson has made analogous observations about European species of plants brought from the Azores to England. There are more examples that I could give. In regard to animals, several authentic instances could be adduced of species having largely extended their range from warmer to cooler latitudes within historical times, and conversely. We do not know with certainty that these animals were strictly adapted to their native climate,

Figure 5.5 Himalayan rhododendrons.

though in all ordinary cases we assume that to be the case. Nor do we know whether they have subsequently become specially acclimatized to their new homes, so as to be better fitted for them than they were at first.

It seems reasonable to assume that our domestic animals were originally chosen by uncivilized man because they were useful in some way and because they reproduced readily under confinement, and not because they were subsequently found capable of being transported large distances. The common and extraordinary capacity of our domestic animals to not only be able to withstand a wide range of climates but of also being perfectly fertile (a far more severe test) under such a wide range of conditions thus suggests that a large proportion of other animals now living in nature could also easily be brought to withstand widely different climates. We must not, however, push the foregoing argument too far, remembering that some of our domestic animals originated from several wild stocks. The blood, for instance, of a tropical and arctic wolf may perhaps be mingled in our domestic breeds. Although the rat and mouse cannot be considered as domestic animals, we certainly have transported them to many parts of the world so that they now have a far wider geographical range than any other rodent; they live under the cold climate of the Faroe Islands in the North Atlantic and of the Falkland Islands in the South Atlantic,, and yet also on many an island in the intensely hot tropics. Thus adaptation to any special climate may be looked at as a quality readily grafted onto an innate flexibility of constitution that is common to most animals. Viewed thus, the ability to endure the most different climates by man himself and by his domestic animals, and the fact of the extinct elephant and rhinoceros having previously endured a glacial climate whereas the living species are now all tropical or subtropical in their habits, ought not to be looked at as anomalies, but rather as examples of a very common flexibility of constitution brought into action under peculiar circumstances.

How much of the acclimatization of species to any peculiar climate is due to mere physiological flexibility of individuals and how much to natural selection acting on varieties with different innate constitutions over time—or to both of these mechanisms combined—is an obscure question. It seems reasonable that both physiological flexibility and experience have had some influence. And as it is not likely that humans should have succeeded in selecting so many breeds and sub-breeds with constitutions specially fitted for their own regions, the result must, I think, be due to physiological flexibility. On the other hand, natural selection would inevitably tend to preserve those individuals that were born with constitutions best adapted to whatever country they inhabited. In treatises on many kinds of cultivated plants, certain varieties are said to withstand certain climates better than others. This is strikingly shown in works published in the United States about fruit trees: certain varieties are habitually recommended for the northern states and others for the southern states; as most of these varieties are of recent origin, they cannot owe their constitutional differences to habit. The case of the Jerusalem artichoke, which is never propagated in England by seed, and of which consequently new varieties have never been produced, has even been advanced as proving that acclimatization cannot be achieved, for it is now as tender as it ever was! The case of the kidney bean has also been cited for a similar purpose, and with much greater weight; but until someone will plant his kidney beans so early that a very large proportion of them are destroyed by frost and then collect seed from the few survivors, with care to prevent accidental crosses, and then again get seed from those seedlings with the same precautions over a dozen or more generations, then the experiment cannot be said to have been tried. Nor let it be supposed that differences in the constitution of seedling kidney beans never appear, for I have read that some seedlings are hardier than others, and in fact I have observed striking instances of this in my own studies. It may well be possible to select for cold tolerance in kidney beans.

On the whole, we may conclude that physiological flexibility and experience along with use and disuse, have, in some cases, all played a considerable part in the modification of the constitution and structure of animals and plants. **But clearly these effects have often been largely combined with—and sometimes overmastered by— natural selection acting on innate variation among individuals within species.**

Correlated Variation

By "correlated variation" I mean that all of the parts of an organism are so tied together during its growth and development that when slight variations in any one part occur,, and are gradually accumulated by natural selection, some other, apparently unrelated parts become modified as well. This is a very important subject, but one that we understand very poorly. We shall presently see that what seems like correlation in many other cases can be easily explained by simple inheritance.

One of the most obvious legitimate cases of correlated variation is when anatomical variations arising in the young or larvae of a species later affect the structure of the mature animal. The several parts of the body that are symmetrical, and which at an early embryonic period are identical in structure and are necessarily exposed to similar environmental conditions, seem eminently liable to vary in similar ways: we see this in the right and left sides of the body varying in the same manner, for example. The front and hind legs of many species also vary together. However, these tendencies may be overruled in some cases more or less completely by natural selection. For instance, a family of stags once existed with an antler only one side of the head. If this situation had been of any great use to the breed it would probably have been rendered permanent by selection, and the left and right sides in that species would no longer be "correlated."

It also seems that hard body parts can affect the form of adjoining soft body parts, and that may account for the apparent correlation of some traits. Some authors believe that the diversity in the shape of the pelvis among birds causes the remarkable diversity in the shape of their kidneys. Others believe that, in humans, the shape of the child's head is influenced, through pressure, by the shape of the mother's pelvis. And in snakes, according to the German naturalist Hermann Schlegel, the form of the body and the manner of swallowing determine the position and form of several of the most important digestive organs.

However, the nature of the relationship is frequently quite obscure. The French authority on developmental abnormalities Isidore Geoffroy Saint-Hilaire[5] is firmly convinced that certain physical malformations tend to appear together, although what causes this connection remains uncertain. What can be more singular than the relationship in cats between completely white fur, blue eyes, and deafness, or between the tortoise-shell color in cats and being female; or with pigeons that have feathered feet also having skin between the toes, or with the amount of down on the skin of young pigeons and the future color of its plumage? And I think it can be hardly accidental that the two orders of mammals that are most unusual in their skin coverings, that is, the Cetacea (whales, dolphins, and porpoises) and the Edentata[6] (armadillos, anteaters, and sloths, for example) are also the most abnormal in their teeth.

Perhaps the best example showing the importance of the laws of correlation and variation—unrelated to usefulness and therefore not affected by natural selection—is that of the difference between the outer and inner flowers in some members of the Asteracea (or Compositae) family of plants (including daisies, asters, and sunflowers) and those in the Apiaceae family (including carrots and parsley). These plants all produce clusters of small flowers that give the appearance of being a single flower. For example, everyone is familiar with the peripheral ray flowers and the central florets of the daisy, where the small clustered flowers form a central disk (Figure 5.6); that is, although the daisy looks like a single flower, it is really composed of many individual flowers, with those on the outside (the peripheral flowers) looking very

---

[5] Isidore Geoffroy Saint-Hilaire is the son of French naturalist Étienne Geoffroy Saint-Hilaire.
[6] These animals are now all placed within the order Xenarthra.

Peripheral flower

Open central flower

Unopened central flower

Figure 5.6 Daisy (*Leucanthemum vulgare*) showing peripheral and central flowers.

different—each peripheral flower produces a single long petal—from those more centrally located. This difference is often associated with the ray flowers being either sterile (having neither male nor female parts) or forming only female parts. But in some of these plants, the seeds of the different flowers also differ in shape and sculpture. These differences have sometimes been attributed to the pressure placed on the florets[7] by the involucra—the group of small green leaflets at the base of each flower cluster; the shape of the seeds in the peripheral florets of some members of the Asteracea (daisies, for example) support this idea. On the other hand, with plants in the Apiaceae family (carrots and parsley), Dr. Hooker tells me that the species with the densest heads do *not* differ most frequently in their inner and outer flowers. It might have been thought that by drawing their nourishment from the reproductive organs, the ray petals cause the abortion of those organs; but this cannot be the sole cause, for in some members of the Asteracea the seeds of the outer and inner florets differ even though there is no difference in the petals. Possibly these several differences may be connected with the different flow of nutrients toward the central versus the peripheral flowers; we know that with irregular flowers, those nearest to the axis are most likely to become abnormally symmetrical. I may add, as an example of this fact, and as a striking correlation, that in many geraniums, the two upper petals in the central flower of the cluster often lose their patches of darker color; when this occurs, the adhering nectary—which of course produces the sugar-rich nectar—is quite aborted. The central flower thus becomes unusually symmetrical. When the color is absent from only one of the two upper petals, the nectary is much shortened, although not fully aborted.

With respect to the development of the corolla of these composite flowers, Christian Konrad Sprengel's idea that the peripheral florets serve to attract insects is highly likely, as insects are highly advantageous, or even essential, for fertilizing

---

[7] One of the small flowers that makes up the head of a composite flower.

these plants; if so, then natural selection has probably been at work. But with respect to the seeds, it seems impossible that their shape differences, which are not always correlated with any difference in the corolla, can in any way benefit the plant; yet, in the carrot/parsley family (the Apiaceae), the seeds are sometimes very straight in the exterior flowers but hollow and curved in the central flowers. Indeed, the Swiss botanist Augustin de Candolle (the father of Alphonse de Candolle, see below), based his main taxonomic divisions for the entire order on such characteristics. Thus we see that modifications of structure, viewed by systematists[8] as of great importance, may be wholly due to the laws of variation and correlation, without being, as far as we can judge, of the slightest benefit to the species or subject to the laws of natural selection.

Although correlated variation is common, certain structures that are common to whole groups of species exist not through correlation but rather to inheritance. An ancient ancestor may have acquired, through natural selection, some particular modification in structure and, after thousands more generations, some other and independent modification. These two modifications, having been transmitted to a whole group of descendants with diverse habits, would naturally be thought to be in some necessary matter correlated, even though they were actually independently evolved.

Some other correlations are apparently due to the manner in which natural selection alone can act. For instance, the younger de Candolle (Alphonse) has noted that winged seeds are never found in fruits that do not open. I should explain this correlation as follows: it would be impossible for seeds to gradually become winged through natural selection unless the capsules surrounding them eventually opened to the outside, releasing the seeds; only in that way could seeds that were a little better adapted to be wafted by the wind gain an advantage over others less well fitted for wide dispersal. In short, through natural selection, we would never expect winged seeds to evolve in plants that never released their seeds to the wind.

Compensation and the Economy of Growth

Both Étienne Geoffroy Saint-Hilaire (Isidore Geoffroy Saint-Hilaire's father) and Johann Wolfgang von Goethe formulated, at about the same time, their "law of compensation." As Goethe expressed it, "in order to spend on one side, nature is forced to economise on the other side." I think this also holds true to a certain extent with our domesticated species: if nourishment flows to one part or organ in excess, it rarely flows to another part, at least to such a high degree. For example, it is difficult to get a cow to give a lot of milk and to fatten it readily at the same time. Similarly, some varieties of cabbage yield an abundant and nutritious foliage but only a few seeds, while

---

[8] Systematists are biologists who try to understand how different groups of organisms are related to each other.

other varieties produce a copious supply of oil-bearing seeds, but a less-developed foliage. And again, when the seeds in our fruits become atrophied, the fruit itself increases considerably in size and quality, again suggesting an increase of nutrients in one part being caused by a decrease in another part. Similarly, a large tuft of feathers on the heads of our poultry is generally accompanied by a smaller comb on the top of the head, and a large beard is similarly accompanied by diminished wattles under the chin.

In nature it is difficult to tell whether this "law of compensation" applies or not. I don't see any way of distinguishing between the effects of a part being largely developed through natural selection—with another and adjoining part being reduced by the same process—on the one hand, and the excess growth of some part directly causing a withdrawal of nutrients from some other part, on the other hand.

I also suspect that some of the presumed cases of compensation that have been proposed in the literature may be combined with some other facts under the more general principle that natural selection is continually trying to economize every part of an organism's organization. If, under changed environmental conditions a structure that was previously useful now becomes less useful, its diminution and eventual loss will be favored, for it will benefit the individual not to have its nutrients wasted in trying to build and maintain a useless structure. This is the only way that I can understand a fact with which I was much struck when examining barnacles some years ago, namely that when one barnacle is parasitic within another, and is thus protected, it more or less completely loses its own outer covering, or shell. This is certainly the case with the males of the barnacle genus *Ibla* and is seen in a truly extraordinary manner within the genus *Proteolepas*. The outer covering in the larvae of all other barnacles consists of three highly important anterior segments of the head enormously developed and furnished with large nerve fibers and muscles; but in members of the parasitic and protected genus *Proteolepas*, the whole anterior part of the head is reduced to the merest rudiment[9] attached to the bases of the prehensile antennae.

Many analogous cases could be given. Clearly it would be a decided advantage to each individual of any species not to produce a large and complex structure that has now become superfluous; in the struggle for life that all animals experience, individuals will have a better chance of supporting themselves by wasting less of what they eat. Thus I believe that natural selection will tend in the long run to reduce any part of the organization that becomes, through changed habits, superfluous, without causing some other part to be largely developed in a corresponding degree. Conversely, natural selection may also perfectly well succeed in increasing the size of an organ if that should prove beneficial, without requiring the compensatory reduction of an adjoining part.

---

[9] A rudiment is the barest beginning of a structure. Think of "rudimentary knowledge." In this case the head only starts to develop, but never gets very far.

Multiple, Rudimentary, and Lowly Organized Structures Are Especially Variable

As the French zoologist Isidore Geoffroy Saint-Hilaire (the son of Étienne, as mentioned earlier in the chapter) has noted, when any part or organ occurs many times in the same individual (e.g., the vertebrae of snakes or the stamens in polyandrous[10] flowers; i.e., those having numerous stamens), their number usually varies among individuals; in contrast, when the same part or organ occurs within an organism in fewer numbers, the number is constant from one individual to the next. Mr. Étienne Geoffroy Saint-Hilaire, together with a number of botanists, has noted that multiple parts in flowers are also extremely likely to vary in structure. This observation fits well with Sir Owen's suggestion that the simple repetition of parts reflects what he calls "low organization"; as many naturalists have suggested, more primitive organisms tend to vary more than do more advanced organisms. This makes sense to me if by "primitive" we mean that the various parts of an organism's organization have not become specialized for particular functions: as long as the same part has to perform many different tasks, we can perhaps see why it should remain variable—that is, why natural selection would not have preserved or rejected each little deviation of form as carefully as when the part has had to serve some specialized purpose. In the same way, a knife that has to cut all sorts of very different things may be almost any shape, whereas a cutting tool used for some specific purpose will be of some particular shape. It should never be forgotten that natural selection can act only through and for the advantage of each being.

It is also widely acknowledged that rudimentary parts of all sorts tend to be highly variable in form. I will return to this topic later. Here let me just say that their great variability among individuals seems to result from their uselessness and thus from natural selection having no power to stop deviations in their structure from persisting.

A Part Developed in Any Species to an Extraordinary Degree Tends to Be Highly Variable, in Comparison with the Same Part in Related (Allied) Species

Several years ago I was much struck by the English naturalist Mr. George Robert Waterhouse's remark that parts developed to an extraordinary degree in some species tend to be far more variable than the same parts developed normally in related species. Sir Owen seems to have come to a similar conclusion in his studies. I am convinced that this is in fact a rule of great generality, although I cannot possibly

---

[10] The Greek derivation of polyandrous: *poly* = many; and *androus* = husbands.

Figure 5.7 A beautiful Indian peacock (*Pavo* sp.) with a fully fanned tail. This is a good example of secondary sexual characteristics, a term which, as used by the Scottish anatomist John Hunter, refers to characteristics associated with only one sex without being directly connected to the physical act of reproduction.

introduce here the long array of facts that I have collected that have led me to this conviction. It should be understood that the rule applies only to parts that are unusually well developed in one species or in a few species in comparison with the same part found in many closely allied species. Thus, although a bat's wing is a most abnormal structure to find among mammals, the rule would not apply here because all bats have wings; it would apply only if one particular bat species had wings that were remarkably different in some way in comparison with those of the other species in the same genus.

The rule applies very strongly in the case of secondary sexual characteristics, a term which, as used by the Scottish anatomist, John Hunter, refers to characteristics associated with only one sex without being directly connected to the physical act of reproduction—such as the long and fantastically colored feathers of the male peacock (Figure 5.7). The rule applies to both males and females, although males are more likely to show such secondary sexual characteristics and in a highly variable manner (Figure 5.7). But our rule is not confined to such characteristics, as is clearly shown in the case of hermaphroditic barnacles; indeed, the rule almost always holds good for these interesting animals, which I have studied in detail for many years. In a future work I will list all the more remarkable cases. Here I will give only one example, but one that illustrates the rule in its largest application.

Rock barnacles have very important outer body parts called "opercular valves," which are calcareous structures (i.e., containing calcium carbonate) that are moved together or apart to open and close a barnacle's shell (Figure 5.8A). Typically, they differ extremely little even among the members of distinct genera. However, in several species of the barnacle genus *Pyrgoma*, these valves present a marvelous

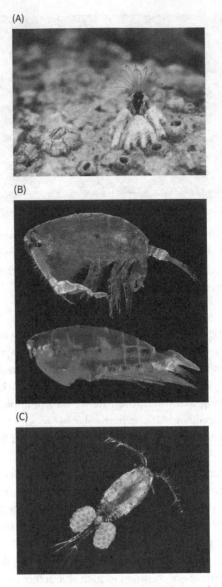

Figure 5.8 (A) Underwater photo of acorn barnacle (*Balanus glandula*). (B) The pontellid copepod (*Pontella securifer*). Various parts glow fluorescent green when viewed under blue light. Top: Note the male's hinged antenna and the claw on its last leg, both used for grabbing females. (C) Freshwater female copepod (*Cyclops* sp.); note the two egg masses attached to her abdomen.

amount of diversification. The comparable valves found among the different species of that genus are sometimes completely different in shape, and the amount of variation among the individuals of even the same species is so great that I can say without exaggeration that *varieties* of the same species differ more from each other in the characteristics derived from these important organs than they do among species belonging to other genera.

Over the years I have also paid particular attention to birds, as individuals of the same species inhabiting a common area generally vary extremely little; I find that the rule seems to hold up very well in this class of animals. I can't determine how well it applies in plants, though; the great variability in plants makes it particularly difficult to compare their relative degrees of variability.

Whenever we see any part or organ of a species developed to a very remarkable degree in any particular species, it is fair to presume that that part has a highly important function. And yet, it is eminently liable to variation among individuals. Why should this be? On the view that each species has been independently created with all of its parts just as we see them now, I can see no explanation. But if we assume that groups of species are descended from other, ancestral species, and have been modified over time by natural selection, then I think the observation makes good sense.

First, let me make some preliminary remarks. If any part in one of our domesticated animals—or even the whole animal—be neglected during selection for breeding future generations, with no selection being applied in any direction, then that part—or the entire breed—will soon cease to have a uniform character: it may be said that the breed is degenerating. We see a nearly parallel case in rudimentary organs[11] and in those that have not been specialized for any particular purpose: in such cases, natural selection has not—or cannot—come into full play, so that the organization is left in a fluctuating condition.

But what especially concerns us here is that the features in our domestic animals that are presently undergoing rapid change due to deliberate selection of traits in breeding are also eminently liable to variation within each breed. Look, for example, at individuals of the same breed of pigeon and see what a prodigious amount of variation there is in the beak of tumblers (see Figure 1.2), in both the beak and the wattle of carriers (see Figure 1.3A), in the carriage and tail of fantails (see Figure 1.3C), and so on, which are all currently features of major concern to English pigeon fanciers (i.e., breeders). Even in the same sub-breed of pigeon, as in that of the short-faced tumbler, it is notoriously difficult to breed nearly perfect birds; many depart widely from the standard. Truly, there is a constant struggle between the tendency to revert to a less perfect state along with an innate tendency to new variations, on the one hand, and the power of steady selection to keep the breed true, on the other. In the long run, selection wins the day: we do not expect to fail so completely as to breed a bird as coarse as a common tumbler pigeon from a good short-faced strain. But as long as active selection is going on, we can always expect to see much variability in the parts undergoing modification.

---

[11] We now refer to "rudimentary" organs as "vestigial," a mere remnant of something formerly better developed.

Now let us turn to nature. Whenever any particular part has been developed to an extraordinary manner in any one species compared with the development of that part in other species in the same genus, we may conclude that this part has undergone an extraordinary amount of modification since the period when the several species within that genus branched off from their common ancestor. This period will seldom be extremely remote, as species rarely endure for more than one geological period. An extraordinary amount of modification implies an unusually large amount of variability continued over a long period of time that has been continually accumulated for the benefit of the species by natural selection. But as the variability of the extraordinarily developed part or organ has been so great in the fairly recent past, we might, as a general rule, still expect to find more variability in such parts than in other parts of the organisms that have remained nearly constant for a much longer time. This is, I am convinced, precisely the case. I see no reason to doubt that the struggle between natural selection and the tendency to reversion and variability will eventually cease, and that the most abnormally developed organs may then be made constant. Thus, when an organ, however abnormal it may seem to us, has been transmitted in approximately the same condition to many modified descendants over many generations— as in the case of the bat's wing—it must, according to our theory, have existed for an immense period of time in nearly the same state; thus it is now no more variable than any other structure. It is only in those cases in which the modification has been extraordinarily great and comparatively recent that we would expect to find considerable variability still present.

Specific Characters Vary More Than Generic Characters

The same principle discussed in the preceding section may also explain the differences in variability in specific and generic characteristics. It is notorious that specific traits vary more than generic ones do. For example, suppose that in a large genus of plants, some species had blue flowers while some had red flowers; flower color would then be only a species-defining characteristic and no one would be surprised at seeing one of the blue species varying into red or one of the red species varying into blue within the genus. But if all the species in that genus had blue flowers, the color would become a generic character and its variation would be a more unusual circumstance. I have chosen this example because most naturalists would argue that specific characteristics are more variable than generic ones simply because they are taken from parts of less physiological importance than those commonly used for classing genera. I believe this explanation is only partly, and indirectly, true; I shall return to this point in the chapter on classification (Chapter 14). With respect to particularly *important* characteristics I have repeatedly noticed in works on natural history that when some important organ or part that is generally very constant throughout a large group of species differs considerably in some closely allied species, it is often variable among individuals of some of the species. This shows that when a trait that is usually used to define genera sinks in value and becomes only useful in defining species, it often becomes variable, even though its physiological importance to the organism may remain the same. **In other words, characteristics that vary among species also vary greatly among individuals of those species, regardless of their physiological importance.**

On the ordinary view of each species having been specially and independently created, why should one part of a structure that differs from the same part in other independently created species of the same genus be more variable than those parts that are very similar in the same species? I do not see that any explanation can be given. **But if we believe that species are only strongly marked and fixed varieties, then we might expect to often find them still continuing to vary in those parts of the structure that have varied in the recent past.** To state the case in a different way, the characteristics in which all the species of a genus resemble each other, and in which they differ from allied genera, are called generic characteristics and are used to define the members of a genus; those characteristics may be attributed to inheritance from a common ancestor, for it can rarely have happened that natural selection will have modified several unrelated species, fitted to more or less widely different habits, in exactly the same way. And as these so-called generic characteristics have been inherited from before the time when the various species first branched off from their common ancestor and subsequently have not come to differ to any large extent, it is not likely that they should vary today. On the other hand, since the characteristics that distinguish one species from another within the same genus (specific characteristics) have varied and come to differ *after* the time when the various species branched off from their common ancestor, it is reasonable that they should still often be somewhat variable today—at least more variable than those parts of the organism that have remained constant for a very long period.

Secondary Sexual Characteristics Are Highly Variable

I think most naturalists will admit that secondary sexual characteristics are highly variable. It will also be admitted that species belonging to any particular group differ from each other more widely in their secondary sexual characteristics than in other parts of their anatomy. Consider, for example, the turkey, chicken, pheasant, and other gallinaceous[12] birds, and compare, for instance, the amount of difference between the males, in which secondary sexual characteristics are strongly displayed, with the far more limited differences between the females. Although we don't know what causes the original variability of those characteristics, we can easily see why they

---

[12] Gallinaceous refers to birds of the order Galliformes.

should not have been made as constant and uniform as other traits: they are accumulated by sexual selection, which is less rigid in its action than ordinary selection since it does not involve death but only gives fewer offspring to the less-favored males. As secondary sexual characteristics are highly variable, sexual selection will have had a wide scope for action and may thus have succeeded in giving to the species of the same group a greater amount of difference in these traits than in others.

Remarkably, the differences between the two sexes of the same species are generally displayed in the very same parts of the anatomy in which species of the same genus differ from each other. Let me just give the first two examples on my long list of examples to illustrate this point. As the differences in these cases are of a very unusual nature, the relation can hardly be accidental.

Among most groups of beetles, the terminal section of the leg is called the tarsus, or foot. For beetles in most groups, all individuals have the same number of segments in the tarsi. But in the beetle family Engidae,[13] as the entomologist John Obadiah Westwood has remarked, the number varies greatly, and even more intriguing, the numbers differ in the two sexes of each species. Again, in the ground-dwelling Hymenoptera (a very large group of species that includes the wasps, bees, and ants), the pattern of nerves in the wings is a taxonomic characteristic of the highest importance, because it is common to large groups; but in certain genera the pattern differs among different species, as it also does between males and females of the same species. The very versatile archaeologist and biologist Sir John Avebury Lubbock has recently remarked that several minute marine crustaceans—in the copepod genus *Pontella*—offer excellent illustrations of this law (see Figure 5.8B,C). "In *Pontella*, for instance, the sexual characters are afforded mainly by the anterior antennae and by the fifth pair of legs: the differences between species are also principally given by these organs."[14] This relation has a clear meaning in my view of things: I look at all the species of the same genus as having certainly descended from a common ancestor, just as have the two sexes of any one species. Consequently, whatever part of the structure of the common ancestor, or of its early descendants, became variable, variations of this part would, mostly likely, be taken advantage of by both normal selection and sexual selection. While natural selection would act to gradually fit the different species to their particular niches, sexual selection would gradually act to fit the two sexes of the same species to each other, or to enable the males to best struggle successfully with other males for possession of the females.

Finally then, here are the key facts before us: (1) characteristics that distinguish species from each other are more variable than are those that distinguish the different genera from each other, or those that are possessed by all the species; (2) characteristics that are developed in a species in an extraordinary manner are frequently far

---

[13] These beetles are now in the family Erotylidae.
[14] In male copepods, the right first antenna is always hinged, to grasp the female in mating, and the right limb of the fifth pair of hind legs is always shaped like a claw, for the same purpose. In some species, both the left and right limbs are so modified.

more variable than the same parts in other species in the same genus; (3) parts that are common to a whole group of species show only a slight degree of variability, however extraordinarily those parts may be developed; (4) secondary sexual characteristics are highly variable and show great differences in closely allied species; and (5) secondary sexual characteristics and ordinary specific differences are generally displayed in the same parts of the anatomy. These principles are all closely connected, all being mainly due to (1) all of the species of a particular group being descended from a common ancestor, from whom they have inherited much in common; (2) parts that have recently and largely varied being more likely to go on varying than parts that have long been inherited and have not varied; (3) natural selection having more or less completely, according to the lapse of time, overmastered the tendency to reversion and to further variability; (4) sexual selection being less rigid than ordinary selection; and (5) variations in the same body parts having been accumulated by both natural selection and sexual selection and having been thus adapted for secondary sexual purposes as well as for ordinary, non-sexual purposes.

Distinct Species Present Analogous Variations, So That a Variety of One Species Often Presents a Trait Typical of a Related Species, or Reverts Back to Some Trait Possessed by an Early Ancestor

These claims will be most readily understood by considering our domestic animals and plants. In widely separated regions, the most distinct breeds of pigeon each sometimes present subvarieties with reversed feathers on the head or with feathers on the feet—characteristics never found in the aboriginal rock pigeon that gave rise to all pigeon breeds long ago. These then are "analogous variations" in two or more distinct races. The frequent appearance of 14 or even 16 tail feathers in some pouter pigeons (see Figure 1.2) may be considered as a variation representing the normal structure of another race of pigeons, the fantail. I presume that no one will doubt that all such analogous variations are due to the several races of the pigeon having inherited from a common parent the same constitution and the same tendency to vary in certain directions when acted on by similar (but presently unknown) influences.

Among plants we have a similar case of analogous variation in the enlarged stems of the Swedish turnip and the rutabaga (Figure 5.9), plants that several botanists rank as mere varieties produced by cultivation from a common parent. If this is not the case, then it represents analogous variation in two so-called distinct species; and to these then, a third species may be added, namely the common turnip (Figure 5.10), which shows the same enlarged stems. If we are to believe that each species was independently created, we should have to attribute this similarity in the enlarged stems of these three plants not to descent from a common ancestor and a consequent tendency to vary in a similar manner, but rather to three separate yet closely related

Figure 5.9 Rutabaga (*Brassica napus*).

Figure 5.10 Common turnip (*Brassica rapa*).

acts of creation. Many similar cases of analogous variation have been observed by the French botanist Charles Victor Naudin in the great gourd family, and by various other authors in wheat and other cereals. Similar cases occurring with insects under natural conditions have recently been discussed by the American entomologist Mr. Benjamin Walsh.

With pigeons, we have a particularly striking case: the occasional appearance of slaty-blue birds with two black bars on the wings, white loins, a bar at the end of the tail, and with the outer feathers externally edged with white near their base; such characteristics occasionally appear in all pigeon breeds. As these marks all characterize the parent rock pigeon (see Figure 1.3D), I presume no one will doubt that this is a clear case of reversion to ancestral characteristics and not of a new analogous variation suddenly appearing in the several breeds. We may, I think, confidently come to this conclusion because, as we have seen earlier, these colored marks are eminently liable to appear in the offspring when two distinct and differently colored breeds are mated; there is nothing in the external conditions of life to cause that reappearance of the slate-blue coloration, or with the several other marks described, beyond the influence of the mere act of crossing on the laws of inheritance, whatever they may turn out to be.

No doubt it is a very surprising fact that characteristics should sometimes reappear after having been lost for many—indeed, probably for many hundreds of—generations. But when a breed has been crossed only once with some other breed, the offspring occasionally show a tendency to revert in character to the foreign breed for many generations—some say for a dozen or even a score of generations. This is quite remarkable; after 12 generations, the proportion of blood, to use a common expression, from one ancestor is only 1 part in 2,048; and yet, as we see, it is generally believed that this remnant of foreign blood somehow retains a tendency to reversion.[15] In a breed that has not been crossed, but in which both parents have lost some particular characteristic that their ancestor possessed, the tendency, whether strong or weak, to reproduce that lost character might, as was formerly remarked, be transmitted to offspring for almost any number of generations. When a characteristic that has been seemingly lost in a breed reappears after a great number of generations, the most probable hypothesis is not that one individual suddenly takes after an ancestor removed by some hundred generations, but rather that in each successive generation the character in question has been somehow lying latent and at last, under unknown favorable conditions, is developed.[16] With the barb pigeon (see Figure 1.2), for example, which very rarely produces a blue individual, there is probably a latent tendency in each generation to produce blue plumage.[17]

If all the species in the same genus are indeed descended from a common ancestor, as I claim, we might expect them to occasionally vary in an analogous manner, so that the different varieties of two or more species would occasionally resemble each other, or that a variety of one species would occasionally resemble a different distinct

[15] We now know that the traits Darwin is talking about are controlled by particular recessive alleles—the forms of a gene that are expressed only when paired with an identical allele, and are not expressed when paired with a dominant allele—not by anything in the blood.

[16] Darwin is so close to the truth here! Surely a reading of Mendel's 1866 paper would have given him the mechanism he was looking for—the idea of discrete alleles that never change, except through mutation. If only one of Darwin's friends had sent him a copy of that paper—who knows what might have happened?

[17] Yes! Because the rare recessive allele for blue plumage never completely disappeared from the population.

species in certain characteristics—after all, this other species would be, according to our view, only a well-marked and permanent variety. But characteristics that are exclusively due to analogous variation would probably be of an unimportant nature since the preservation of all functionally important characteristics will have been fixed through natural selection, in accordance with the different habits of the different species. We might further expect that the various species within any one genus would occasionally exhibit reversions to long lost characteristics that represent those of the ancient ancestor of all current members of that genus. As, however, we do not know the common ancestors of *any* natural groups, we cannot distinguish such reversionary characteristics from those that are instead analogous. If, for instance, we did not know that the ancestral rock pigeon was not feather-footed (Figure 5.11A) or turn-crowned (Figure 5.11B), we could not know whether the appearance of such characteristics in our domestic breeds were reversions to an ancestral state or only analogous variations. Perhaps we would have inferred that the blue color was a case of reversion, based on the number of distinctive markings that are correlated with this tint, and that would probably not have all appeared together from simple variation. More especially we might have inferred reversion from the blue color and from the several marks appearing so often when differently colored breeds were crossed. Under nature, certainly, we cannot usually know which cases are reversions to formerly existing characteristics and which ones are new but analogous variations; even so, according to our theory, we can expect to find some of the varying offspring of a species assuming characteristics that are already present in at least some other members of the same group. And this is undoubtedly the case.

The difficulty in characterizing variable species is largely due to the varieties within that species mocking, as it were, other species in the same genus. A considerable list could also be given of forms intermediate between two other forms, which themselves can only doubtfully be ranked as species; this shows, unless all these closely allied species are assumed to have been independently created, that some species have, by varying, assumed some of the characteristics of the others.

Figure 5.11 (A) A feather-footed pigeon. (B) Turn-crowned pigeons are characterized by the feathers on the head projecting backward.

But the best evidence of analogous variation is afforded by organs or other body parts that are generally constant in character but which occasionally vary so as to resemble, in some degree, the same part or organ in an allied species. I have collected a long list of such cases—including plant, insect, crustacean, reptile, bird, and mammalian examples; but here, as before, I lie under the great disadvantage of not being able to give them all. I can only repeat that such cases certainly occur, and seem to be very remarkable.

I will, however, give one curious and complex case, one that does not affect any important character but that occurs in several species of the same genus, partly under domestication and partly in nature. It is a case almost certainly of reversion to an ancestral state. I am talking here about donkeys and horses. The donkey (*Equus africanus asinus*, also commonly known as the "ass"; Figure 5.12A) sometimes has very distinct transverse bars on its legs, like those on the legs of the zebra. It has been said that these bars are plainest in the foals, and from the inquiries that I have made, I believe this to be true. They also have a stripe on the shoulder, sometimes a double stripe that varies considerably in both length and outline. A white donkey (but not an albino) has been described that has neither a spinal nor a shoulder stripe; these stripes are sometimes very obscure, or even completely lost, as is the case for dark-colored donkeys. The wild horse from the central Asian plains, the koulan, first documented by the German biologist Peter Simon Pallas, is said to have been seen with a double shoulder stripe. The English zoologist Mr. Edward Blyth, now in India, has seen a specimen of the onager (*Equus hemionus*)—the "Asiatic wild ass" of central Asia—with a distinct shoulder stripe, though it normally has none; I have been informed by that expert on the horses of India, Colonel Skeffington Poole, that the foals of this species are generally striped on the legs and faintly on the shoulder as well. The quagga (*Equus quagga quagga*) (Figure 5.12B) of South Africa, though so plainly barred like

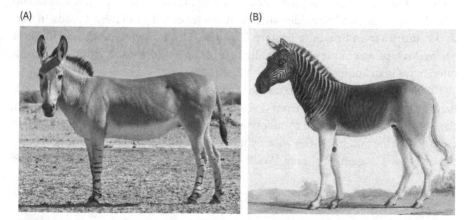

(A)   (B)

Figure 5.12 (A) A donkey (*Equus africanus asinus*); donkeys were first domesticated about 5,000 years ago. (B) The quagga (*Equus quagga quagga*) was a subspecies of the plains zebra. It became extinct in the wild by 1878, and the last zoo animal died in 1883.

a zebra over its body, has no bars on its legs; but Dr. Asa Gray of Harvard University has illustrated one specimen with very distinct zebra-like bars on the animal's hocks, a region of the animal's hind limb.

In England, I have recorded cases of the spinal stripe in horses of the most distinct breeds and of all colors. Transverse bars on the legs are fairly common among some breeds (duns and mouse duns[18]) and I have also seen this in one instance in a chestnut; duns sometimes also have a faint shoulder stripe, and I have seen a trace of one in a bay horse as well. The term "dun" includes a large range of coat colors, from one between brown and black to something closely approaching cream-colored. My son made a careful sketch for me of a dun Belgian cart horse with a double stripe on each shoulder and with leg stripes as well, and I myself have seen a dun Devonshire pony with three parallel stripes on each shoulder. A small dun Welsh pony with the same three parallel stripes has also been carefully described to me.

In northwest India, the Kathiawari breed of horses is so often striped that, according to Colonel Poole, who examined this breed for the Indian government, a horse without stripes is not considered as purebred. The spine is always striped; the legs generally show bars; they commonly have a shoulder stripe, which is sometimes double and sometimes triple, and even the side of the face is sometimes striped. The stripes are often plainest on the foals, and sometimes they disappear in older horses. Colonel Poole has seen both gray and bay Kathiawari horses striped when first foaled. I also have good cause to suspect, from information given to me by Mr. W. W. Edwards, that with the English racehorse, the spinal stripe is also much commoner in foals than in full-grown animals. I myself have recently bred a foal from a bay mare (the offspring of a Turkoman horse and a Flemish mare) mated with a bay English racehorse. When this foal was a week old, it had many very narrow, dark, zebra-like bars on its hind quarters and on its forehead, and its legs were feebly striped; all the stripes soon disappeared completely as the horse grew. Without here entering still further details, let me just state that I have collected cases of leg and shoulder stripes in horses of very different breeds in various countries from Britain to Eastern China, and from Norway in the north to the Malay Archipelago in the south. In all parts of the world, these stripes occur most often in duns and mouse duns.

Charles Hamilton Smith, who has written on this subject, believes that the several breeds of horse are descended from several different aboriginal species, one of which, the dun, was striped, and that the above-described appearances are all due to ancient crosses with the dun stock. But we may safely reject this view. It is highly improbable that the heavy Belgian cart horse, Welsh ponies, cobs, the lanky Kathiawari race, and

---

[18] Dun horses have a lighter hair coloration than normal. Mouse dun refers to a dark brownish gray color. We now know that these color patterns are caused by a simple dominant gene called "dun," which affects the deposition of both red and black pigment equally.

so forth, inhabiting the most distant parts of the world, should all have been crossed with one supposed aboriginal stock.

Now let us turn to the effects of cross-breeding among the several species in the horse genus (*Equus*). Rollins asserts that the common mule—the offspring of matings between two different species (a male donkey and a female horse)—is particularly likely to have bars on its legs; indeed, according to the English zoologist Mr. Philip Henry Gosse, in certain parts of the United States about nine out of ten mules have such striped legs. I once saw a mule with its legs so much striped that anyone might have thought it was a hybrid zebra; Mr. William Charles Martin's excellent treatise (*History of the Horse*, 1845) includes a figure of another similar mule. In four colored drawings that I have seen of hybrids from matings between donkeys and zebras, the legs were much more plainly barred than the rest of the body, and in one of them there was a double shoulder stripe. Finally, and this is another most remarkable case, a hybrid from matings between the donkey and the onager has been illustrated by Dr. Gray; this hybrid had three short shoulder stripes, like those on the dun Devonshire and Welsh ponies, and also had some zebra-like stripes on the sides of its face, even though the donkey only occasionally has stripes on its legs and the onager has none; it doesn't even have a shoulder stripe. With respect to the zebra-like stripes on the hybrid's face, I was so convinced that stripes of color never appear from what is commonly called chance that I was led to ask Colonel Poole whether such face stripes ever occurred in the eminently striped Kathiawari (also called Kattywar) breed of horses; as we have seen, the answer was yes.

How on Earth can we explain all of these facts? We see several distinct species of the horse genus becoming, by simple variation, striped on the legs like a zebra or striped on the shoulders like a donkey. In the horse we see this tendency very strongly whenever a dun tint appears—a tint that approaches that of the general coloring of the other species of the genus. The appearance of the stripes is not accompanied by any change of form or by any other new characteristic. We see this same tendency to become striped most strongly displayed in hybrids from between several of the most distinct species.

To understand these facts, let us consider the several breeds of pigeons as a parallel case: all of the breeds are descended from a pigeon of a bluish color, with certain bars and other distinctive marks (the rock pigeon; see Figure 1.3D); when any breed assumes by simple variation a bluish tint, these bars and other marks invariably reappear,, and do so without any other change of form or character. When the oldest and truest different pigeon breeds of various colors are crossed, we see a strong tendency for the blue tint and bars and marks to reappear in the mixed offspring. As I have previously stated, the most probable hypothesis to account for the reappearance of these very ancient characteristics is that there is a tendency in the young of each successive generation to produce the long-lost character, and that this tendency, from unknown causes, sometimes prevails. And we have just seen that in several species of the horse genus, the stripes are either plainer or appear more commonly in the young than in the old. If we were to think of the various pigeon breeds, some of which have bred true for centuries, as separate species, then the case would be exactly parallel with that of species in the horse genus: we see striping patterns periodically among members of the genus *Equus* because the common ancient ancestor was striped! I am confident that were we to look back thousands on thousands of generations, we would see in the common parent of the donkey, the onager, the quagga, the zebra, and our domestic horse an animal striped like a zebra but perhaps otherwise very differently constructed. **All of these animals must have descended from the same striped ancestor long, long ago.**

Anyone who believes that each equine species was independently created will, I presume, assert that each species has been created with a tendency to vary in this particular manner, both under nature and under domestication, so as often to become spontaneously striped like the other species of the genus, and that each has been created with a strong tendency, when crossed with species inhabiting distant quarters of the world, to produce hybrids resembling in their stripes, not their own parents, but other species of the genus. To admit this view, it seems to me, is to reject a real cause for an unreal cause—or at least an unknown one. It makes the works of God a mere mockery and deception; I would almost as soon believe the medieval assertion that fossil shells had never actually contained living animals but had instead been created in stone so as to mock the shells now living on the seashore.

Summary

Our ignorance of the laws of variation is profound. Not in one case out of a hundred can we pretend to explain why this or that part has varied as it has. **But whenever we have the means of making a comparison, the lesser differences between varieties of the same species and the greater differences between species of the same genus all seem to have been produced through the same laws.** Here are some of the things we think we know. Changed environmental conditions may somehow induce variability, habit and flexibility (plasticity) may produce constitutional peculiarities, and use and disuse may strengthen or weaken organs, respectively. Symmetrical parts tend to vary in the same ways and to the same degree, and some other parts are somehow related to each other so that variations in one part cause those others parts to become modified as well. Modifications in hard parts and in external parts can sometimes affect the form of softer and internal parts. When one part is largely developed, perhaps it draws nourishment from adjoining parts; but every part of the structure that can be saved without detriment will be saved. Changes of structure at an early age may affect parts developed later in life, and many such cases of "correlated variation," the mechanism behind which we are unable to understand, undoubtedly do occur.

In addition, multiple parts are especially variable both in number and in structure, perhaps because such parts have not been closely specialized for any particular function, so that their modifications have not been closely checked by natural selection. It follows, probably from this same cause, that so-called lower organisms are more variable than those standing higher on the scale and which have their whole organization more specialized. Rudimentary organs, such as the eyes in some cave animals, from being useless, are not regulated by natural selection and thus are free to vary—and vary they do. Species-defining characteristics (i.e., specific characteristics) are more variable than those that define genera (generic characteristics) or those that have long been inherited and have not differed from this same period.

In this chapter, I have sometimes referred to special parts or organs being still variable because they have recently varied and thus come to differ; but we have also seen in Chapter 2 that the same principle applies to the whole individual: for in a region where many species of a particular genus are found—that is, where there has been much former variation and differentiation, or where the production of new specific forms has been actively at work—in that district and among those species, we now find, on average, the most varieties.

Secondary sexual characteristics are highly variable, and such characteristics differ much among species in the same group. Variability in the same anatomical features has generally been taken advantage of by selection in giving secondary sexual differences to the two sexes of the same species as well as specific differences to various species within the same genus.

Any part or organ developed to an extraordinary size or in an extraordinary manner, in comparison with the same part or organ in allied species within the genus, must have gone through an extraordinary amount of modification since the genus arose. Thus we can understand why it should often still be variable in a much higher degree than other parts, for variation is a long-continuing and slow process, and natural selection will in such cases not as yet have had time to overcome the tendency to further variability and reversion to a less modified state. But when a species with any extraordinarily developed organ has given rise to many modified descendants—which in my view must be a very slow process, requiring a great lapse of time—in this case, natural selection has succeeded in giving a fixed character to the organ, in however extraordinary a manner it may have been developed.

Species inheriting nearly the same constitution from a common parent and exposed to similar influences naturally tend to present analogous variations, or these same species may occasionally revert to some of the characteristics of their ancient ancestors. Novel modifications, however they arise, will add to the beautiful and harmonious diversity of nature.

Whatever the cause may be of each slight difference between offspring and their parents—and a cause for each *must* exist!—we have reason to believe that it is the steady accumulation of beneficial differences over a great many generations that has given rise to all the more important modifications of structure in relation to the habits of each species.

Key Issues to Talk and Write About

1. Explain Darwin's argument that the occasional appearance of a blue color and certain distinctive markings when different varieties of pigeons are bred together tells us much about the ancient ancestor of the modern horse and donkey.
2. Based on the arguments given in this chapter, for which of the following do you think Darwin makes the most convincing case, and why does that argument seem most convincing to you?
   a. Use and disuse of parts
   b. Correlated variation
   c. Compensation
3. What does Darwin mean by "analogous variation"? Give an example.
4. Find out two interesting things about one of the people that Darwin mentions in this chapter. Choose from the following:
   Louis Agassiz
   Edward Blyth
   Augustin de Candolle
   Asa Gray
   Richard Owen
   Peter Simon Pallas
   Isidore Geoffroy Saint-Hilaire
   Étienne Geoffroy Saint-Hilaire
   Benjamin Silliman
   Charles Hamilton Smith
   Christian Konrad Sprengel
   Thomas Vernon Wollaston
   Johann von Goethe
   George Robert Waterhouse
   Hewett Cottrell Watson
   John Obadiah Westwood
5. Explain what Darwin means by the term "secondary sexual characteristics."
6. Following the guidelines given in Chapter 1 (see page 28, Key Issue 10), choose one of the following paragraphs from Chapter 5 and write an informative but standalone one-sentence summary of that paragraph. Choose either of the following two paragraphs:
   a. Page 123, the paragraph that begins, "I also suspect that some of the presumed cases of compensation that have been brought forward...." or
   b. Page 136, the paragraph that begins, "In northwest India, the Kathiawari breed of horses is so often striped that...."

7. Try your hand at revising the following sentences, to make them clearer and more concise:
   a. "The same number of joints in the tarsi is a character common to very large groups of beetles, but in the Engidae the number varies greatly."
   b. "They also have a stripe on the shoulder that is very variable in length and outline, and is sometimes double."

Bibliography

Kirby, W. F., and W. Spence. 1815–1826. *An Introduction to Entomology: Or Elements of the Natural History of Insects*. London.
Martin, W. C. 1845. *History of the Horse*. London.

6

Difficulties with the Theory

If modern animals and plants have evolved to their present condition slowly, over long periods of time, why don't we now see clear intermediate forms between all of the existing species? And how could incredibly complicated organs like the vertebrate eye possibly have developed gradually by small steps? If the intermediate stages weren't functional (what good is half an eye?), how could natural selection possibly account for the evolution of such structures to their present state of great complexity? And how could natural selection be at all involved in causing the development of body parts that appear to play no important role in an organism's life? And how can anyone argue that flowers, diatoms, and other such beautiful organisms weren't created especially for us to admire? Darwin answers all of these questions patiently, logically, and brilliantly and with one marvelous example after another.

Long before readers have arrived at this part of my work, a crowd of difficulties will have occurred to them. Some of the difficulties are so serious that to this day I can hardly think of them without being in some degree staggered. But, to the best of my judgment, most of the imagined difficulties pose no real problems, and those that are real are not, I think, fatal to the theory.

Each of these difficulties and objections may be placed under one of the following four headings:

1. First, if all species have descended from other species by very small steps, then why do we not everywhere see innumerable transitional forms between species? Why is not all of nature in confusion, instead of each species being, as we see them, so well defined?
2. Second, is it really possible that an animal having, for instance, the structure and habits of a bat, could have been formed by the modification of some other animal with very different habits and a very different structure? And can we honestly believe that natural selection could produce, on the one hand, an organ of trifling importance such as a giraffe's tail, which serves merely as a flyswatter, and, on the other hand, an organ so wonderfully complex as the eye?
3. Third, can instincts also be acquired and modified through natural selection? What about, for example, the instinct that leads bees to make beehives, whose inner cells are so complex as to practically anticipate the discoveries of profound mathematicians?

The Readable Darwin. Second Edition. Jan A. Pechenik, Oxford University Press. © Oxford University Press 2023.
DOI: 10.1093/oso/9780197575260.003.0007

4. Fourth, how can we account for the fact that when members of different species are mated together, they either produce no offspring or produce offspring that are themselves sterile, whereas when distinct varieties of the same species are crossed, they have no trouble producing viable offspring?

I will discuss the first two points in this chapter and the other points in succeeding chapters.

On the Absence or Rarity of Transitional Varieties

As natural selection acts solely by preserving advantageous modifications, each new superior form will tend, in the midst of intense competition or predation pressure, to take the place of and finally to exterminate its own less improved predecessors. Extinction and natural selection go hand in hand. Thus, if we look at each species as being descended from some unknown ancestral form, both the parent and all of the transitional varieties will generally have been exterminated by the very process of forming and gradually perfecting the new form.

But there is a problem: If innumerable transitional forms have existed in former times, why do we not now find them embedded in countless numbers in the Earth's crust? I will discuss this question in detail later, in the chapter "On the Imperfection of the Geological Record" (Chapter 10). Here I will simply say that I believe the answer lies mainly in the geological record being incomparably less perfect than is generally supposed. Although the Earth's crust is indeed a vast museum, the natural collections have been imperfectly preserved and only at long intervals of time. I will talk more about this later.

It may also be argued that when several closely related species inhabit the same territory at the same time, we surely ought to find at the present time many transitional forms filling in the gaps between them. Let us a take a simple case: in traveling from north to south over a continent, we generally encounter at successive intervals a series of closely allied or representative species, **each evidently filling nearly the same ecological niche where it lives**. These representative species often meet and overlap, and, as the one becomes rarer and rarer, the other becomes more and more frequent, until the one fully replaces the other as we continue our travels. But if we compare these species where they intermingle, they show no intermediate traits: they are generally as absolutely distinct from each other in every detail of structure as are specimens taken from the centers of their distribution. How can we explain this?

According to my theory of natural selection, these allied species are descended from a common ancestor, so that during the process of evolution each has gradually become adapted to the conditions of life in its own region , and has supplanted and exterminated its original ancestral form and all the transitional varieties between its past and present states. Thus we should not expect to see numerous transitional varieties in each region now, though they must have existed there at one time in the past and may in fact be embedded there still as fossils. But in the regions between where we now find the two related species having intermediate conditions of life, why do we not now find closely linking intermediate varieties? For a long time this problem quite confounded me. But I think I can now in large part explain it.

In the first place, just because an area of land is now continuous, we should be extremely cautious in assuming that it has always been so. Geology leads us to believe that most continents have in fact been broken up into a number of separate islands even within the past several million years. On such isolated islands, distinct species might have gradually evolved without the possibility of intermediate varieties existing in intermediate zones.

But I will pass over this way of escaping from the difficulty; although I do not doubt that the *formerly* isolated condition of lands that are now continuous has played an important part in forming new species—most especially with freely crossing and widely wandering animals—I believe that many perfectly defined species have also formed within strictly continuous areas.

Here is my reasoning. In looking at species as they are now distributed over a wide area, we generally find that the individuals of each species are fairly numerous over a large territory and then become somewhat abruptly rarer and rarer at the extremes of their range, finally disappearing altogether. Thus the neutral territory *between* two representative species is generally small and narrow compared with the territory dominated by each. We see the same thing when we climb mountains: sometimes it is quite remarkable how abruptly a common alpine tree species disappears as we continue to climb upward, as the Swiss botanist Alphonse de Candolle has observed. The same sharp boundaries between species have been noticed by the British naturalist Professor Edward Forbes in dredging up animals from the depths of the sea.

To those who look at climate and the other physical conditions of life as the all-important elements controlling the distributions of animals and plants, these facts ought to cause surprise as climate, elevation, and water depth graduate away almost imperceptibly. But distributions are not controlled only by physical conditions. Remember that almost every species, even where it is the most abundant, would increase immensely in numbers were it not for the impact of other species; the members of nearly all species either prey on other species or serve as prey for others (i.e., each living organism is either directly or indirectly related in the most important manner to other living organisms). **Thus we see that the area occupied by any given species in any given country by no means depends exclusively on barely perceptible, gradually changing physical conditions, but rather in large part on the presence of the other species on which it feeds, or by which it is destroyed, or with which it competes for food or space.** And, as all of these other species are already well-defined objects, not blending into one another by small, barely perceptible graduations, the range of any one species, depending as it does on the range of other species, will tend to be sharply defined.

Moreover, on the edges of its range, where each species exists in smaller numbers, the species will be extremely liable to utter extermination during fluctuations in the number of its enemies, or of its prey, or in climate. The geographical range of the species will then become even more sharply defined.

We can probably apply this same rule—that two or more allied or representative species occupying a continuous area generally have wide individual ranges with a much narrower neutral territory between them and become rarer and rarer within that narrow zone of overlap—to the distribution of *varieties within* a given species. Thus, if we take a varying species inhabiting a very large area, we shall have to adapt two varieties to two large areas and a third variety to a narrow zone between them. The intermediate variety in that narrow intermediate zone must exist in smaller numbers because it inhabits a narrow and smaller area. As far as I can tell, this rule indeed holds good with varieties in nature. I have in fact met with striking examples of this rule in the case of individuals intermediate between well-marked varieties in the barnacle genus *Balanus* (see Figure 5.8). Furthermore, it would seem from information given to me by the botanists Mr. Hewett Watson and Dr. Asa Gray, and the entomologist Mr. Thomas Vernon Wollaston, that when varieties intermediate between two other forms occur, they are indeed generally found in much smaller numbers than the forms that they connect.

If we trust these facts and inferences and thus conclude that intermediate varieties linking two other varieties together generally have existed in smaller numbers than the forms that they connect, then we can understand why intermediate varieties should not last very long in nature—why, in fact, they should generally be exterminated and disappear, and do so sooner than the forms that they originally linked together. For any form existing in fewer numbers would, as already remarked, run a greater chance of being exterminated than one existing in large numbers. In this particular case the intermediate form would be readily susceptible to the pressures imposed by the closely related forms existing on both sides of it. But the really important point here is that, during the gradual process of further modification by which two varieties will presumably develop into two distinct species, the two forms that exist in larger numbers, from inhabiting larger areas, will have a great advantage over the intermediate variety that exists in small numbers and lives in a narrow zone in between the other two. **Forms existing in larger numbers will always have a better chance of presenting further favorable variations for natural selection to work with than will the rarer forms that are fewer in number.**

In the never-ending race for life, then, the more common forms will always tend to beat and supplant the less common forms, for the less common forms will be modified and improved over time much more slowly, simply because there are fewer individuals involved. I believe it is this very same principle that accounts for the most common species in each country presenting on average more well-marked varieties than do the rarer species (see Chapter 2). Let me illustrate: suppose we have three varieties of sheep: the first is adapted to an extensive mountainous region; the second is adapted to a comparatively narrow, hilly tract of land; and the third is adapted to the wide plains at the base of the mountain. Suppose, too, that the sheep herders in these areas are all trying with equal steadiness and skill to improve their stocks by artificial selection. The great sheep holders on the mountains or on the plains will surely improve their breeds more quickly than the small holders on the intermediate, narrow, hilly tract. Consequently, either the improved mountain breed or the improved plains breed will soon replace the less improved hill breed. Thus, the two breeds that originally existed in greater numbers will now come into close contact with each other, without the interposition of the now-supplanted, intermediate hill variety.

To summarize, I believe that all species come to be reasonably well-defined, and to never present an inextricable chaos of varying and intermediate links, for several reasons. First, because variation is a slow process, new varieties can form only very slowly. Remember, natural selection can do nothing until favorable individual differences or variations occur and until a niche in some particular part of the country can be better filled by some modification of some one or more of its inhabitants. The advent of new niches in an area will depend on slow changes of climate or on the occasional introduction of new inhabitants (e.g., through immigration) and probably—and even more importantly—on some of the old inhabitants becoming slowly modified themselves over many generations. The new forms thus produced, along with the old ones, would then act on and react to each other, so that, in any one region and at any one time, we ought to see only a few species presenting slight modifications of structure that are in some degree permanent. And this is in fact what we see.

Second, areas that are now continuous must often have existed as isolated pockets in which many forms may have separately become sufficiently distinct to rank as representative species of the area. This is especially likely for animals that congregate for mating and that wander far and wide over the entire area. In this case, varieties that were intermediate between the several representative species and their common parent must formerly have existed within each isolated portion of the land—but those intermediate varieties will have been supplanted and exterminated over time by natural selection and thus will no longer be living.

Third, when two or more varieties have been formed in different parts of a strictly continuous area, intermediate varieties will probably have formed in the zones between them; but those intermediate varieties will generally have been short-lived. Why? Because the intermediate varieties will, for reasons that I have already discussed, exist in the intermediate zones in fewer numbers than the varieties that they connect. For this reason alone, the intermediate varieties will be readily vulnerable to accidental extermination. Moreover, organisms that exist in greater numbers will, on average, present more varieties and are consequently more likely to become further improved over time through natural selection, thereby gaining further advantages over any intermediates. Thus, over time, the intermediates will almost certainly be beaten and supplanted by the forms that they once connected.[1]

---

[1] We now also have evidence that there can be selection against interbreeding between individuals belonging to adjacent varieties. Such selection will discourage the formation of intermediate varieties in the first place.

And finally, looking not to any one time but to all time, if my theory is correct, then countless intermediate varieties that linked all the species within any particular group close together must assuredly have once existed. However, the process of natural selection tends to constantly exterminate both the ancestral forms and the intermediate links. Thus, evidence of their former existence will only be found among their fossil remains; however, these remains are preserved in an extremely imperfect and intermittent record, as I shall attempt to show in Chapter 10.

On the Origin and Transitions of Organic Beings with Peculiar Habits and Structure

Opponents of my views have asked how could a terrestrial meat-eating (i.e., carnivorous) animal have been converted into one with aquatic habits? How could such an animal have survived in its transitional state? Well, a number of existing carnivorous animals do in fact show very clear intermediate grades from strictly terrestrial to aquatic habits; as each experiences a severe struggle for life, it is clear that each must be quite well adapted to its place in nature.

Look, for example, at the American mink (*Neovison vison*) of North America (Figure 6.1A). It has webbed feet, like a duck, but resembles an otter in its fur, short legs, and in the form of its tail. During the summer this animal dives for and feeds on fish, but during the long winter it leaves the frozen water and feeds, like other polecats, on mice and other land animals. If a different case had been taken, and I had been asked how an insect-eating four-legged animal could possibly have been converted into a flying bat, the question would have been far more difficult to answer. Yet I think that such difficulties have little weight.

Here, as on other occasions, I lie under a heavy disadvantage: out of the great many striking cases that I have collected, I have space here to give only one or two examples of transitional habits and structures found in related species and of diversified habits among members of the same species.

Look at the family of squirrels, for instance, which includes at least 200 distinct species. Here we have the finest gradations, from some animals with their tails only slightly flattened to others with, as the naturalist Sir John Richardson has noted, the posterior part of their bodies rather wide and with the skin on their flanks rather full, and finally to the so-called flying squirrels, which have their limbs—and even the base of their tails—united by a broad expanse of skin that serves as a parachute and allows them to glide through the air to an astonishing distance from tree to tree (Figure 6.1B). We cannot doubt that each modification is of some use to each kind of squirrel in its own habitat by enabling it to escape birds or other beasts of prey, or to collect food more quickly, or to lessen the danger from occasional falls. But it does not follow from this fact that the structure of each squirrel is the *best* that it could possibly be under all possible conditions. Let the climate and vegetation change over time, or let other competing rodents or new beasts of prey immigrate into the

Figure 6.1 (A) The American mink (*Neovison vison*). (B) Flying squirrel (*Glaucomys volans*). (C) The flying lemur (colugo) (*Galeopterus variegatus*) from the rainforests of Malaysia.

squirrel's territory, or let old ones within the territory become gradually modified, and we would expect at least some of the squirrels to decrease in numbers or become exterminated—unless they also gradually became modified and improved in structure in a corresponding manner in subsequent generations. Therefore I can see no difficulty, especially under changing environmental conditions, in the continued preservation of individuals with fuller and fuller flank membranes, with each modification being useful and each therefore being propagated to future generations, until, by accumulated effects of this process of natural selection over long periods of time, a perfect so-called flying squirrel was produced.

Now let's consider the Southeast Asian flying mammals known as colugos (Figure 6.1C), a group formerly ranked among the bats but that is now believed to belong to a separate order.[2] An extremely wide flank membrane stretches from the corners of

---

[2] This is now known as the order Dermoptera. Only two colugo species exist today, both still restricted to Southeast Asia.

the jaw to the tail of these animals and includes limbs with elongated fingers. This flank membrane is furnished with an extensor muscle. Although no graduated links of structure fitted for gliding through the air now connect the colugos with any other members of the order, there is no difficulty in supposing that such links existed in the past or that each was developed in the same manner as with the less perfectly gliding squirrels, with each modification of structure having been useful to its possessor. Nor can I see any insuperable difficulty in further believing that the membrane-connected fingers and forearm of the colugos might have been greatly lengthened by natural selection, and this, as far as the organs of flight are concerned, would have converted the animal into a bat. In certain bats the wing membrane extends from the top of the shoulder to the tail and includes the hind legs; here, perhaps, we see traces of an apparatus originally suited for gliding through the air rather than for flight.

If about a dozen genera of birds had become extinct before our time, who would have ever thought that some birds might have once used their wings solely as flappers, like the flightless steamer duck (see Figure 5.1) does today; or as fins in the water and as front legs on land, like the penguin; or as sails, like the ostrich; or functionally for no purpose at all, as in the kiwi (Figure 6.2). Yet the structure of each of these birds is good for it under the conditions of life in which it lives, for each has to live by a struggle. But it is not necessarily the *best* structure possible under all possible conditions. Please do not infer from my remarks that any of the grades of wing structure

Figure 6.2 An immature and adult male kiwi (*Apteryx haastii*).

here alluded to indicate the steps by which birds actually acquired their perfect power of flight; but they do serve to show what diversified means of transition from state to state are at least possible as animals adapt to different conditions.

Let us continue this argument. Seeing that some members of such primarily water-breathing groups as the Crustacea (a group that includes the crabs and hermit crabs) and the Mollusca (the phylum that includes the snails) are adapted instead to life on the land, and seeing that we have both flying birds and flying mammals, and flying insects of the most diversified types, and that we formerly had flying reptiles, it is conceivable that flying fish (Figure 6.3), which now glide far through the air, slightly rising and turning by the aid of their fluttering fins, might have become modified by now into perfectly winged animals. If this had occurred, who would have ever imagined that in an early transitional state they had inhabited the open ocean and had used their incipient organs of flight exclusively, as far as we know, to escape being devoured by other fish?

When we see any structure highly perfected for any particular habit, such as the wings of a bird for flight, we should bear in mind that animals displaying early transitional grades of the structure will seldom have survived to the present day; rather, they will have been supplanted by their successors, which were gradually rendered more perfectly adapted for their lifestyle through natural selection over a great many generations. Furthermore, we may conclude that transitional states between structures suited for very different habits of life will rarely, when first developed, have appeared in great numbers and under many subordinate forms.

Thus, to return to our imaginary illustration of the flying fish, it seems unlikely that fishes capable of true flight would ever develop under many subordinate forms—for

Figure 6.3 Flying fish shortly after take-off.

Figure 6.4 A great kiskadee (*Pitangus sulphuratus*) standing on a tree branch.

taking prey of many kinds in many ways, for example, on the land and in the water—until their organs of flight had come to a high degree of perfection so as to have given them a decided advantage over other animals in the battle for life. Thus the chance of discovering species with transitional grades of structure in fossils will always be less than in the case of species with fully developed structures, simply because the transitional forms will have existed in smaller numbers.

I will now give a few examples of changed and diversified habits in individuals of the same species. In either case, it would be easy for natural selection to adapt the structure of the animal to its changed habits or exclusively to just one of its several habits. It is, however, difficult to decide whether habits generally change first and structure afterward, or whether slight modifications of structure lead to changed habits; both probably occur almost simultaneously in many cases. But really, this problem is immaterial for us.

For cases of changed habits, I need only remind readers of the many British insects that now feed on non-native plants, or even feed exclusively on artificial substances. Of diversified habits, I could give innumerable examples. I have, for example, often watched a bird in South America known as the great kiskadee (*Pitangus sulphuratus*, also called the tyrant flycatcher) (Figure 6.4) hovering over one spot and then proceeding to another, just like a kestrel does, and at other times standing stationary at the water's edge and then dashing into it, like a kingfisher behaves toward a fish. In our own country, the great tit (*Parus major*) (Figure 6.5) may be seen climbing branches,

Figure 6.5 The great tit (*Parus major*).

Figure 6.6 The common treecreeper (*Certhia familiaris*).

almost like a treecreeper (Figure 6.6); but then sometimes it kills small birds by blows on the head, like a shrike. And I have many times seen and heard it hammering away at the seeds of the yew tree on a branch and thus breaking them like a nuthatch does. In North America, the black bear (Figure 6.7) was seen by the English explorer, fur trader, and naturalist Samuel Hearne swimming for hours with its mouth open wide, thus catching insects in the water almost like a baleen whale catches krill.

Figure 6.7 A black bear (*Ursus americanus*).

As we sometimes see individuals following habits different from those normal for their species and even for other species in the same genus, we might expect that such individuals would occasionally give rise to new species having such anomalous habits and with their structure either slightly or considerably modified from that of their normal fellows.

Such instances do indeed occur in nature. Can a more striking instance of adaptation be given than that of a woodpecker, for climbing trees and seizing insects in the chinks of the bark? Yet, in North America, there are some woodpeckers that feed largely on fruit and others with elongated wings that they use to chase insects in flight. And on the plains of La Plata, Argentina, where hardly a trees grows, there is a woodpecker species (*Colaptes campestris*) (Figure 6.8) that has two toes before and two behind, a long pointed tongue, pointed tail feathers sufficiently stiff to support the bird in a vertical position on a post (but not as stiff as those of typical woodpeckers), and a straight, strong beak. The beak, however, is not as straight or as strong as in typical woodpeckers, but it is strong enough to bore into wood. This animal is, in all the essential parts of its structure, clearly a woodpecker. Even in such trifling characters as its coloring, the harsh tone of its voice, and its undulatory flight pattern, its close blood relationship to our common woodpecker is plainly declared. Yet, as I can assert from my own observations as well as those of the South American explorer and naturalist Félix de Azara, in certain large districts this bird does not climb trees; indeed, it makes its nest in holes in the banks of rivers! In certain other districts, this very same woodpecker frequents trees and bores holes in the trunk for its nest, as described by the ornithologist Mr. William Henry Hudson. I may mention as another illustration of the varied habits found within this genus that a Mexican *Colaptes* species has been said by the Swiss zoologist, Henri Louis Frédéric de Saussure to bore holes into hard wood in order to lay up a store of acorns. The habits and lifestyles of woodpeckers are indeed diverse.

Figure 6.8 The South American woodpecker, Pica-pau-do-campo (*Colaptes campestris*).

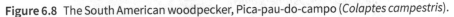

Petrels (Figure 6.9A,B) offer another example of diverse habits. They are generally the most aerial and oceanic of birds, but in the quiet waters off Tierra del Fuego, at the southern tip of South America, the petrel, in its general habits, in its astonishing power of diving, in its manner of swimming and of flying when made to take flight, would be mistaken by anyone for an auk or a grebe (Figure 6.9C); it is essentially a petrel, but with many of its structures profoundly modified over time to match its new habits of life. Compared with that bird, the woodpecker of La Plata has had its structure modified only slightly. In the case of dippers (members of the genus *Cinclus*) (Figure 6.10), even the most careful observer would never suspect its sub-aquatic habits merely by examining its dead body. Yet this bird, which is allied with the thrush family, subsists by diving, using its wings to move under the water and grasping stones with its feet.

Last, all of the hymenopterous[3] insects (class Insecta, order Hymenoptera) are terrestrial except for members of a single genus, *Proctotrupes*, which the biologist Sir John Lubbock has discovered to be aquatic in its habits. It often enters the water and dives about by the use of its wings rather than its legs, and remains as long as 43 hours beneath the surface before coming up again for air. And yet despite its abnormal habits, it exhibits no corresponding modifications in structure.

He who believes that each being has been created as we now see it must occasionally have felt surprise when meeting an animal having habits not in agreement with its structure. What can be plainer than that the webbed feet of ducks and geese are formed for swimming? And yet there are web-footed upland geese that rarely go near the water. And, as described by the French-American ornithologist John James

---

[3] Members of the insect order Hymenoptera, which includes at least 150,000 species of bees, ants, sawflies, and wasps. In comparison, there are fewer than 5,500 mammal species.

Figure 6.9 (A) The northern giant petrel (*Macronectes halli*). (B) A diving-petrel (*Pelecanoides urinatrix*). (C) A pied-billed grebe (*Podilymbus podiceps*). These are expert divers: "part bird, part submarine." They are common in freshwater marshes, sluggish rivers, and estuaries.

Figure 6.10   A Eurasian dipper (the water ouzel in Darwin's day), genus *Cinclus*.

Audubon, the frigate bird (Figure 6.11A) alights on the surface of the ocean using its toes, all four of which are webbed. On the other hand, grebes and coots are eminently aquatic, even though their feet are not webbed: their toes are only bordered by membrane. What seems plainer than that the long toes found in wading birds, which are not furnished with any membrane between the toes, are formed for walking over swamps and floating plants? And yet the purple gallinule (Figure 6.11B), the moorhen (Figure 6.11C), and corncrake (Figure 6.11D) are members of a single family (Rallidae), even though the moorhen is nearly as aquatic as the coot, and the corncrake is nearly as terrestrial as the quail or partridge. In such cases—and many others could be given—habits have changed without a corresponding change of structure. The webbed feet of the upland goose may be said to have become almost rudimentary in function, though not in structure. In the frigate bird, the deeply scooped membrane between the toes shows that that structure has begun to change.

He who believes in separate and innumerable acts of creation may say that in all of these cases it has pleased the Creator to cause a being of one type to replace a being of another type. But this seems to me to be only restating the facts in dignified language. **He who believes in the struggle for existence and in the principle of natural selection will acknowledge that every organic being is constantly endeavoring to increase in numbers, and that if one individual varies in a useful way ever so little, either in habits or in structure, and in doing so gains an advantage over some other inhabitant of the same region, it will seize on the place of that inhabitant, however different that may be from its own home.** Such a person will not be surprised to learn of geese and frigate birds living on the dry land and rarely alighting on the

Figure 6.11 (A) A male great frigatebird (*Fregata minor*). (B) Purple gallinule, water walker (*Porphyrio martinica*). (C) Moorhen (*Gallinula chloropus*). (D) Corncrake (*Crex crex*).

water despite their having webbed feet; or that there are long-toed corncrakes living in meadows instead of swamps; or that there are woodpeckers living where hardly a tree grows; or that there are diving thrushes, and diving hymenopteran insects, and birds such as petrels that live always at sea (except to breed).

Organs of Extreme Perfection and Complication

To suppose that the eye, with all its inimitable contrivances for adjusting the focus to different distances, for admitting different amounts of light, and for correcting spherical and chromatic aberration, could have been formed by natural selection, seems, I freely confess, absurd in the highest degree. But when it was first said that the Sun stood still and that the world revolved around it, the common sense of mankind mistakenly declared that doctrine to be false. It seemed obvious that the Sun revolved around the Earth. As every philosopher knows, however, the old Latin saying of *vox populi vox Dei* ("The voice of the people is the voice of God") cannot be trusted in science. Reason tells me the following:

- That if we can show the existence of numerous graduations from a simple and imperfect eye to one that is complex and perfect, with each grade being useful to its possessor, as is certainly the case; and
- if further, the eye's characteristics can vary among individuals and that those variations are inherited, as is also certainly the case; and
- if such variations should be useful to an animal under changing conditions of life,

then the difficulty of believing that a perfect and complex eye could eventually be formed by natural selection, though perhaps almost impossible for our imaginations to grasp, should not be considered to doom the theory. How a nerve comes to be sensitive to light hardly concerns us any more than how life itself originated, but as some of the lowest organisms can perceive light even though no nerves can be detected, it is clearly possible that certain sensitive elements in their cytoplasm should have eventually become aggregated and developed into nerves endowed with this special sensibility.

In searching for the stages through which any particular organ in any particular species has been perfected, we should look exclusively to its direct ancestors. But this is scarcely ever possible, and so we are forced to look to other species and genera belonging to the same group—that is, to the collateral descendants, the nieces and cousins descended from the same ancestor—in order to see what gradual stages are possible, hoping for the chance of finding some gradations that have been transmitted in an unaltered or little-altered condition. The state of the same organ in distinct classes of organisms may incidentally throw light on the steps by which it has been perfected. Let us take this approach in trying to understand the evolution of eyes.

The simplest organ that can be called an eye consists merely of an optic nerve surrounded by pigment cells and covered by translucent skin that lets in light, but without any lens or other refractive body. We may, however, descend even a step lower and find, according to the French biologist Monsieur S. Jourdain, clusters of pigment cells lacking nerves and resting merely in cytoplasm, apparently serving as crude organs for the perception of light and dark. In certain sea star species, small depressions in the layer of pigment surrounding the nerve are filled, as described by Monsieur Jourdain, with transparent gelatinous matter projecting with a convex surface like the cornea in higher animals. Monsieur Jourdain suggests that this serves not to form an image, but only to concentrate the luminous rays and facilitate their perception. In this concentration of the light rays we gain the first and by far the most important step toward the formation of a true, picture-forming eye; for if we place the naked extremity of the optic nerve—which in some of the lower animals lies deeply buried in the body and in others near the surface—at the right distance from the light-concentrating apparatus, an image will be formed on it.

Among the arthropods[4] and annelids we may start from an optic nerve simply coated with pigment, the latter sometimes forming a sort of pupil but without any lens or other optical contrivances. With insects we now know that the numerous facets on the cornea of their great compound eyes form true lenses (Figure 6.12) and that their photosensitive receptors cells, the cones, include curiously modified nervous filaments. But these organs are so incredibly diversified among arthropods that the German biologist Fritz Müller formerly divided these animals into three main classes with seven subdivisions, in addition to establishing a fourth major class of arthropods with simple eyes.

When we reflect on the wide, diversified, and graduated range of structures that we find in the eyes of lower animals, and when we bear in mind how small the number of all living forms must be in comparison with all those that have become extinct, it becomes easier to believe that natural selection may well have gradually converted the simple apparatus of an optic nerve, coated with light-absorbing pigment and covered with a transparent membrane, into an optical instrument as perfect as is possessed by any members of the Arthropoda.

Anyone who has come this far with my argument ought not to hesitate in going one step further: if you find upon finishing this volume that large bodies of facts, otherwise inexplicable, can now be explained satisfactorily by the theory of gradual modification through natural selection, then you should admit that even a structure as perfect as an eagle's eye might thus be formed, even without knowing any of the transitional steps to the final product. Some have objected that in order to modify the eye and still preserve it as a perfect instrument, many changes would have to have happened simultaneously, which, it is assumed, could not be done through natural selection. But, as I have attempted to show in my work on the variation of domestic

---

[4] This group, members of the phylum Arthropoda, includes animals as diverse as insects, spiders, crabs, lobsters, barnacles, and copepods.

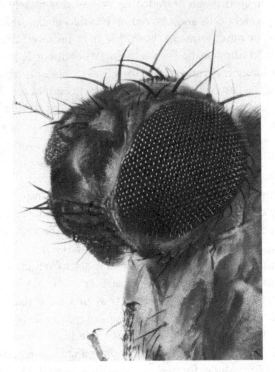

Figure 6.12 Compound eye of a fly. Note the hundreds of lenses at the eye's surface.

animals, we need not suppose that the modifications were all made simultaneously, as long as they were extremely slight and the process was gradual. Also, different kinds of modification can serve for the same general purpose. As my fellow naturalist Mr. Alfred Russel Wallace has remarked, "if a lens has too short or too long a focus, it may be amended either by an alteration of curvature, or an alteration of density; if the curvature be irregular, and the rays do not converge to a point, then any increased regularity of curvature will be an improvement. So the contraction of the iris and the muscular movements of the eye are neither of them essential to vision, but only improvements that might have been added and perfected at any stage of the construction of the instrument."

Even within the Vertebrata,[5] the highest division of the animal kingdom, we can start from an eye so simple that it consists—as in the lancelets (see Figure 4.16)— of a little sack of transparent skin, furnished with a nerve and lined with pigment, but destitute of any other apparatus. In fishes and reptiles, as Sir Richard Owen of the British Museum has remarked, "the range of gradations of dioptric (or refractive) structures is very great," while in humans, according to the high authority of

---

[5] This subclass of the phylum Chordata includes all animals with vertebrae (e.g., birds, fish, reptiles, amphibians and mammals—including humans).

the German doctor and biologist Rudolf Virchow, even our own beautiful crystalline lens is formed during embryonic development by an accumulation of epidermal cells lying in a sack-like fold of the skin, and the vitreous body is formed from embryonic subcutaneous tissue.

To arrive at a valid conclusion regarding eye formation, with all its marvelous yet not absolutely perfect characters, it is crucial that reason should conquer the imagination. But I have felt the difficulty far too keenly myself to be surprised at others hesitating to extend the principle of natural selection to so startling a length. Still, let us continue the argument.

It is scarcely possible to avoid comparing the eye with a telescope. We know that the telescope has been perfected by the long-continued efforts of the highest human intellects; naturally we infer that the eye has been formed by a somewhat analogous process. But might this inference be presumptuous? Have we any right to assume that the Creator works by intellectual powers like ours? If we must compare the eye to an optical instrument, we ought to imagine a thick layer of transparent tissue containing fluid-filled spaces, with a nerve sensitive to light beneath. Then suppose every part of this layer to be continually changing very slowly in density over time, so as to separate into layers of different densities and thicknesses, placed at different distances from each other, and with the surfaces of each layer slowly changing in form. Further we must suppose that there is a power, represented by natural selection or the survival of the fittest, always intently watching each slight alteration in the transparent layers and carefully preserving any of these variants that tend to produce a more distinct image. We must suppose each new state of the instrument to be multiplied by the millions, each to be preserved until a better one is eventually produced. Then we must suppose the old ones to be all destroyed, probably through competition with individuals possessing the better ones.

In living organisms, natural variation will cause the slight alterations, reproduction through the generations will multiply them almost infinitely, and natural selection will then pick out each improvement with unerring skill. Let this process go on for millions of years, and during each year on millions of individuals of many kinds.[6] May we not then believe that a living optical instrument might eventually be formed that is as superior to one made of glass as are the works of the Creator to those of Man?

Modes of Transition

If someone could demonstrate the existence of even just one complex organ that could not possibly have been formed by numerous successive, slight modifications, my theory would absolutely break down. But I can find no such case. No doubt many

---

[6] Actually, the process may occur much more quickly than that: See Nilsson, D-E. and S. Pelger. 1994. A pessimistic estimate of the time required for an eye to evolve. *Proceedings of the Royal Society of London Series B* 256: 53–58.

organs exist of which we do not know the transitional forms, particularly if we look to much-isolated species around which, according to my theory, there has been much extinction. Or again, let's take an organ common to all the members of a class. In this case, the organ must have been originally formed very long ago, and all the many current members of the class must have been developed after that time. In order to discover the early transitional grades through which the organ in question has passed, we should have to look to very ancient ancestral forms, forms that have long been extinct.

We should be extremely cautious in concluding that an organ could not have been formed by transitional gradations of some kind. Many cases could be given among the lower animals of the same organ performing several wholly distinct functions at the same time. In dragonfly larvae and in the fish genus *Cobites*, for example, the digestive tract does three different things: it digests, it respires, and it excretes. And if the freshwater *Hydra* (Figure 6.13) is turned inside out, the inner and outer surfaces of the animal will exchange tasks: the former exterior surface will now digest, and the former stomach will now exchange gases.

In such cases, natural selection might slowly lead to specialization, so that the whole organ or one part of an organ that had previously performed two or more functions might now perform one function alone, if any advantage would be gained from doing so. Thus, by many insensible steps, its nature would be greatly changed.

Consider a similar example with plants. Many plants regularly produce several differently constructed flowers; if such plants were to produce only one kind of flower, a great change in the character of the species would be brought about with comparative suddenness. It is, however, likely that the two sorts of flowers now borne by the same plant were originally differentiated by finely graduated steps, which may still be followed in a few cases.

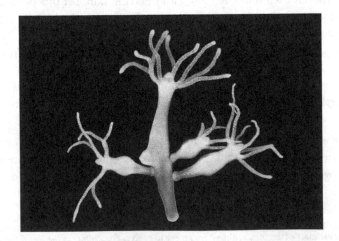

Figure 6.13 This freshwater hydra is reproducing asexually by budding.

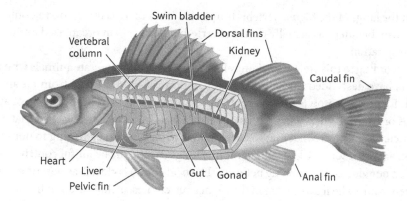

Figure 6.14 Fish swim bladder (shown in blue). Most fish use this organ to become neutrally buoyant in the water, after which they need not expend energy to stay at that particular depth.

We also know of instances in which two distinct organs, or the same organ in two very different forms, may simultaneously perform the same function in a single individual. This provides an extremely important means of future transition. Let me give one example. Some fish with gills take up the oxygen dissolved in water at the same time that they also take up oxygen in gaseous form from the air in their swim bladders (Figure 6.14), this latter organ being divided by highly vascularized partitions and being provided with air through a specialized pneumatic duct. In addition, the swim bladder of certain fishes has come to also function as an accessory to the organs of hearing. To give a similar example from the plant kingdom, plants climb by three distinct means—by spirally twining, by clasping a support with their sensitive tendrils, and by emitting aerial rootlets. These different mechanisms of climbing are usually found in distinct groups of plants; however, a few species exhibit two of the above mechanisms or even all three combined in the same individual. In all such cases, one of the two organs might readily be modified and perfected so as to perform all the work, being aided by the other organ during the progress of modification. This other organ—now freed of its responsibilities—might later become gradually modified for some other quite distinct purpose or be wholly obliterated.

The illustration just given of the fish swim bladder is a good one because it clearly shows us the highly important fact that an organ originally constructed for one purpose (in this case for adjusting an animal's buoyancy) can in fact be converted into one for a widely different purpose (respiration). Indeed, all physiologists admit that the swim bladder is homologous[7] (that is, "ideally similar in position and structure")

---

[7] Today that term is used more clearly to mean similarity by descent from a common ancestor. The contrasting term "analogous" refers to structures that are similar in structure or function, but without descent from a common ancestor.

with the lungs of the higher vertebrate animals. Thus there is no reason to doubt that the swim bladder has actually been converted into lungs—an organ used exclusively for respiration.

According to this view, then, it may be inferred that all vertebrate animals with true lungs have descended through many generations of reproduction from an ancient and unknown prototype animal that was furnished with a buoyancy-control apparatus or swim bladder. We can thus understand the otherwise strange fact that every particle of food and drink that we swallow has to pass over the opening to our windpipe, and with some risk of falling into the lungs despite the beautiful contrivance by which our glottis (the opening between the vocal cords) is closed, as described by Sir Owen. And while it is true that in the higher vertebrates, including adult humans, the gills have wholly disappeared, yet we still see gill slits on the sides of the neck during embryonic development, along with the loop-like course of the arteries marking their former positions. Conceivably, the now utterly lost gills might instead have been gradually worked on by natural selection for some other distinct purpose. For instance, the German zoologist Hermann Landois has shown that the wings of insects develop from the tracheae;[8] thus it is highly probable that in this large and successful group of animals, organs that once served for respiration have actually been converted into organs for flight. What a magnificent thought!

In considering the evolutionary modification of organs, it is so important to bear in mind the probability of an organ being converted from one function to another that I will give one more example, this time from the world of barnacles. Stalked barnacles (Figure 6.15) have two minute folds of skin, which I have called "ovigerous frena." These folds produce a sticky secretion that holds the embryos in place within the parent's nursery chamber until they hatch from their egg membranes. These barnacles have no gills; gas exchange is accomplished very simply, across the general body surfaces and those of the nursery chamber, together with the small frena (folds of membrane that restrict motion). On the other hand, the acorn barnacles, like members of the genus *Balanus*, for example (see Figure 5.8A) have no ovigerous frena; their embryos lie loose at the bottom of the nursery chamber, within the well-enclosed outer shell of the adult. They do have, however, in the same relative position as the frena of stalked barnacles, a set of large, much-folded membranes that freely communicate with the circulatory pouches of the animal's nursery chamber and outer body, and which have been considered by all naturalists to act as gills. Now I think that no one will dispute that the ovigerous frena in the one family of barnacles are strictly homologous with the gills of the other family; indeed, they graduate into each other. Therefore it need not be doubted that the two little folds of skin that had originally served as ovigerous frena, but which also very slightly aided in gas exchange, have been gradually converted by natural selection into gills simply through an increase in their size and the obliteration of their adhesive glands. If all stalked barnacles had

---

[8] These are major components of the insect (and spider) gas exchange system, guiding air from small openings on the outside of the body to individual cells within the animal's tissues.

Figure 6.15 A stalked ("gooseneck") barnacle (*Pollicipes pollicipes*).

before now gone extinct—and they have indeed suffered far more extinction than have the sessile barnacles—then who would ever have imagined that the gills of the acorn barnacle had originally existed as organs whose role was to prevent the eggs from being washed out of the nursery chamber?

There is another possible mode of evolutionary transition to consider: namely, reproducing earlier or later in life. This idea has lately been promoted by the American paleontologist and comparative anatomist Professor Edward Drinker Cope and others in the United States. We now know that some animals can reproduce at a very early age, before they have acquired their perfect adult characters. If this power became thoroughly well-developed in a species, it seems probable that the presently existing morphologically distinct adult stage would sooner or later be lost from the life cycle. In that case, especially if the larva was very different-looking from the adult form, the character of the species would be greatly changed and simplified.[9]

Finally, we know that many animals, after arriving at maturity, go on changing in character during nearly their whole lives. With mammals, for instance, the form of the skull typically changes a great deal with age. The Scottish pathologist and naturalist Dr. James Murie has given some striking instances of this with seals. And, of course, everyone knows how the horns of stags become more and more branched with age and that the plumes of some birds become more finely developed as each grows older. Professor Cope writes that the teeth of certain lizards also change much in shape with advancing years; with crustaceans, some parts—both important ones and trivial ones—assume a new character after maturity, as recorded by Fritz Müller. In such cases—and many more could be given—if the age for reproduction were

---

[9] Indeed, this may well explain the evolution of planktonic "larvaceans" (solitary free-swimming organisms in the subphylum Urochordata, phylum Chordata) from immobile sea squirt ancestors.

(A)   (B)

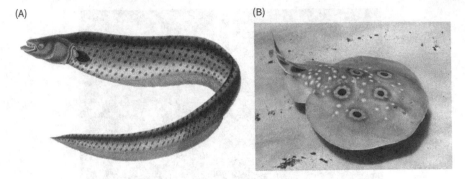

Figure 6.16 (A) The electric knife fish (*Gymnotus inaequilabiatus*). (B) The electric ray (*Torpedo torpedo*).

delayed, the character of the species, at least in its adult state, would be modified. Nor is it improbable that the earlier stages of development would in some cases be hurried through and finally lost. Whether species have often—or even ever—been modified through this comparatively sudden mode of transition, I can form no opinion. But if this has in fact occurred, the differences between the young and the mature, and between the mature and the old, were probably acquired long ago by gradual steps.

Special Difficulties of the Theory of Natural Selection

Although we must be extremely cautious in concluding that any organ could not have been produced by successive, small, transitional gradations, several cases seem to point in that direction. But I believe that all of these cases can ultimately be explained through the power of natural selection and survival of the fittest acting over very long periods of time.

One of the most difficult situations to explain is that of neuter, nonreproductive insects, which often look very different from both the males and the fertile females of the same species. I will discuss that case in the next chapter. Here I will instead discuss another case of special difficulty for my theory: the electric organs of fishes.

It is impossible to conceive by what gradual steps these wondrous organs have been produced. But this is not surprising, for we do not even know with certainly of what use these organs are to their bearers. In the Amazonian electric knife fish (*Gymnotus* spp.) (Figure 6.16A), and the electric ray *Torpedo* (Figure 6.16B), the electric organs no doubt serve as a powerful means of defense and perhaps also for securing prey; yet, in the ray, as noted by the Italian neurophysiologist Carlo Matteucci, a similar organ in the tail manifests very little electricity, even when the animal is greatly irritated—so little, in fact, that it can hardly be of any use in either defense or prey capture. Moreover, the Irish surgeon and anatomist Dr. Robert McDonnell has shown that, besides the organ just referred to, there is another organ near the ray's head that is also not known to be electrical but which appears nevertheless to be truly homologous with the electric battery in the electric ray. It is generally believed that there is a close similarity between these organs and ordinary muscles—in the fine structural details, in the distribution of the nerves, and in the manner in which they are acted on by various reagents. Remember, too, that muscular contraction is accompanied by electrical discharge. Moreover, as the English physician Dr. Charles Bland Radcliffe insists, "in the electrical apparatus of the torpedo during rest, there would seem to be a charge in every respect like that which is met with in muscle and nerve during rest, and the discharge of the torpedo, instead of being peculiar, may be only another form of the discharge which depends upon the action of muscle and motor nerve." Beyond this we presently know little. But as we know so little about the uses of these organs, and as we know nothing about the habits and structure of the ancestors of the existing electric fishes, it would be extremely bold to maintain that there is no mechanism by which these organs might have been gradually developed over time.

Electric organs appear at first to offer another and far more serious difficulty: they occur only in about a dozen kinds of fish, of which several are only distantly related to each other and to any of the others. When we find the same organ in several members of the same group, especially in members having very different habits of life, we may generally attribute its presence to inheritance from a common ancestor and its absence in some of the members to loss through disuse or natural selection. If all electric organs had in fact been inherited from some one ancient ancestor, we might have expected that all existing electric fishes would be closely related to each other. But this is far from the case. Nor does the geological record suggest that most fishes formerly possessed electric organs, which most of their modified descendants have now lost.

So how can we account for this intriguing distribution of electric organs among fish? In looking at the subject more closely, we find that the electric organs of the different fish species are situated in different parts of the body. They also differ in construction among the different species, as well as in the arrangement of the disc-like plates that are stacked one on another like batteries, and, according to Italian anatomist Filippo Pacini, even in the mechanism through which the electricity is produced. Last, the electric organs of the different species are supplied with nerves proceeding from different sources: this is perhaps the most important of all the differences. Thus, the electric organs found in the various fish species cannot be considered as homologous, but only as analogous in function—they seem not to have arisen only once in a common ancestor; if they had, they would have closely resembled each other in all respects. Thus there is no need to explain how an identical organ arose separately in several remotely related species: the organs are *not* identical. But we still have a lesser yet still great difficulty to resolve: By what gradual steps have these organs been developed in each separate group of fishes?

The light organs that occur in a few insect species belonging to widely different families, and which are situated in different parts of the body in different species, offer, under our present state of ignorance, a difficulty almost exactly parallel to that of the electric organs of fishes. Other similar cases could be given. For instance, in plants we find what seems to be the same curious contrivance in both orchids and milkweeds (genus *Asclepias*), namely a mass of pollen grains borne on a footstalk with an adhesive gland, which sticks the pollen to the backs of insects for transport to other flowers. And yet these two groups are about as remotely related as is possible among flowering plants. But, here again, the parts are not homologous (i.e., they do not come from a common ancestor).

In all cases of beings far removed from each other in the scale of organization but which are furnished with similar and peculiar organs, it will be found that although the general appearance and function of those organs may be the same, we can always detect fundamental differences between them. For instance, the eyes of cuttlefish, squid, and octopus (class Cephalopoda, in the phylum Mollusca) seem remarkably similar to those of vertebrate animals. In such distantly related groups no part of this resemblance can be due to inheritance from a common ancestor. The English biologist Mr. St. George Jackson Mivart has advanced this case as one of special difficulty, but I am unable to see the force of his argument. Any organ intended for vision must be formed of transparent tissue and must include some sort of lens for projecting an image at the back of a darkened internal chamber. Beyond this superficial resemblance, there is hardly any real similarity between the eyes of cephalopods and vertebrates, as may be seen by consulting Christian Victor Hensen's admirable memoir on these organs in cephalopods. I can't discuss this in detail here, but I may specify a few of the points of difference. The crystalline lens in the eyes of the most advanced cuttlefish (Figure 6.17) consists of two parts, placed one behind the other like two lenses, both having a very different structure and disposition to what we find among vertebrates. Moreover, the cuttlefish retina is wholly different, with a very clear inversion of the elemental parts and with a large mass of nervous tissue included within the membranes of the eye. The relations of the muscles are also as different from those

Figure 6.17 Cuttlefish belong to the same animal group that includes the octopus and the squid, and to the phylum that also includes the snails and oysters.

of the vertebrate eye as it is possible to conceive, and so on in many other points. It's really not clear whether the same terms should even be used in describing the eyes of cephalopods and vertebrates. Anyone may, of course, deny that the eye in either case could have been developed through the natural selection of slight variations in many successive generations. But if one admits that this process produced the eye in one case, it is clearly possible as well in the other.

Fundamental structural differences in the visual organs of the two groups might have been anticipated, in accordance with this view of their manner of formation. Just as two people have sometimes independently hit on the same invention, so in the several foregoing cases it appears that natural selection, working for the good of each being and taking advantage of all favorable variations, has produced functionally similar organs independently in distinctly different groups, with none of the similarity owing to inheritance from a common ancestor.

Fritz Müller, in order to test the conclusions reached in this book, has carefully constructed a nearly similar line of argument. Several families of crustaceans—most of which are aquatic—include a few species that are physically modified to breathe air and live on land. In carefully studying the members of two of these families that are clearly closely related to each other, Mr. Müller found that the members of both species agree very closely in all important characters: in their sense organs and circulatory systems, in the position of the tufts of filtering hairs within their complex stomachs, and in the whole structure of the water-breathing gills—even to the microscopic hook by which they are cleansed. Thus, we might have expected that the equally important air-breathing apparatus would have looked the same in the land-dwelling species of both groups as well: For why should this one apparatus, which has an identical purpose in both groups of animals, have been made to differ, while all the other important organs are closely similar or even essentially identical?

Mr. Müller argues that this close similarity in so many points of structure must, in keeping with my views, be accounted for by inheritance from a common ancestor. But as the vast majority of the species in the above two families, as well as in most other crustaceans, are fully aquatic in their habits, it is improbable in the highest degree that their distant common ancestor should have been adapted for breathing air. Mr. Müller, thus being led to carefully examine the apparatus in the air-breathing species, found clear species-specific differences in several important points, including the position of the openings, the manner in which they are opened and closed, and in some other related details. Such differences are fully understandable and might even have been expected, if we suppose that several species belonging to several distinct families had slowly and independently become adapted over many generations to live more and more out of water and to breathe the air. For, by belonging to different families, the different species would have differed to a certain extent; in keeping, then, with the principle that the nature of each variation depends on both the nature of the organism and that of the surrounding conditions, the different species would not have been expected to vary in exactly the same ways. Consequently, natural selection would have had different materials or variations to

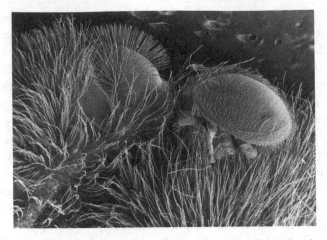

Figure 6.18 The parasitic varroa mite (*Varroa destructor*) on a honeybee host. Mites are arachnids, a group that also includes the spiders.

work on, so that even when eventually arriving at the same functional result, the structures thus acquired would almost necessarily have differed. In contrast, on the hypothesis of separate acts of creation, the whole case would remain unintelligible. This line of argument seems to have had great weight in leading Mr. Müller to accept the views that I have put forth in this book.

Another distinguished zoologist, the late Professor Jean Louis René Antoine Edouard Claparède of Switzerland, has argued in the same manner and has arrived at the same result. He shows that there are parasitic mites (Figure 6.18) (members of the arachnid order Acari) that are all furnished with hair claspers, even though they belong to several distinct families and subfamilies. These organs must have been independently developed in the different species, as they differ too much from each other to have been inherited from a common ancestor: in some species they are formed by the modification of the forelegs; in others, through modifications of the hind legs; in others, through modification of the maxillae or lips, or even of the appendages on the underside of the hind part of the body.

In all of these cases we see the same end gained and the same function performed in organisms that are not at all, or only remotely, related, and performed by organs that look similar but that develop quite differently in the different species. Indeed, it is a common rule throughout nature that, even in closely related species, the same end is often achieved by very different means. The feathered wing of a bird is so differently constructed from the membrane-covered wing of a bat, and still more so are the differences between the four wings of a butterfly and the two wings of a fly, and the two wings with the forewings (called elytra) of a beetle. Bivalve shells are made to open and shut, but they do so through an amazing variety of patterns in hinge construction—from the long row of neatly interlocking teeth in the small marine clam *Nucula*, to the simple ligament of a blue mussel (Figure 6.19). Similarly, seeds

Figure 6.19 Blue mussel (*Mytilus edulis*) showing how shells are joined along the edge by a simple strip of tissue called the ligament (pink).

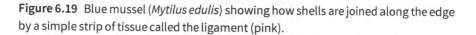

are disseminated in such a remarkable variety of ways: by their minuteness; by their capsule being converted into a light balloon-like envelope; by being embedded in pulp or flesh, formed of the most diverse parts and rendered nutritious, as well as conspicuously colored so as to attract and be devoured by birds; by having hooks of many kinds and even thin, serrated spines that adhere to the fur of passing dogs and other quadrupeds; or by being furnished with little wings and plumes, as different in shape as they are elegant in structure, so as to be wafted by every breeze.

I will give just one more example, as this subject of the same end being gained by the most diversified means well deserves our attention. Some authors maintain that the various animals and plants on this planet have been formed in many ways for the sake of mere variety, almost like toys in a toy shop, but such a view is not credible. For plants with separate sexes, the pollen clearly needs help in reaching another flower for fertilization; the same is also true for hermaphroditic plants (i.e., the flowers have both sexes) in which the pollen does not simply fall onto the stigma of that same flower (see Figure 4.7). In several species, the pollen grains themselves are light and incoherent and may simply be blown by the wind through mere chance onto the stigma of another flower; indeed, this is the simplest plan one can imagine.

However, an almost equally simple, though very different plan occurs in many plants in which a symmetrical flower secretes a few drops of nectar and is consequently visited by insects; the visiting insects then carry the pollen from the anthers to the stigma. From this simple stage of pollination we may pass through an inexhaustible number of contrivances, all functioning for the same purpose and effected in essentially the same manner but entailing changes in every part of the flower. The nectar may be stored in variously shaped receptacles: for example, with the flower's stamens and pistils modified in many different ways, sometimes forming trip-like contrivances and sometimes capable of neatly adapted movements through irritability or elasticity.

From such structures we may advance until we come to a case of truly extraordinary adaptation, as recently described by Dr. Hermann Crüger (Director of the Botanic Garden on the island of Trinidad) for the bucket orchids (genus *Coryanthes*) (Figure 6.20A). This orchid has part of its lower lip (the labellum, which is really a modified petal) hollowed out into a great bucket, into which drops of almost pure water continually fall from two secreting horns that stand above it; when the bucket is half full, the water overflows by a spout on one side. The base of the labellum lies over the bucket and is itself hollowed out into a sort of chamber with two entrances at the sides; there are a number of curious fleshy ridges within this chamber. Now the most ingenious person, if he or she had not witnessed what takes place, could never have imagined what purpose all these parts serve. But Dr. Crüger saw crowds of large bumblebees visiting the gigantic flowers of this orchid. They visited not to suck nectar, but rather to gnaw off the ridges within the chamber above the bucket. While focused on their gnawing, they frequently pushed each other into the bucket. With their wings being thus wetted they could not fly away, but were instead compelled to crawl out through the passage formed by the spout or overflow. Dr. Crüger saw a "continual process" of bees thus crawling out from their involuntary bath. Remarkably, the passage out is narrow and is roofed over by the column, so that a bee, in forcing its way out, cannot help but rub its back against the sticky stigma and then against the sticky

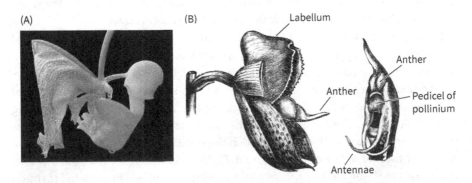

Figure 6.20 (A) A flower of the remarkable bucket orchid (*Coryanthes verrucolineata*). (B) Illustration of *Catasetum*.

glands of the pollen masses. The pollen masses are thus glued to the back of whatever bee first happens to crawl out through the passage of a lately expanded flower and are thus carried away by the bee. Dr. Crüger sent me a flower preserved in spirits of wine, along with a bee he had killed before it had quite crawled out; the dead bee still had a pollen mass fastened to its back. When the thus-encumbered bee flies to another flower, or to the same flower a second time, and is pushed by its comrades into the bucket and then crawls out through that narrow passage, the pollen mass must necessarily first contact the flower's sticky stigma and adhere to it—and the flower is thus fertilized.

Now at last we see the full use of every part of the flower: of the water-secreting horns and of the bucket half full of water, which not only prevents the bees from flying away but also forces them to crawl out through the spout and rub against the precisely placed sticky pollen masses and sticky stigma. What an incredible system!

The construction of the flower in another closely related orchid in the genus *Catasetum* (Figure 6.20B) is widely different, although it serves the same end and is equally curious. Bees visit these flowers, like those of *Coryanthe*, in order to gnaw the labellum. In doing so they inevitably touch a long, tapering, touch-sensitive projection that I have called the antenna. When touched, this "antenna" transmits a sensation or vibration to a certain membrane, which is instantly ruptured. This in turn sets free a spring by which the pollen mass is shot forth, like an arrow, in precisely the right direction, so that it adheres by its sticky extremity to the bee's back. The pollen mass of the male plant (for the sexes are separate in this orchid) is thus carried by the bee to the flower of the female plant, where it is brought into contact with the stigma. The stigma is sticky enough to break certain elastic threads: retaining the pollen thereby released, fertilization is thus accomplished.

How, it may be asked, in the foregoing and in innumerable other instances, can we understand the graduated scale of complexity and the multifarious means for gaining the same end in such a wide variety of organisms? The answer no doubt is, as I have already remarked, that when two forms—which already differ from each other in some slight degree—continue to vary, the variability will not be of the same exact nature; the results obtained through natural selection for the same general purpose will consequently not be the same in both instances. We should also bear in mind that every highly developed organism that we see has already passed through many changes, and that each modified structure tends to be inherited; thus each modification will not readily be lost and may be again and again further altered. The structure of each part of each species, then, for whatever purpose it may serve, is the sum of many inherited changes through which the species has already passed during its successive adaptations to changed habits and changed conditions of life over many generations.

Even though in many cases it is most difficult even to conjecture about the transitions by which various organs have arrived at their present state, I have been astonished at how rarely we find any organ toward which no transitional grade is known to lead, especially considering how small the number of living species is compared with the number of those that have gone extinct in the past. It is certainly true that we never, or rarely, see new organs in any being appearing suddenly, as if they were created for some specific purpose; this reminds me of that well-noted old, but somewhat exaggerated canon of natural history, *Natura non facit saltum* ("Nature does not make jumps"). Indeed, we meet with this admission in the writings of almost every experienced naturalist. As the eminent French zoologist Henri Milne-Edwards has well expressed it, nature is prodigal in variety but stingy in innovation. **Why, based on the theory of Creation, would there be so much variety and yet so little real novelty?** Why should all the parts and organs of so many independent beings, each supposed to have been separately created for its proper place in nature, be so commonly linked together by graduated steps? Why should not nature take a sudden leap from structure to structure? On the theory of natural selection, we can clearly understand why nature should not do so, and in fact *cannot* do so: natural selection acts only by taking advantage of slight successive variations; she can never take a great and sudden leap, but must advance slowly by short and sure steps, over many, many generations.

Organs of Little Apparent Importance, as Affected by Natural Selection

As natural selection acts by life and death—by the survival of the fittest and by the destruction of the less well-fitted individuals—I have sometimes felt great difficulty in understanding the origin or formation of parts of little importance, parts whose value seems too small to cause their preservation in successively varying individuals. It seems almost as great a problem, though of a very different kind, as in the case of the most perfect and complex organs such as eyes.

In the first place, however, we are much too ignorant with regard to the detailed functioning and ecological interactions of any one organic being to say what slight modifications might or might not be important to that organism. In Chapter 4, I gave instances of what seemed like very trifling characteristics, such as the down on the outside of fruit, the color of its flesh, and the color of the skin and hair of quadrupeds, which from being correlated with constitutional differences or from discouraging or encouraging the attacks of insects, might assuredly be acted on by natural selection. But the tail of the giraffe certainly looks like an artificially constructed flyflapper! At first it seems incredible that this tail could have been adapted for its present purpose by successive slight modifications, each better and better fitted for so trifling an object as to drive away flies! Yet we should pause before being too certain even in this case. In particular, we know that the distribution and existence of cattle and many other animals in South America absolutely depends on their ability to resist insect attacks; thus, individuals that could by any means defend themselves from these small enemies would be able to range into new pastures, thereby gaining a great advantage over others. It is not that the larger quadrupeds are actually destroyed (except in some rare cases) by flies, but that they are incessantly harassed and their strength reduced, so that they are more vulnerable to disease, or less able to search for food during a future famine, or to escape from predators.

Moreover, organs now of trifling importance may well have been of great importance to an early ancestor, and, after having been slowly perfected in some former period, were then transmitted through the generations to existing species in nearly the same state, even though they may now be of little use; any injurious changes to their structure would of course have been checked by natural selection. We may, for example, perhaps account for the general presence of tails and their use for so many purposes in many land animals—which in the structure of their lungs or modified swim bladders betray their aquatic origins—by noting how important the tail is as an organ of locomotion in most aquatic animals. Once a well-developed tail has been formed in some aquatic animal, it might subsequently come to be modified for all sorts of purposes—as a flyflapper, for example, or as an organ for grasping, or as an aid in turning, as in the case of dogs, although the degree to which the tail helps a dog turn must be slight, for a rabbit can double about even more quickly than a dog with hardly any tail at all.

On the other hand, we may easily be mistaken in attributing importance to certain characteristics and in believing that they have been developed through natural selection. We must by no means overlook the effects of so-called spontaneous variations, which seem to depend little on the nature of the surrounding conditions, or the tendency of organisms to sometimes revert to long-lost characteristics, or the effects of the complex "laws of growth," such as correlation, compensation, and the effects of pressure of one part on another,[10] as discussed in Chapter 5. Nor should we overlook the role of sexual selection, through which characteristics of use to one sex are often gained and then transmitted more or less perfectly to the other sex, even though they are of no use to that other sex. But structures thus indirectly gained, although at first of no advantage to a species, may subsequently have been taken advantage of by its modified descendants under new conditions of life and with newly acquired habits.

For example, if all the woodpeckers we knew of were green, and we did not know that there were also many black or black and white kinds (often with a patch or two of red), I dare say that we should have thought that the green color was a beautiful adaptation to conceal these tree-frequenting birds from their enemies, and, consequently, that it was a characteristic of considerable importance, one that had been acquired through natural selection; but in fact, the color probably results mostly from sexual selection. Similarly, a trailing palm plant in the Malay Archipelago climbs the loftiest trees by using exquisitely constructed hook clusters around the ends of its branches. No doubt this contrivance is extremely useful to the plant, but the fact remains that we see very similar hooks on many plants that are not climbers and which seem to serve instead as a defense against browsing quadrupeds. Thus, the spikes on the palm plant may at first have been developed for this defensive function and subsequently

---

[10] Once again, we see that Darwin knows nothing about the genetic basis of variation—poor fellow!

have been improved and taken advantage of by the plant over many thousands of generations as it underwent further modification and became a climber.

As a third example, the naked skin on the head of a vulture (Figure 6.21A) is generally considered as a direct adaptation for wallowing in putridity. Or perhaps it is due to the direct action of the decomposing and putrifying matter on which it feeds. But we should be very cautious in drawing any such inference; after all, the skin on the head of the clean-feeding male turkey is also naked (Figure 6.21B), even though these birds feed mostly on grains, seeds, nuts, leaves, and small insects.

(A)

(B)

Figure 6.21 (A) The head of a turkey vulture (*Cathartes aura*). (B) The turkey (*Meleagris gallopavo*).

Last, the sutures in the skulls of young mammals have been proposed to be beautiful adaptations for easing the movement of babies through the birth canal at birth. No doubt they do facilitate that process and may even be indispensable for it, but sutures also occur in the skulls of young birds and reptiles, which have only to escape from a broken egg; thus we may infer that this structure has arisen originally only from the natural laws of growth and has simply then been taken advantage of by mammals for childbirth.

We are profoundly ignorant of the cause of each slight variation or individual difference. We are immediately made aware of this by reflecting on the differences between the breeds of domesticated animals found in different countries, particularly in the less civilized countries where there has been but little methodical selection for any particular traits. Animals kept by savages in different countries often have to struggle to find food and are thus to some extent exposed to the forces of natural selection; individuals with slightly different constitutions would succeed best under different climates. With cattle, susceptibility to the attacks of flies is correlated with body color, as is the likelihood of being poisoned by certain plants; thus even color would be subjected to the action of natural selection. Some observers are convinced that a damp climate affects the growth of the hair, and that with the hair the animal's horns are correlated. Mountain breeds always differ from lowland breeds; a mountainous country probably affects the hind limbs by exercising them more, and possibly even affects the form of the animal's pelvis. If so, then by the law of homologous variation, the front limbs and the head would probably also be affected. The shape of the pelvis might also affect, by pressure, the shape of certain parts of the young in the womb. The laborious breathing necessary in high-altitude regions tends to increase the size of the chest; again, the correlation with other body parts would come into play. The effect of lessened exercise together with abundant food on the whole organization is probably still more important, and this, in fact, as the German animal breeder Hermann von Nathusius has lately shown in his excellent treatise, is apparently one chief cause of the great modification that the various pig breeds have undergone.

But we are far too ignorant to speculate on the relative importance of the several known and unknown causes of variation. I have made the preceding remarks only to show that if we are unable to account for the characteristic differences in our several domestic breeds, which nevertheless are generally admitted to have arisen through ordinary reproductive processes from one or at most a few parent stocks, we ought not to lay too much stress on our ignorance of the precise cause of the slight analogous differences between true species.

Utilitarian Doctrine, How Far True? Beauty, How to Explain It?

The previous remarks lead me to say a few words about the protest lately made by some naturalists against my idea that every detail of structure has been produced for the good of its possessor. They believe instead that many structures have been created for the sake of beauty, to delight man or the Creator—a point beyond the scope of scientific discussion—or for the sake of mere variety, a view that I have already discussed. Such doctrines, if true, would be absolutely fatal to my theory.

I fully admit that many structures are now of no direct use to their possessors, and may never have been of any use to the ancestors of those individuals, but this does not prove that they were formed solely for beauty or variety. No doubt the definite action of changed conditions, and the various causes of modifications, which I have already discussed, have all produced an effect—probably a great effect—independently of any advantage thus gained. **But a still more important consideration is that most of the organization of every living creature is due to inheritance.** Consequently, though each being is assuredly well fitted for its place in nature, many structures now have no very close and direct relation to present habits of life. Thus, we can hardly believe that the webbed feet of the upland goose of South American grasslands, or of the frigate bird, which cannot swim or even walk well and which takes most of its food in flight, are of special use to these birds. And it seems so unlikely that the similar bones in the arm of the monkey, in the foreleg of the horse, in the wing of the bat, and in the flipper of the seal are of special use to these animals; we may safely attribute these structures to inheritance. But webbed feet no doubt were as useful to the distant ancestor of the upland goose and of the frigate bird as they now are to the most aquatic of living birds.

Thus it is logical to believe that the ancient ancestor of the seal did not possess a flipper but rather a foot with five toes fitted for walking or grasping, and we may further venture to believe that the several bones in the limbs of the monkey, horse, and bat were originally developed on the principle of usefulness, probably through the reduction of more numerous bones in the fin of some ancient fish-like ancestor of the whole class. It is scarcely possible to decide how much allowance ought to be made for such causes of change as the definite action of external conditions, so-called spontaneous variations, and the complex laws of growth;[11] but with these important exceptions, we may conclude that the structure of every living creature either now is, or was formerly, of some direct or indirect use to its owner.

With respect to the belief that organisms have been created beautiful for the delight of people, let me first note that the sense of beauty obviously depends on the nature of the mind, irrespective of any real quality in the admired object: the idea of what is beautiful is not innate and is not unalterable. We see this, for instance, in the men of different races admiring an entirely different standard of beauty in their women. If beautiful objects had been created solely for our gratification, it ought to be shown that there was less beauty on the face of the Earth before we appeared than since we came on the stage. Were the beautiful snail shells, such as the volutes and cone snails, from tens of millions of years ago in the Eocene epoch, and the gracefully sculpted ammonites (Figure 6.22), which went extinct some 65 million years ago, all created

---

[11] Again, Darwin would have so much enjoyed, and profited from, an introductory course in genetics. Imagine the look on his face if he sat in on such a class and first learned about genes.

(A)   (B)

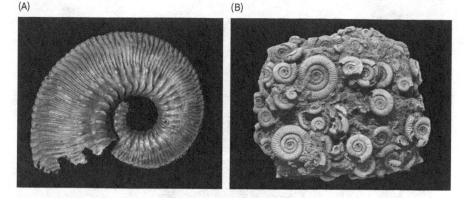

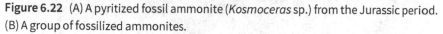

Figure 6.22 (A) A pyritized fossil ammonite (*Kosmoceras* sp.) from the Jurassic period. (B) A group of fossilized ammonites.

so that we might—ages afterward—admire them in our curio cabinets? Similarly, few objects are more beautiful than the microscopic silica cases of diatoms (Figure 6.23); were these created so that they might someday be examined and admired by us under the higher powers of microscopes?

And what about flowers? Flowers must be among the most beautiful productions of nature; but they have been made conspicuous in contrast to the green leaves around them, and consequently also beautiful, only so that they may be easily observed by the insects that are needed to complete their life cycles. I have come to this conclusion from finding it an invariable rule that flowers that are routinely fertilized by the wind are never gaily colored. Moreover, several plants habitually produce two kinds of flowers: one that is open and colored, to attract insects, and the other closed, not colored, destitute of nectar, and never visited by insects. Thus we may conclude that if insects had never appeared on the face of the Earth, our plants would not now be decked with beautiful flowers but would have produced only such poor flowers as we see on fir, oak, nut, and ash trees, and on grasses, spinach, docks (*Rumex obtusifolius*), and nettles, all of which are fertilized by the wind.

A similar line of argument holds for fruits. Everyone can agree that a ripe strawberry or cherry is as pleasing to the eye as to the palate and that the gaily colored fruit of the spindle-wood tree and the scarlet berries of the holly (Figure 6.24) are beautiful objects. But this beauty serves merely as a guide for birds and beasts, in order that the fruit may be eaten and the matured seeds then disseminated by the eater. Indeed, I infer that this is the case from having as yet found no exception to the rule that seeds embedded within a fruit of any kind (that is, within a fleshy or pulpy envelope) are always disseminated in this way as long as the fruit is colored in any brilliant tint or rendered conspicuous by being white or black.

On the other hand, I willingly admit that a great number of male animals—as with all our most gorgeous birds, some fishes, reptiles, and mammals, and a host of

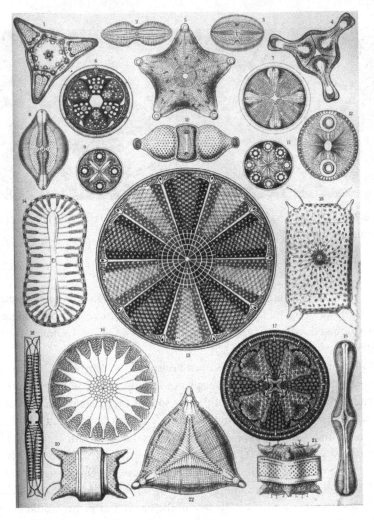

Figure 6.23 An assortment of diatoms.

magnificently colored butterflies—have been rendered beautiful for beauty's sake. But this has resulted through sexual selection, not for our delight; that is, by the more beautiful males having been continually preferred by the females. So it is with the music of birds. We may infer from all this that a nearly similar taste for beautiful colors and musical sounds runs through a large part of the animal kingdom. When the female is as beautifully colored as the male, which is often the case with birds and butterflies, the cause apparently lies in the colors acquired through sexual selection having been transmitted to both sexes instead of to the males alone. How the sense of beauty in its simplest form—that is, the reception of a peculiar kind of pleasure from perceiving certain colors, forms, and sounds—was first developed in the mind of humans and of the lower animals is a very obscure subject. The same

Figure 6.24 Holly berries.

sort of difficulty is presented if we ask how it is that certain flavors and odors give us pleasure, while others displease us. Habit appears to play a role in all these cases, but there must be some fundamental cause in the constitution of the nervous system of each species.

Natural selection cannot possibly produce any modification in a species exclusively for the good of another species, even though throughout nature one species incessantly takes advantage of and profits by the structures of others. But natural selection can, and does, often produce structures for the direct injury of other animals, as we see, for example, in the fang of the adder and in the ovipositor of the ichneumon wasps (Figure 6.25) (insect family Ichneumonidae), which it uses to deposit its eggs into the living bodies of other insects.

If it could be proven that any part of the structure of any one species had been formed for the exclusive good of another species, it would annihilate my theory, for such could not have been produced through natural selection. Although many works on natural history claim that some structures in one species do indeed serve for the exclusive benefit of a different species, I cannot find even one example that seems to me to hold any weight. Most authors admit that the rattlesnake has a poison fang for its own defense and for killing its prey. But some of these same authors also believe that the rattlesnake is furnished with a rattle to deliberately harm itself by warning its prey. I would almost as soon believe that the cat curls the end of its tail when preparing to spring in order to warn the doomed mouse of its coming fate!

It is much more likely that the rattlesnake uses its rattle, the cobra expands it frill, and the puff adder swells while hissing loudly and harshly in order to alarm the many birds and beasts that are known to attack even the most venomous species. Snakes act on the same principle that makes the hen ruffle her feathers and expand her wings when a dog approaches her chickens—but I do not have space here to say more about the ways that animals endeavor to frighten away their enemies.

Figure 6.25 Ichneumon wasp (*Xorides praecatorius*) with prominent ovipositor (bottom).

Natural selection will never produce—in any organism—any structure that injures that organism more than it provides a benefit to that organism, for natural selection acts solely by and for the good of each individual. No organ will be formed, as Reverend William Paley has remarked, for the purpose of causing pain or for injuring its owner. If a fair balance be struck between the good and evil caused by each part, each part will be found on the whole to be advantageous to its possessor. Over long periods of time, under changing conditions of life, if any part comes to be injurious it will be modified; if not, the organism will become extinct, as indeed myriads have become extinct in the past.

Natural selection tends only to make each organism as perfect as, or slightly more perfect than, the other inhabitants of the area in which it competes. And we see that this is the standard of perfection attained under nature. The endemic native plants and animals of New Zealand, for instance, are perfect when compared with each other; but their populations are now rapidly dwindling before the advancing legions of plants and animals that have been introduced from Europe.

No, natural selection will not produce absolute perfection, and we do not often see, as far as we can judge, such a high standard met under nature. According to Mr. Müller, for example, the correction for the aberration of light is not perfect even in that most perfect organ of sight, the human eye. The German physician and physicist, Hermann von Helmholtz, whose judgment no one will dispute, describes in the strongest terms the wonderful powers of the human eye, but then adds these

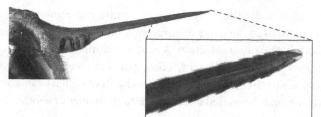

Figure 6.26 A bee stinger removed from a honeybee, lateral view. Inset, barbs are more apparent when viewed from above.

remarkable words: "That which we have discovered in the way of inexactness and imperfection in the optical machine and in the image on the retina, is as nothing in comparison with the incongruities which we have just come across in the domain of the sensations. One might say that nature has taken delight in accumulating contradictions in order to remove all foundation from the theory of a preexisting harmony between the external and internal worlds." If our reason leads us to admire with enthusiasm a multitude of inimitable contrivances in nature, this same reason tells us, though we may easily err on both sides, that some other contrivances are less perfect. Can we consider the sting of the bee to be perfect, considering that when used against many kinds of enemies it cannot be withdrawn, because of the backward-pointing teeth along its length, and so causes the death of the bee by tearing out its entire digestive tract?

But if we look at the bee's stinger (Figure 6.26) as having existed in a remote ancestor as a boring and serrated instrument,[12] like that seen in so many other hymenopterans, and that it has since been modified—but not perfected—for its present purpose, with the poison originally adapted for some other purpose, such as to produce galls and since intensified, we can perhaps understand how it is that the use of the sting should so often cause the insect's death: for if the power of stinging is generally useful to the social community, it will fulfill all the requirements of natural selection even though it may cause the death of some few members.

And again, if we admire the truly wonderful power of scent by which the males of many insects find their females for mating, can we also admire the production for this single purpose of thousands of male drones, which are utterly useless to the community for anything other than to fertilize the eggs of a receptive queen and which are ultimately slaughtered by their industrious and sterile sisters? This is certainly not evidence of perfection in nature. It may be difficult, but we ought also to admire the savage instinctive hatred of the queen bee, which urges her to destroy the young queens—her own daughters—as soon as they are born, or to perish herself in the

---

[12] Actually, the stinging structure was probably used for egg laying originally, which would explain why only female bees can sting.

combat, for undoubtedly this is good for the entire community; maternal love or maternal hatred, though the latter fortunately is most rare, is all the same to the inexorable principle of natural selection. And if we admire the several ingenious contrivances by which orchids and many other plants are fertilized by insects, can we consider as equally perfect the release of dense clouds of pollen by our fir trees, so that just a few granules may be wafted by chance onto the ovules of another tree?

Summary: The Law of Unity of Type and of the Conditions of Existence Embraced by the Theory of Natural Selection

In this chapter, I have discussed some of the difficulties and objections that may be brought against my theory of natural selection. Although many of them are serious, I think that in the preceding discussion I have thrown light on several facts that a belief in numerous independent acts of special creation are utterly unable to explain. We have seen that species at any one period are not indefinitely variable and are not linked together by a multitude of intermediate gradations, partly because the process of natural selection is always very slow and acts only on a few forms at a time, and partly because the very process of natural selection implies the continual supplanting and extinction of the preceding and intermediate steps. Closely allied species, now living throughout a large, continuous area, must often have been formed when the area was broken up into isolated sections and when the conditions of life did not sensibly graduate away from one part to another. When two varieties of a species are formed in different parts of a continuous area, an intermediate variety will often be formed as well, fitted for life in the intermediate zone between the two varieties; but, as I have explained, the intermediate variety will usually exist in smaller numbers than the two forms it connects. In consequence, during the course of further modification, the two major forms, because they exist in larger numbers, will have a great advantage over the less numerous intermediate variety and will thus generally succeed in eventually supplanting and exterminating it.

We have seen in this chapter how cautious we should be in concluding that the most different habits of life could not graduate from one to the other—that a bat, for instance, could not have been formed by natural selection from an animal that at first only glided through the air.

We have also seen that a species under new conditions of life may change its habits. Or it may have a variety of habits with some very unlike those of its closest relatives in the same family. We can understand, then, bearing in mind that each living being is *trying* to live wherever it *can* live, how it has come about that there are fully terrestrial upland geese with webbed feet, and ground-living woodpeckers, and diving thrushes, and petrels with the habits of open ocean auks.

The belief that an organ as perfect as the human eye could ever have been formed gradually by natural selection is enough to stagger anyone. And yet, in the case of the vertebrate eye or any other organ, if we know of a long series of gradations in complexity, each good for its possessor, then, under changing conditions of life there is no logical obstacle to eventually acquiring any conceivable degree of perfection through natural selection. Even in those cases in which we don't know of any intermediate or transitional states, we should be extremely cautious in concluding that none existed; the transformation of many organs shows what wonderful changes in function are at least possible. The fish's swim bladder, for example, has apparently been converted into an air-breathing lung. In some ancient variety, the same organ must have simultaneously performed two very different functions (buoyancy control and gas exchange) in the same individual and then gradually become specialized for just one of those functions—gas exchange. In many other cases, two distinct organs simultaneously performed the same function, opening the path, through redundancy, for one of the organs to then transition to a very different function.

We have seen that in two distantly related organisms, organs having the same function and closely resembling each other in external appearance may have been separately and independently formed; if so, essential differences in their structure can almost always be detected when such organs are examined closely. This naturally follows from the principle of natural selection. The common rule throughout nature is "Infinite diversity of structure for gaining the same end," which again follows from the same great principle of natural selection.

In many cases we are far too ignorant to claim that any particular part or organ is so unimportant for the welfare of a species that modifications in its structure could not have been slowly accumulated through natural selection. In many other cases, modifications probably result directly from the laws of variation or of growth, independently of any benefits having been thus gained. But even such structures have probably often been further modified by natural selection for the good of individuals under new conditions of life. It seems also likely that a part formerly of high importance to an organism has frequently been retained (such as the tail of an aquatic animal by its now-terrestrial descendants), even though it has now become of such small importance that it could not, for its present function, have been acquired by means of natural selection.

Natural selection can produce nothing in one species for the exclusive good or injury of another, although it may well produce parts, organs, and excretions (such as the sugar-containing honeydew of aphids) highly useful or even indispensable to other species, or even injurious to another species, as long as those parts, organs, or secretions are at the same time useful to the possessor. Wherever population sizes are large, natural selection acts through competition between the inhabitants and consequently leads to successes and failures in the battle for life in that particular region or area. Thus the inhabitants of one region, generally the smaller one, often yield to the inhabitants of another region (generally the larger one), and for this simple reason: the larger region will be home to more individuals and more diversified forms, and the competition will thus have been more severe there, so that the standard of perfection will have been rendered higher. Natural selection will not necessarily lead to absolute perfection, nor, as far as we can judge by our limited faculties, can absolute perfection be everywhere expected.

On the theory of natural selection we can clearly understand the full meaning of that old rule in natural history, *Natura non facit saltum*—"Nature does not make jumps." This rule, if we look only at the present inhabitants of our world, is not strictly correct. But if we include all those ancestors from past times, whether known or unknown, the rule must, on the theory of natural selection, in fact be strictly true. Organisms evolve gradually over long periods of time.

It is generally believed that all living beings have been formed on two great laws: the Unity of Type and the Conditions of Existence. Unity of Type, as promoted by Étienne Geoffroy Saint-Hilaire and Sir Richard Owen, refers to the fundamental similarities in structure that we see in organisms found within the same taxonomic group, regardless of their lifestyles. My theory of natural selection explains unity of type quite simply, by descent from a common ancestor.

The second law, Conditions of Existence, as promoted so often by the French anatomist Georges Cuvier, refers to the way in which many structures—the hands of people, the wings of birds, the flippers of a seal, for example—are so marvelously adapted for the particular lifestyles of their owners, as though their lifestyle (i.e., their "conditions of existence") directly controlled their anatomy. This law is again fully explained by natural selection: natural selection acts by either now slowly adapting the varying parts of each being to its lifestyle and passing those adaptations along to offspring, or by the ancestors having adapted those parts to that lifestyle in the past. Thus, in fact, the Conditions of Existence law is the higher law, as it includes, through the inheritance of former variations and adaptations that were selected for in ancestors, that of Unity of Type.

Key Issues to Talk and Write About

1. Find out two interesting things about one of the people that Darwin mentions in this chapter. Choose from the following:
   Antoine Edouard Claparède
   Edward Drinker Cope
   Georges Cuvier
   Félix de Azara
   Alphonse de Candolle
   Henri Louis Frédéric de Saussure
   Edward Forbes
   Asa Gray
   Samuel Hearne
   John Avebury Lubbock
   Carlo Matteucci
   Henri Milne-Edwards

St. George Jackson Mivart
Fritz Müller
Filippo Pacini
Jean Louis René
Thomas Vernon Wollaston
Rudolf Virchow
Hermann von Helmholtz
Hermann von Nathusius
Hewett Cottrell Watson

2. Picture a cow: slow moving, feeding on grass, and occasionally mooing. Based on your reading of this chapter, try to come up with a scenario in which a cow-like animal in the wild might eventually, over many thousands of generations, become adapted for feeding on frogs.

3. According to Darwin, why do we rarely see intermediate stages between an ancestral form and a modern form adapted for a very different lifestyle (e.g., winged birds and their presumably non-winged ancestors?)

4. Darwin argues that although we can't directly trace the exact steps through which a complex organ like the eye progressed from some ancient ancestor of today's eyed species, we can get a good sense of what those steps may have been by looking at how that organ is developed today in different species. How convincing do you find this argument? Explain your reasoning.

5. How might redundant parts or organs have played a role in the gradual evolution of novel structures and novel functions, according to Darwin?

6. According to Darwin, how should we determine whether similar organs in very different species were independently evolved, or passed down and evolved from a common ancestor? Give one example of his argument.

7. Using Darwin's comparison of the cephalopod and vertebrate eyes, tell a classmate or friend how natural selection can explain the appearance of similarly specialized organs in two completely different groups of animals.

8. Briefly summarize Darwin's argument about why we should not expect natural selection to bring about perfection in any of the traits that it shapes.

9. What are the male parts of a flower called? What are the female parts of a flower called? Which of Darwin's descriptions of fertilization among orchids do you find most interesting, and why?

10. List all of the animals and plants that Darwin uses as examples in this chapter.

11. Write a stand-alone, accurate, and informative sentence that summarizes the essential content of one of the following paragraphs:
    a. "We also know of instances in which two distinct organs...." (see page 163)
    b. "For example, if all the woodpeckers we knew of were green...." (see page 175)
    c. "And what about flowers? Flowers must be among the most beautiful productions of nature...." (see page 179)

12. Try rewriting these sentences to make them clearer and more concise:
   a. "In the foregoing cases, we see the same end gained and the same function performed, in beings not at all or only remotely allied, by organs in appearance, though not in development, closely similar."
   b. "But in fact, the color is probably mostly the result of sexual selection."
   c. "The pollen does not spontaneously fall on the stigma of the flower."

Bibliography

See the following papers for fascinating information about the remarkable eyes of brittle stars:

Aizenberg, J., and G. Hendler. 2004. Designing efficient microlens arrays: Lessons from Nature. *Journal of Materials Chemistry* 14: 2066–2072.

Aizenberg, J., A. Tkachenko, S. Weiner, L. Addadi, and G. Hendler. 2001. Calcitic microlenses as part of the photoreceptor system in brittlestars. *Nature* 412: 819–822.

7

Miscellaneous Objections to the Theory of Natural Selection

In this chapter, Darwin addresses a number of major criticisms that had been thrown at his theory of evolution by natural selection and survival of the fittest, particularly those delivered by zoologist St. George Jackson Mivart, who had initially been a strong supporter of the theory. In doing so, Darwin discusses in detail about some remarkable examples that he hasn't mentioned before, including the evolution of baleen whales from toothed ancestors, the evolution of climbing in plants, and the evolution of breasts in mammals. This chapter appeared in *The Origin of Species* for the first time in the sixth edition, on which this book is based.

In this chapter I will consider various miscellaneous objections against my views, to make some of the previous discussions a little clearer. But it would not be useful to discuss all of the objections, as many have been made by writers who have not taken the trouble to understand the subject.

For example, a distinguished German naturalist has asserted that the weakest part of my theory is that I consider all organisms to be imperfect. However, what I really said is that all are not as perfect as they might have been in relationship to the sorts of lives they lead. This is clearly shown to be the case just from the number of native forms in many parts of the world that have yielded their places to intruding organisms—invasive or introduced species—from other countries; if the native species had been perfectly adapted to their surroundings, then the invading species wouldn't have been able to get a foothold. Nor can any organisms—even if they had been perfectly adapted to their niches at one time—have remained so perfectly adapted when environmental conditions changed unless they themselves also changed; no one will dispute that the physical conditions of each country, as well as the numbers and kinds of its inhabitants, have undergone many such changes.

Another critic has recently insisted, with some parade of mathematical accuracy, that since longevity must be a great advantage to all species, anyone who believes in natural selection "must arrange his genealogical tree" so that all descendants have longer lives than any of their ancestors did! Can our critic not conceive that a biennial plant or one of the lower animals might range into a cold climate and perish there every winter, and yet, owing to the advantages gained through natural selection, survive as a species from year to year by means of its seeds or its eggs? The English

The Readable Darwin. Second Edition. Jan A. Pechenik, Oxford University Press. © Oxford University Press 2023.
DOI: 10.1093/oso/9780197575260.003.0008

biologist Mr. Edwin Ray Lankester has recently discussed this subject, concluding that longevity probably varies with the relative structural and physiological complexity of each species, as well as with the amount of energy expended in its reproduction and general activity. These conditions have probably been determined largely through natural selection.

Some have also argued that, as none of the animals and plants of Egypt that we know about have changed during the last 3,000 or 4,000 years, probably none have changed in any other part of the world either. But as the English philosopher Mr. George Henry Lewes has remarked, although the ancient domestic races illustrated on various Egyptian monuments, or embalmed, are indeed closely similar or even identical with those now living, all naturalists nevertheless admit that such races have been produced through the prior modification of their original types. Moreover, Egypt is not typical of other parts of the world: in Egypt, the conditions of life seem to have remained absolutely uniform during the last several thousand years. The many animals that have remained unchanged since the end of the most recent glacial period would have been an incomparably stronger case to offer, for those organisms have been exposed to great changes of climate and have migrated over great distances. The fact of little or no modification having taken place in those organisms since the glacial period would have been of some use against those who believe in an innate and continuous law of development but has little to do with the doctrine of natural selection or survival of the fittest. **Natural selection doesn't imply continuous change, only that when appropriate variations or individual differences of a beneficial nature happen to arise within a population, those will be preserved under certain favorable circumstances.**

The celebrated paleontologist Heinrich Georg Bronn, at the close of his translation of *The Origin* into German, asks, "How can the principle of natural selection allow a variety to lie side by side with the parent species?" I can answer this easily: If both have become fitted for slightly different habits of life or conditions, they might live together perfectly well, since they will not be directly competing for the same resources. And if we put "polymorphic species" on one side (in which several distinctly different body types are found within a single population), and all mere temporary variations on the other (such as differences in size, degree of albinism and so forth), the more permanent varieties are generally found inhabiting very distinct habitats—such as high land versus low land, or dry versus moist districts. Moreover, in the case of animals that wander about a good deal and cross freely when mating, their varieties seem to be generally confined to distinct regions so that interactions between them are minimal.

Modifications Not Necessarily Simultaneous

Mr. Bronn also insists that distinct species never differ from each other in just single characteristics, but rather in many parts. He then asks, how is it possible that so many parts of the organization should have been modified simultaneously through variation and natural selection? But why must we assume that all parts of an organism have been modified simultaneously? As I have noted earlier, the most striking modifications, excellently adapted for some particular function, might well be acquired by successive slight variations, first in one part and then in another. And as those variations would be transmitted all together to the next generation, it would look to us as though they had been simultaneously developed, although that was not actually the case. The best answer to the above objection, however, is afforded by those domestic races that people have modified for some special purpose, chiefly through our powers of selection over time. Look at the racehorse and the cart horse (see Figure 4.10) for example, or at the greyhound and the mastiff: over many generations we have modified their whole frames and even their mental characteristics, but if we could trace each step in the history of their transformation—and the latter steps can indeed be so traced—we would not see great and simultaneous changes, but first just one part and then another slightly modified and improved. Even when we have applied artificial selection to some one characteristic—of which our cultivated plants offer the best examples—we invariably find that although this one part—whether it be the flower, the fruit, or the leaves—has been greatly changed over many generations, almost all the other parts have been slightly modified as well. These modifications have presumably taken place through the principle of correlated growth, along with so-called spontaneous variation—mechanisms that, as I have explained earlier, we understand very little about.

Modifications That Are Apparently of No Direct Service

A much more serious objection has been put forth by Mr. Bronn, and even more recently by the French anatomist Pierre Paul Broca: many characteristics appear to be of no use whatever to their owners and therefore could not have been selected for. Mr. Bronn gives as examples the length of the ears and tails in the different species of hares and mice, along with the complex folds of the enamel in the teeth of many animals and a multitude of analogous cases. With respect to plants, the subject has been admirably discussed by the Swiss botanist Carl Wilhelm von Nägeli. Although Mr. von Nägeli admits that natural selection has accomplished much, he insists that the families of plants differ from each other mainly in morphological characteristics that seem to be quite unimportant for the welfare of the species. He consequently believes that plants show an innate tendency toward an automatic, progressive, and more perfect development over time. As examples of cases in which natural selection could not have acted, he offers the arrangement of cells in the plant's tissues and of the leaves on the axis. To these we might add such items as the numerical divisions in the parts of the flower, the position of the ovules (the structure that produces and contains the female reproductive cells), and the shape of the seeds when they seem of no use for dissemination.

There is much force in the above objections. Even so, we ought to be extremely cautious in pretending to know what structures are now, or have been, of use to any species. There must be some efficient cause for each slight individual difference, as well as for the more prominent variations that occasionally arise; if the unknown cause were to act persistently, it is almost certain that all the individuals of the species would be similarly modified. With respect to the assumed uselessness of various body parts and organs, it is hardly necessary to point out that even in the higher and best-studied animals we see many structures that are so highly developed that nobody doubts that they are of importance, even though their use has not been, or has only recently been, determined. Mr. Bronn gives the length of the ears and tail in the several species of mice as instances, though trifling ones, of differences in structure that can be of no special use to the owner. However, Dr. Joseph Schöbl, a Czech anatomist, notes that the external ears of the common mouse are supplied in an extraordinary manner with nerves, so that they no doubt serve also as tactile organs; thus the length of the ears can hardly be unimportant. Moreover, as I will presently clarify, we now find that tails are highly useful prehensile organs for the members of some species; thus the tail's use would certainly be influenced by its length.

Since Mr. von Nägeli focuses his arguments on plants, I will do so as well. I certainly admit that the flowers of orchids present a great many curious structures, which a few years ago would have been considered to be mere morphological differences without any special function. But we now know that these odd structures are of the highest importance for achieving fertilization through the aid of insects and have probably been gained through natural selection. No one until recently would have imagined that in dimorphic and trimorphic plants—plants with either two or three different types of flowers—the different lengths of the stamens (which produce pollen) and pistils (which produce eggs and seeds) (Figure 7.1) and their arrangement on the flower could possibly have been of any practical use, but we now know that they are. For example, in certain whole groups of plants the ovules stand erect, and in others they are suspended; but in a few other species one ovule stands erect while a second ovule is instead suspended, all within the same ovarium. These positions seem at first to have no functional significance, but the Director of the Royal Botanical Gardens Dr. Joseph Hooker informs me that within the same ovarium, only the upper ovules are fertilized in some cases, and in other cases only the lower ones are fertilized. He suggests that this probably depends on the direction in which the pollen tubes enter the ovarium. If so, then the exact position of the ovules, even when one is erect and the other is suspended within the same ovarium, would follow from the selection of any slight deviations in position that favored fertilization and seed production.

Several plants belonging to distinct orders routinely produce flowers of two kinds—one that is open and with an ordinary structure, and the other that is closed and imperfect. These two kinds of flowers sometimes differ wonderfully in structure and yet may be seen to graduate into each other on the same plant. The ordinary and open flowers can be successfully intercrossed, with obvious benefits. However, the closed and imperfect flowers are manifestly of great importance as well, as they yield

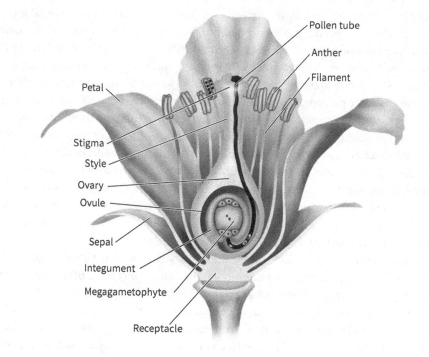

Figure 7.1 Schematic drawing of the parts of a flower.

with the utmost safety a large stock of seed with the expenditure of wonderfully little pollen. As just stated, the two kinds of flowers often differ much in structure. The petals in the imperfect flowers almost always consist of mere undeveloped rudiments, and the pollen grains are reduced in diameter. Five of the alternate stamens are rudimentary in the flowering plant *Ononis columnae*; in some species within the violet genus *Viola*, three of the stamens are in this same condition, with two retaining their proper function but being of very small size. In 6 out of 30 of the closed flowers that I examined in an Indian violet (I can't tell the species, as none of the plants ever produced perfect flowers under my care), the sepals[1] were reduced from the normal number of five to only three. In one section of the Malpighiaceae (a family of exclusively tropical and subtropical plants), the closed flowers, according to the French botanist Antoine Laurent de Jussieu, are still further modified, for the five stamens that stand opposite the sepals are all aborted, with a sixth stamen standing opposite a petal being the only one developed; this stamen is not present in the ordinary flowers of these species. Moreover, the style is aborted, and the ovaria are reduced from three to two. Now although natural selection may well have had the power to prevent some of the flowers from expanding and to reduce the amount of pollen—since pollen

---

[1] Sepals make up the calyx of flowering plants, which encloses the developing flower bud.

would be superfluous in a closed flower—yet hardly any of the above special modifications can have been thus determined but must have followed from what I have called "the laws of growth." These laws would include the functional inactivity of parts as pollen production was gradually reduced and the flowers were opened less and less. Also owing to these rather mysterious "laws of growth," we often see differences in the same parts or organs caused by differences in their relative positions on the same plant. In the Spanish chestnut tree, for example, and in certain fir trees, the leaves on the nearly horizontal branches diverge at different angles from those on the upright branches. And in the British plant *Adoxa*, which gives off a decidedly musk-like scent, the uppermost flower generally has two calyx lobes with the other organs being tetramerous (i.e., divided into fours), while the surrounding flowers generally have three calyx lobes, with the other organs being pentamerous (i.e., divided into five parts). As far as we can tell, such modifications follow from the relative position and interaction of the parts and not as a result of natural selection.

In many other cases we find modifications of structure that botanists generally consider to be highly important even though they affect only some of the flowers on a given plant, or occur on distinct plants that grow close together under the same conditions. As these variations seem of no special use to the plants, they cannot have been influenced by natural selection. As before we are quite ignorant of the causes of these variations and can't even invoke any proximate agency, such as their relative position on the organism. Here I will give only a few instances of many that could be given. We commonly see that some flowers on a given plant are four rayed, others three rayed, and so forth. The Swiss botanist Mr. Augustin Pyramus de Candolle, for example, notes that the flowers of the Iranian poppy (*Papaver bracteatum*) offer either two sepals with four petals, which is typical of most poppies, or three sepals with six petals. The French botanist Augustin Saint-Hilaire gives the following similar cases: the genus *Zanthoxylon* belongs to a division of the citrus family Rutaceae with a single ovary; but in some species, flowers with either one or two ovaries may be found on the same plant and even within the same cluster of flowers (i.e., panicle). Mr. Saint-Hilaire found for *Gomphia olivaeformis* toward the southern extreme of its range, two flower forms of which he was initially certain were distinct species; but he later saw both forms growing on the same bush.

We thus see that, with plants, many morphological changes may be governed by factors independent of natural selection. But with respect to Mr. von Nägeli's doctrine of an innate tendency toward perfection or progressive development, can it be said in the case of these strongly pronounced variations that the plants have been caught in the act of progressing toward a "higher state of development"? On the contrary, the fact that the parts in question differ or vary greatly on the same plant suggests that such modifications are of extremely small importance to the plants themselves, although they may be helpful to us for our classifications. The acquisition of a useless part can hardly be said to raise an organism in the natural scale of complexity; in the case of the imperfect, closed flowers described above, if any new principle has to be invoked, it must be one of retrogression rather than of progression, as it must be as well with the many parasitic animals that have become secondarily simplified for their new lifestyles.

If the above characteristics are unimportant for the welfare of the species, any slight variations that occurred in these parts would not have been accumulated and augmented through natural selection. When a structure that has been developed through long-continued selection ceases to be of service to a species, it generally becomes variable, for it will no longer be regulated by this same power of selection; we see this with rudimentary organs. And when modifications have been induced that do not affect the success of the species, they may be, and apparently often have been, transmitted in nearly the same state to numerous, otherwise modified descendants. It cannot have been of much importance to the greater number of mammals, birds, or reptiles whether they were clothed with hair, feathers, or scales, and yet hair has been transmitted to almost all mammals, feathers to all birds, and scales to all true reptiles. Even so, any structure that is common to many allied forms is ranked by us as of high importance for classifying organisms, and consequently is often assumed to be of vital importance to the species. I am inclined to believe, however, that many morphological differences that we consider to be important—such as the arrangement of the leaves, the divisions of the flower or of the ovarium, the position of the ovules, and so forth—first appeared in many cases as simple fluctuating variations, which sooner or later became constant not through natural selection, but rather through the nature of the organisms and of the surrounding conditions, as well as through the interbreeding of distinct individuals. For as these morphological characteristics do not affect the welfare of the species, any slight deviations in them could not have been governed or accumulated through selection. It is a strange result that we thus arrive at: characteristics that are of the most importance to the systematist are only of slight importance to the lives of the organisms they study. Later, when I talk about the genetic principle of classification (Chapter 14), we will see that this is by no means as paradoxical as it may at first appear.

Although we have no good evidence of the existence in living organisms of any innate tendency toward progressive development, such a tendency would not necessarily be expected to follow from the actions of natural selection, as I have tried to show in Chapter 4. Indeed, the best definition that has ever been given of a high standard of organization is the degree to which various body parts have been specialized or differentiated; natural selection tends toward specialization, inasmuch as the parts are thus enabled to perform their functions more efficiently.

Supposed Incompetence of Natural Selection to Account for the Incipient Stages of Useful Structures

The distinguished zoologist Mr. St. George Mivart has recently collected all the objections that have been advanced to date against the theory of natural selection, as put forward by both Mr. Alfred Russel Wallace and myself, and has illustrated them with admirable art and force. When thus marshaled, they make a formidable array. However, as Mr. Mivart does not give any of the various facts and considerations that go against his conclusions, readers have no chance to weigh the evidence on both sides. When discussing special cases, Mr. Mivart passes over the effects of the increased use and disuse of parts, which I have always maintained to be highly important, and which I have addressed in my book *The Variation of Animals and Plants Under Domestication* at, I believe, greater length than any other writer. He likewise often assumes that I attribute nothing to variation independently of natural selection, whereas in *The Variation of Animals and Plants Under Domestication* I have collected more well-established cases than can be found in any other work that I know of. My judgment may not be trustworthy, but after reading Mr. Mivart's book with considerable care and comparing each section with what I have said on the same topic, I never before felt so strongly convinced of the general truth of the conclusions that I have arrived at here.

All of Mr. Mivart's objections will be, or have already been, considered in this book. The one new point that appears to have struck many readers is, "that natural selection is incompetent to account for the incipient stages of useful structures." This subject is intimately connected with that of the gradation of characteristics, often accompanied by a change of function—for instance, the conversion of a swim bladder into lungs—points that I discussed at length in the previous chapter under two headings. Nevertheless, I will here consider in some detail several of the cases advanced by Mr. Mivart, selecting those that are the most illustrative, as want of space prevents me from considering them all. In every case I think you will agree that natural selection prevails in accounting for what we see among organisms in nature today.

First let us consider the giraffe. This animal, by its lofty stature and its much elongated neck, forelegs, head, and tongue, has its whole frame beautifully adapted for browsing on the higher branches of trees. It can thus obtain food well beyond the reach of the other hoofed animals (i.e., ungulates) inhabiting the same lands; this must be a great advantage to it during times of famine. Indeed, the Niata cattle in South America show us how important a small change in structure could be in preserving an animal's life during such periods. These cattle can browse on grass as well as other animals do, but, during droughts, because of the way their lower jaw projects, they cannot browse on the twigs of trees, reeds, and so forth on which the common cattle and horses are driven and able to feed. At such times, without the ability to feed on other foods, and unless they are fed by their owners, the Niata cattle perish in large numbers.

Before coming to Mr. Mivart's specific objections, let me just summarize once again how natural selection will act in all ordinary cases. We have modified some of our domestic animals without having necessarily paid much attention to special structural features simply by preserving and breeding from the fleetest individuals, as with the racehorse and the greyhound, or by simply preserving and breeding solely from the victorious birds after staged fights, as with the gamecock. Likewise, with the ancestral giraffe in nature, those individuals that did their browsing higher up than others and were able during famines to reach even just an inch or two above the others, will often have been more likely to feed better and thus survive and reproduce as they roamed over the whole country in search of food. That individuals of the same species often differ slightly in the relative lengths of all their parts may be seen in the many works of natural history in which careful measurements are given. Such differences will have been of great importance to the ancestral giraffe, considering its probable habits of life; for those individuals that had some one of several body parts rather more elongated than usual would generally have been most likely to survive when food was scarce. These individuals will have been more likely to have mated and left offspring, which then would either have inherited the same body peculiarities or at least had a tendency to vary again in the same manner. In contrast, those individuals that were less favored in the same respects will have been most likely to have perished.

Note that when we methodically try to improve a breed, we separate all the superior individuals, allowing only those individuals to freely interbreed. In contrast, natural selection will preserve all the superior individuals, allowing them to interbreed freely, and will destroy all the inferior individuals. This process corresponds exactly with what I have called unconscious selection by man. Continuing the process in nature for many, many generations could easily, it seems to me, convert an ordinary hoofed quadruped[2] into a giraffe.

Mr. Mivart offers two objections to this conclusion. One is that an increased body size would obviously require an increased supply of food; he suggests that it is "very problematical whether the disadvantages thence arising would not, in times of scarcity, more than counterbalance the advantages." But as giraffes do in fact exist in large numbers in Africa, and as some of the largest antelopes in the world—taller in fact than an ox—abound there, why should we doubt that, as far as size is concerned, intermediate gradations in height could formerly have existed there, even when subjected as now to severe famine? Surely, being able to reach, at each increase in size, a supply of food unreachable by other hoofed quadrupeds in the area would have been of some advantage to the nascent giraffe. In addition, the increased bulk of the animals would help to protect them from almost all predators other than the lion. And against that animal, the ancestral giraffe's increasingly tall neck—and the taller the better—would, as the American philosopher and mathematician Mr. Chauncey Wright has remarked, serve as a valuable watchtower for seeing potential predators from afar. Indeed, as the British explorer and naturalist Sir Samuel White Baker remarks, no animal is more difficult to stalk than a giraffe. This animal also uses its long neck as a means of offense or defense, by violently swinging its head—which is armed with stump-like horns—back and forth. The preservation of each species will rarely be controlled by any one advantage, but rather by the combination of them all, both great and small.

---

[2] Any four-legged animal, such as dogs, cats, horses, and cows.

(A)   (B)

Figure 7.2 (A) Guanacos (*Lama glama*). (B) South American llamas (*Macrauchenia patachonica*), which went extinct 10,000–20,000 years ago.

Mr. Mivart's second objection here is that if natural selection is so potent, and if browsing on vegetation higher up is such a great advantage, then why haven't any other hoofed quadrupeds acquired a long neck and lofty stature other than the giraffe, and to a lesser extent the camel, guanaco (Figure 7.2A), and the now-extinct, long-necked llama, *Macrauchenia* (Figure 7.2B)? Or again, why haven't any members of the group acquired a long proboscis? The answer is not difficult when considering South Africa, which was formerly inhabited by numerous herds of giraffe, and can best be given through the following illustration. In every meadow in England in which trees grow, we see the lower branches trimmed to an exact level by the browsing of horses and cattle; so what advantage could there be for sheep, for example, living in that same area, to acquire slightly longer necks when there is no food for them to eat higher up? In every district, some one kind of animal will almost certainly be able to browse higher than the others; it is almost equally certain that this one kind alone could have its neck elongated for this purpose, through natural selection and the effects of increased use. In South Africa, the competition for browsing on the higher branches of the acacias (Figure 7.3) and other trees must be between giraffe and giraffe, and not with the other ungulate animals.

We don't know why various animals belonging to this same category of animals in other parts of the world have not acquired either an elongated neck or a proboscis. But it is as unreasonable to expect a distinct answer to such a question as to ask why some event in the history of humankind occurred in one country but not in another. We don't know what conditions determine the numbers and range of each species, and we cannot even guess what anatomical changes would favor the increase of any species in some new country. We can, however, see in a general manner that various causes might have interfered with the development of a long neck or proboscis in some animals. To reach the foliage at a considerable height implies a substantial increase in the bulk of the body (unless, of course, the animals can climb trees, something that hoofed animals are particularly ill-equipped to do). We know that some

Figure 7.3 Giraffes (*Giraffa camelopardalis*) eating from an acacia tree.

very luxuriant areas, such as South America, support remarkably few large quadrupeds, while South Africa abounds with them to an unparalleled degree. Why this should be so, we don't know. Nor do we know why the later Tertiary period, which ended about 66 million years ago, should have been so much more favorable for their existence than the present time. Whatever the causes might have been, we can see that certain regions and times would have been much more favorable than others for the development of so large a quadruped as a giraffe.

It seems almost certain that an animal cannot acquire some special, well-developed anatomical modification without also having several other body parts modified and coadapted in some way. Although every part of the body varies slightly among individuals, it does not follow that the necessary body parts should always vary in the right direction and to the right extent. With the different species of our domesticated animals, we know that the various parts vary in different ways and to different degrees and that some species show more variability than others. But even if suitable variations arise, it does not follow that natural selection would always be able to act on them to produce a structure that would be clearly beneficial to the species. For instance, if the number of individuals existing in a region is determined chiefly by the extent of predation or by the impact of internal or external parasites, as often seems to be the case, then natural selection will be able to do little in modifying any particular structure for obtaining food.

Last, natural selection is a slow process, and the same favorable conditions must endure for a considerable time before any marked effect can be produced. These reasons are very general and very vague, but that's all we can say at the moment about why, in many quarters of the world, most hoofed quadrupeds have not acquired greatly elongated necks or other means for browsing on the higher branches of trees.

Objections similar to those just mentioned have also been advanced by many other writers. In each case, a variety of factors, in addition to the general ones just discussed, have probably prevented the gradual acquisition of particular structures through natural selection, even though such modifications would seem useful for those organisms to have had. One writer asks, "Why has not the ostrich acquired the power of flight?" But think about the enormous supply of food that would be needed to give this desert bird the ability to move its huge body through the air. Similarly, oceanic islands are inhabited by bats and seals, but not by any terrestrial animals. Yet some of those bats are very distinctive, indicating that they must have occupied those islands for a very long time. Thus, Sir Charles Lyell asks, "Why haven't seals and bats given birth on such islands to forms that can live well on land?" But seals would first have to be converted into terrestrial carnivores of a considerable size and bats into terrestrial insectivorous animals. That would be a great problem, since for seals there would be nothing for them to prey on and for the bats, the ground insects that would be available for food would already be largely preyed on by the reptiles and birds that first colonized and now abound on most oceanic islands.

Gradations of structure, with each stage beneficial to a changing species, will be favored only under certain peculiar conditions. A strictly terrestrial animal, by occasionally hunting for food in shallow water, then in streams or lakes, might eventually be converted into an animal so thoroughly aquatic as to brave the open ocean. But seals would not find the conditions on oceanic islands favorable for their gradual reconversion into a terrestrial form. Bats, as formerly shown, probably acquired their wings by at first gliding through the air from tree to tree, like the so-called flying squirrels (see Figure 6.1C) of today, for the sake of escaping from their enemies or for avoiding falls. But when the power of true flight had once been acquired it would never be reconverted back, at least not for the above purposes, into the less efficient power of simply gliding through the air. Like many birds, bats might have had their wings greatly reduced in size or completely lost through disuse. In such a case, however, they would first have had to acquire the ability to run quickly on the ground using their hind legs alone, so as to compete successfully with birds or other ground animals; bats seem singularly ill-fitted for such a change.

I have made these conjectural remarks only to show that a transition of structure, with each step beneficial to its owner, is a highly complex affair and that there is nothing strange in a transition not having occurred in any particular case.

Finally, more than one writer has asked, "Why have some animals had their mental powers more highly developed than others? Wouldn't such development be advantageous to all? Why haven't apes acquired the intellectual powers of man?" Various explanations could be given, but they are all conjectural and their relative probabilities cannot be weighed. A definite answer to the question should not be expected, seeing that no one can solve the simpler problem of why, of two races of primitive peoples, one has risen higher in the scale of civilization than the other, something that presumably implies increased brainpower.

Let us return to more of Mr. Mivart's objections. Insects often resemble various other objects for the sake of protection—green or decayed leaves, for example, or dead twigs, bits of lichen, flowers, spines, and even the excrement of birds. The resemblance is often wonderfully close, not just in color but in form as well, and even to the manner in which the insects hold themselves. The caterpillars that stick out motionless from the bushes on which they feed, like dead twigs, offer an excellent example of a resemblance of this kind. Cases in which insects imitate the excrement of birds are rare and exceptional. On this point, Mr. Mivart remarks, "As, according to Mr. Darwin's theory, there is a constant tendency to indefinite variation, and as the minute incipient variations will be in *all directions*, they must tend to neutralize each other, and at first to form such unstable modifications that it is difficult, if not impossible, to see how such indefinite oscillations or infinitesimal beginnings can ever build up a sufficiently appreciable resemblance to a leaf, bamboo, or other object for natural selection to seize upon and perpetuate."

But in all the foregoing cases, the insects in their original state no doubt presented some rude and accidental resemblance to an object commonly found in the habitats that they frequented. Nor is this at all improbable, considering the almost infinite number of surrounding objects and the diversity in form and color of the various existing insects. As some rude resemblance is necessary for the first start, we can understand how it is that the larger and higher animals do not (with the single exception, as far as I know, of one fish) gain protection by resembling special objects, but only the surface that commonly surrounds them, and this chiefly through color. Assuming that an insect originally happened to resemble in some degree a dead twig or a decayed leaf, and that it varied slightly in many ways, then all the variations that rendered the insect at all more like any such object and thus favored its escape from predation, would be preserved, while other variations would be neglected and ultimately lost. Or, if they rendered the insect at all less like the imitated object, they would be eliminated through predation. Mr. Mivart's objection would indeed be valid if we were attempting to account for the above resemblances independently of natural selection, through mere fluctuating variability, but as that is not at all what we are trying to do, there is no problem to be resolved.

Nor can I see any force in Mr. Mivart's difficulty with respect to "the last touches of perfection in the mimicry," as in the case given by Mr. Wallace of a walking stick insect (Figure 7.4) found on Borneo that resembles "a stick grown over by a creeping moss or jungermannia." So close was this resemblance, in fact, that a Borneo native maintained that the foliaceous excrescences (leafy outgrowths) really were moss! Insects are preyed on by birds and other enemies whose sight is probably sharper than ours, and every grade in resemblance that aided an insect in escaping notice or detection would help its survival; the more perfect the resemblance, the safer the insect. Considering the nature of the differences between the species in the group that includes the above-mentioned stick insect, there is nothing improbable in this insect showing variations in the irregularities on its surface and in those irregularities having become more or less green-colored over the generations; for within every group, the characteristics that differ among the several species are the most apt to vary, while the generic characteristics, or those common to all the species in the group, are the most constant.

Figure 7.4 A walking stick insect.

The Greenland whale is one of the most wonderful animals in the world, and the baleen (Figure 7.5) or whalebone, one of its greatest peculiarities. The baleen consists of a row of about 300 thin plates ("laminae") on each side of the upper jaw. These plates stand close together transversely to the longer axis of the mouth, and within the main row there are some subsidiary rows. The extremities and inner margins of all the plates are frayed into stiff bristles that clothe the whole gigantic palate and serve to strain or sift particles from the water and thus to concentrate and retain the minute prey on which these huge animals subsist. The middle and longest lamina in the Greenland whale is 10, 12, or even 15 feet long. In other cetacean species we also see graduations in length: in one species, the middle lamina is 4 feet long, according to

(A)   (B)

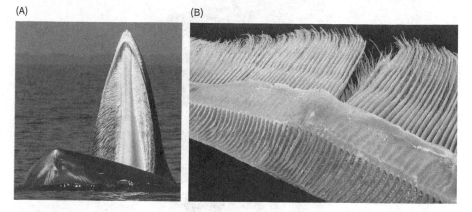

Figure 7.5 (A) A whale's open mouth showing baleen. (B) One plate of baleen from a baleen whale.

the Arctic explorer William Scoresby, in another it is 3 feet long, in another 18 inches long, and in *Balaenoptera acutorostrata*, one of the Rorqual or Minke whales, only about 9 inches long. The quality of the whalebone also differs among the different species.

With respect to the baleen itself, Mr. Mivart remarks that once it "had attained such a size and development as to be at all useful, then its preservation and augmentation within serviceable limits would be promoted by natural selection alone. But how to obtain the beginning of such useful development?" In answer, can't we imagine that the early ancestors of the baleen whales possessed a mouth constructed something like the lamellated beak of a duck,[3] with its rows of comb-like structures ("pectin") along the beak? Ducks, like whales, subsist by sifting the mud and water for food; indeed, the family has sometimes been called Criblatores, or "sifters." I hope I may not be misconstrued into saying that the progenitors of whales actually did possess mouths lamellated like the beaks of ducks. I only wish to show that this is not incredible and that the immense plates of baleen seen in the Greenland whale might have been developed from such lamellae by finely graded steps, each of use to its owner.

Indeed, the beak of a shoveler duck (*Anas clypeata*) (Figure 7.6A) is a more beautiful and complex structure than the mouth of a whale. In the specimen that I examined, the upper mandible is furnished with a row of 188 thin, elastic lamellae on each side, obliquely beveled so as to be pointed, and placed transversely to the longer axis of the mouth. They arise from the palate and are attached to the sides of the mandible by flexible membranes. Those lamellae standing toward the middle are the longest, being about one-third of an inch long, and they project fourteen-hundredths of an inch beneath the edge. At their bases there is a short subsidiary row of obliquely

---

[3] Some duck beaks have a series of tooth-like plates or ridges (see Figure 7.6), used for sieving food particles from the water.

(A)

(B)

(C)

(D)

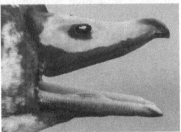

(E)

Figure 7.6 (A) A male shoveler duck (*Anasclypeata*), showing baleen-like projections. (B) Two male torrent ducks (*Merganetta armata*). (C) A male wood duck (*Aix sponsa*). (D) The Egyptian goose (*Alopchen aegyptiacus*). (E) A male merganser, showing its serrated beak (*Mergus merganser*).

transverse lamellae. In these several respects they resemble the plates of baleen in the mouth of a whale. But toward the end of the beak they differ quite a bit, as they project inward instead of straight downward. The entire head of the shoveler, though incomparably less bulky, is about one-eighteenth of the length of the head of a moderately large *Balaenoptera acutorostrata*, the Minke whale, in which species the baleen is only 9 inches long, as noted earlier. Thus, if we were to make the head of the shoveler duck as long as that of *Balaenoptera*, the lamella would be 6 inches long—that is, two-thirds of the length of the baleen found in this species of whale. The lower mandible of the shoveler duck is furnished with lamellae equal in length with those above, but finer; in being thus furnished it differs conspicuously from the lower jaw of a whale, which is destitute of baleen. On the other hand, the extremities of this duck's lower lamellae are frayed into fine bristly points, curiously resembling the plates of baleen. In the bird genus *Prion*, a member of the distinct family of petrels, only the upper mandible is furnished with lamellae. These are well developed and project beneath the margin; thus the beak of this bird resembles, in this respect, the mouth of a whale.

From the highly developed structure of the shoveler's beak we may then proceed, without any great break as far as fitness for sifting is concerned, through the beak of the torrent duck (*Merganetta armata*) (Figure 7.6B), and in some respects through that of the wood duck (*Aix sponsa*) (Figure 7.6C) to the beak of the common duck. I am basing my remarks on information and specimens sent to me by the ornithologist Mr. Osbert Salvin. The lamellae found in the common duck are much coarser than those of the shoveler and are firmly attached to the sides of the mandible. There are only about 50 on each side, and they do not project at all beneath the margin. They are all square-topped and are edged with translucent hardish tissue, as if for crushing food. The edges of the lower mandible are crossed by many fine ridges that project very little. Compared with the beak of the shoveler, the common duck's beak is thus very inferior for sifting, and yet this bird, as everyone knows, constantly uses it for exactly this purpose! There are other species, as I hear from Mr. Salvin, in which the lamellae are considerably less developed than in the common duck, but I do not know whether or not the members of those species use their beaks for sifting food from the water.

Let us turn now to another group of the same family. The beak of the Egyptian goose (*Alopochen*) (Figure 7.6D) closely resembles that of the common duck, except that the lamellae are not so numerous and not so distinct from each other, and do not project so much inward. Yet this goose, I am told by the English ornithologist Mr. Edward Bartlett, "uses its bill like a duck by throwing the water out at the corners." Its chief food, however, is grass, which it crops like the common goose does. In that latter bird (the common goose), the lamellae of the upper mandible are much coarser than they are in the common duck and almost confluent, with about 27 on each side, and they terminate upward in teeth-like knobs. The palate is also covered with hard rounded knobs. The edges of the lower mandible are serrated with teeth that are much more prominent, coarser, and sharper than those of the duck. The common goose does not sift the water at all but instead uses its beak exclusively for tearing or cutting herbage, for which purpose it is so well fitted that it can crop grass closer to the ground than almost any other animal. There are other species of geese, as I hear from Mr. Bartlett, in which the lamellae are less developed than those of the common goose.

Thus we see that a member of the duck family, with a beak constructed like that of the common goose and adapted solely for grazing, or even a member with a beak having less well-developed lamellae, might be converted by small changes into a species like the Egyptian goose, and this into one like the common duck, and this, finally, into one like the shoveler, a bird that is provided with a beak almost exclusively adapted for sifting food from the water; indeed, this bird is so specialized that it could hardly use any part of its beak, except the hooked tip, for seizing or tearing solid food. Let me also add that a goose's beak might be converted by small gradual changes into one with prominent, recurved (curved backward or inward) teeth, like those of the merganser (*Mergus merganser*) (Figure 7.6E), which belongs to the same family.

Returning to the whales now, the northern bottlenose whale (*Hyperoodon ampullatus*, formerly *H. bidens*) (Figure 7.7) lacks true, functional teeth, according to the French naturalist Bernard Germain de Lacépède, but its palate is roughened with small, unequal, hard points of horn. There is, therefore, nothing improbable in supposing that some ancestral cetacean was provided with similar but more regularly spaced points of horn on its palate that aided it in seizing or tearing its food, just as the knobs on a goose's beak do now. If so, it will hardly be denied that the points might then have gradually been converted through variation and natural selection into lamellae that are as well developed as those of the Egyptian goose; in that case they would have been used both for seizing objects and for sifting food from the water. Then we can imagine them being gradually converted into lamellae like those of the domestic duck, and from there onward until they were as well constructed as those of the shoveler, in which case they would have served exclusively as a sifting apparatus.

Figure 7.7 Northern bottle-nose whale (*Hyperoodon ampullatus*).

From this stage, in which the lamellae would be two-thirds the length of the plates of baleen found in *Balaenoptera rostrata*, gradations—which may still be observed in some modern cetaceans—lead us onward to the enormous plates of baleen now seen in the Greenland whale. Nor is there the least reason to doubt that each step in this scale might have been as useful to certain ancient cetaceans, with the functions of the parts slowly changing over the generations, as the various gradations in the beaks of the different modern members of the duck family are now. We should bear in mind that each species of duck is subjected to a severe struggle for existence, so that the structure of every part of its frame must be well adapted to its conditions of life.

The flatfish, members of the fish family Pleuronectidae (Figure 7.8A), are remarkable for their asymmetrical bodies. They rest on one side—usually the left—and occasionally we find reversed adult specimens. At first sight the lower, resting surface resembles the ventral surface of an ordinary fish: it is white in color, less developed in many ways than the upper side, and with the lateral fins often of smaller size. But the eyes offer the most remarkable peculiarity in adults: both eyes are found on the upper side of the head! During the juvenile stage, however, the eyes are positioned on opposite sides of the head just like those of a normal fish; moreover, both sides of the juvenile's body are pigmented equally. As development proceeds, however, the eye that is located on the side of the body that will eventually be facing downward against the substrate begins to glide slowly around the head to the upper side of the body. Obviously, if this lower eye did *not* move to the other side of the body during development, it would end up pointing into the substrate and would in fact be abraded by the sandy bottom. That flatfish are admirably adapted by their flattened and asymmetrical structure for their habits of life is evident from the fact that the various flatfish species (soles, flounders, and halibut, for example) are extremely common. The chief advantages thus gained seem to be protection from their enemies and facility for feeding on the sediment. However, the different members of the family present, as the Danish biologist Jørgen Matthias Christian Schiødte remarks, "a long series of forms exhibiting a gradual transition from *Hippoglossus pinguis*, which does not in any considerable degree alter the shape in which it leaves the ovum, to the soles, which are entirely thrown to one side."

(A)   (B)

Figure 7.8 (A) A flatfish. Note that both eyes are on the same side of the head in adults. (B) The American plaice (*Hippoglossoides platessoides*).

Mr. Mivart, in considering this case, remarks that a sudden spontaneous transformation in the position of the eyes is hardly conceivable, in which I quite agree with him. He then adds. "If the transit was gradual, then how such transit of one eye a minute fraction of the journey toward the other side of the head could benefit the individual is, indeed, far from clear. It seems, even, that such an incipient transformation must rather have been injurious." But he might have found a satisfying answer to this objection in the excellent observations published in 1867 by the Swedish zoologist August Wilhelm Malm. Flatfish, it turns out, when very young and still fully symmetrical, and with their eyes located on opposite sides of the head, cannot long retain a vertical position, owing to the excessive depth of their bodies, the small size of their lateral fins, and to their lacking a swim bladder. Thus, soon growing tired, they fall to the bottom on one side. Mr. Malm has observed that while the fish are thus resting, they often twist the lower eye upward, trying to see above them. Indeed, they do this so vigorously that the eye is pressed hard against the upper part of the orbit. The forehead between the eyes consequently becomes temporarily contracted in breadth. On one occasion Mr. Malm actually saw a young fish raise and depress the lower eye through an angular distance of about 70 degrees.

We should remember that at this early age the skull is cartilaginous and flexible, so that it readily yields to muscular action. We also know that the skull of higher animals can be altered in shape if the skin or the muscles are permanently contracted through disease or some accident, even in young adulthood. Indeed, with long-eared rabbits, if one ear lops forward and downward, its weight drags forward all the bones of the skull on the same side. Mr. Malm states that the newly hatched young of perch, salmon, and several other symmetrical fishes have the habit of occasionally resting on one side at the bottom; of considerable interest to us, he has observed that they often then strain their lower eyes so as to look upwards, so that their skulls are thus rendered rather crooked. These fishes, however, are soon able to hold themselves in a vertical position, so that no permanent effect is thus produced.

Mr. Schiødte believes, in opposition to some other naturalists, that members of the family Pleuronectidae (righteye flounders) are not quite symmetrical even as embryos; if so, we could understand how it is that certain species, while young, habitually fall over and rest on their left side, and other species on their right side. Mr. Malm adds, in confirmation of this view, that adults of the dealfish *Trachipterus arcticus* (a type of ribbonfish), which is not a member of the family Pleuronectidae, rests at the bottom on its left side and swims diagonally through the water; in this fish, the two sides of the head are said to be somewhat dissimilar. Our great authority on fishes, Dr. Albert Günther, concludes his abstract of Mr. Malm's paper by remarking that "the author gives a very simple explanation of the abnormal condition of the Pleuronectoids."

We thus see that the first stages of the transit of the eye from one side of the head to the other, which Mr. Mivart assumes would be injurious, may be attributed to the habit of endeavoring to look upward with both eyes while resting on one side at the sea's bottom—something that no doubt benefits both the individual and the species.

We may also attribute to the inherited effects of use the fact that the mouth in several kinds of flatfish is shifted toward the lower surface, with the jaw bones stronger and more effective on this side of the head—the eyeless side—probably, as the Scottish flatfish authority Dr. Ramsay Traquair supposes, for feeding with ease on the bottom. On the other hand, disuse[4] probably accounts for the less developed condition of the whole inferior half of the body, including the lateral fins, though the English zoologist William Yarrell thinks that the reduced size of these fins is advantageous to the fish, as "there is so much less room for their action, than with the larger fins above." Perhaps the fact that there are 25–30 teeth in the lower halves of the two jaws of the plaice (Figure 7.8B), but only 4–7 in the upper halves, may likewise be accounted for by disuse. From the colorless state of the ventral surface of most fishes and of many other animals, we may reasonably suppose that the absence of color in flatfish on whichever side faces downward is due to the exclusion of light. But it cannot be supposed that the action of light is responsible for the peculiar speckled appearance of the upper side of the sole, so closely resembling the sandy bed of the sea, or for the ability of individuals of some species to alter their body color to closely match that of the surrounding surface (as recently shown by Charles Henri Georges Pouchet at the Muséum National d'Histoire Naturelle in Paris), or for the presence of bony tubercles (nodules) on the upper side of the turbot. Instead, natural selection has probably come into play here, as it probably has in adapting the general shape of the body of these fishes, and many other peculiarities, to their habits of life. We should keep in mind, as I have insisted already several times, that the inherited effects of the increased use of parts, and perhaps of their disuse, will be strengthened by natural selection; all spontaneous variations in the right direction will thus be preserved, as will individuals who strongly inherit any beneficial effects of increased use. How much to attribute in each particular case to the effects of use and disuse, and how much to natural selection, it seems impossible to decide at present.

Let us now consider the mammals. By definition, all mammals have mammary glands, which are indispensable to their existence. Such glands must, therefore, have been developed a very long time ago, and we can know nothing definite about the details of their origins. Mr. Mivart asks: "Is it conceivable that the young of any animal was ever saved from destruction by accidentally sucking a drop of scarcely nutritious fluid from an accidentally hypertrophied cutaneous gland[5] of its mother? And even if one was so, what chance was there of the perpetuation of such a variation?" But Mr. Mivart does not put the case fairly. Most evolutionists admit that mammals are descended from a marsupial form. If so, the mammary glands will have initially developed within the marsupial sack. Even in the case of seahorses (members of the fish genus *Hippocampus*) (Figure 7.9), the eggs are hatched—and the young are then reared for a time—within a sack of exactly this nature. Indeed, an American

[4] As we saw earlier, particularly in Chapter 5, Darwin thought—erroneously, as it turns out—that the effects of use and disuse in the parent could in some cases be directly inherited by offspring.

[5] An increased volume of a gland in the skin due to an increase in the size of its component cells.

Figure 7.9 A seahorse (*Hippocampus* sp.).

naturalist, the Reverend Dr. Samuel Lockwood, believes from what he has seen of the development of the young that they are nourished by a secretion from the cutaneous glands of that sack. Now with the early mammalian ancestors, almost before they deserved to be thus designated, is it not at least possible that the young might have been similarly nourished? And in this case, any individuals that secreted a fluid that was in some degree or manner especially nutritious, so as to take on some of the nutritious characteristics of milk, would in the long run have reared a larger number of well-nourished offspring than would individuals that secreted a fluid poorer in nutrients. Thus the cutaneous glands, which are the homologues[6] of the mammary glands, would have been improved and made more effective over the generations. It fits well with the widely extended principle of specialization that some glands over a certain space of the sack should have become more highly developed than the rest; they would then have formed a breast, but at first one without a nipple, as indeed we see in the platypus (*Ornithorhynchus anatinus*) (see Figure 4.9A), an animal that sits at the base of mammalian evolution. The development of the mammary glands would have been of no use, and thus could not have been achieved through natural selection, unless the young at the same time were able to feed on the secretion. Understanding how young mammals have instinctively learned to suck their mother's breast is no harder than understanding how unhatched chickens have learned to break the eggshell that encloses them by tapping against it with their specially adapted beaks, or how, a few hours after leaving the eggshell, they have learned to pick up grains of food. In such cases, the most probable solution seems to be that the habit was first acquired by practice at a more advanced age and afterward transmitted to the offspring earlier in their development. But the young kangaroo is said not to suck, only to cling to the nipple of its mother, who is then able to actively inject milk into the mouth of her

---

[6] Homologues are parts that have a common ancestry. Here Darwin is suggesting that the mammary glands have their evolutionary origins in these cutaneous glands.

helpless, half-formed offspring. On this point, Mr. Mivart remarks that without some special provision, "the young one must infallibly be choked by the intrusion of the milk into the windpipe." Mr. Mivart then goes on to note that indeed, there is a special provision: the infant kangaroo's larynx "is so elongated that it rises up into the posterior end of the nasal passage, and is thus enabled to give free entrance to the air for the lungs while the milk passes harmlessly on each side of this elongated larynx, and so safely reaches the gullet behind it." Mr. Mivart then asks, how did natural selection remove in the adult kangaroo (and in most other mammals, on the assumption that they have descended over time from a marsupial form), "this at least perfectly innocent and harmless structure?" Let me suggest that the voice, which is certainly of high importance to many animals, could hardly have been used with full force as long as the larynx entered the nasal passage. Indeed, the well-known comparative anatomist Sir William Henry Flower has suggested to me that this structure would have greatly interfered with an animal swallowing solid food.

Let us now turn to the lower divisions of the animal kingdom. Members of the Echinodermata (the phylum containing such animals as sea stars, sea urchins, and brittle stars) are furnished with remarkable organs called pedicellariae (Figure 7.10A), which consist, when well-developed, of a three-part forceps—that is, each pedicellaria is formed of three serrated, moveable fingers, neatly fitting together and placed at the top of a flexible stem, which is moved by muscles. These forceps can firmly seize hold of any object: the comparative anatomist Louis Agassiz at Harvard University has even seen a sea urchin rapidly passing particles of excrement from forceps to forceps down certain lines of its body in order to keep its shell from being fouled. But there is no doubt that besides removing dirt of all kinds from the outside of the shell, they also perform other functions, one of which, apparently, is defense.

With respect to these remarkable structures, Mr. Mivart, as on so many previous occasions, asks, "What would be the utility of the first *rudimentary beginnings* of such structures, and how could such incipient buddings have ever preserved the life of a single *Echinus* [sea urchin]?" He adds, "Not even the *sudden* development of the snapping action could have been beneficial without the freely moveable stalk, nor could the latter have been efficient without the snapping jaws, yet no minute merely indefinite variations could simultaneously evolve these complex coordinations of structure; to deny this seems to do no less than to affirm a startling paradox."

Well, as paradoxical as this may appear to Mr. Mivart, the members of some sea star species in fact do have three-part forceps that are immovably fixed at the base but that are nevertheless capable of a snapping action; this makes sense if they serve, at least in part, as a means of defense. Mr. Agassiz, to whose great kindness I am indebted for much information on the subject, informs me that there are other sea stars in which one of the three fingers of the forceps is reduced to a support for the other two fingers, and that in some genera the third finger is completely lost. In *Echinoneus*, the shell is described by the French echinoderm expert Jean Octave Edmond Perrier as bearing two kinds of pedicellariae, one resembling those of *Echinus* (Figure 7.10B), and the other those of the heart urchin *Spatangus*. Such cases are always interesting as

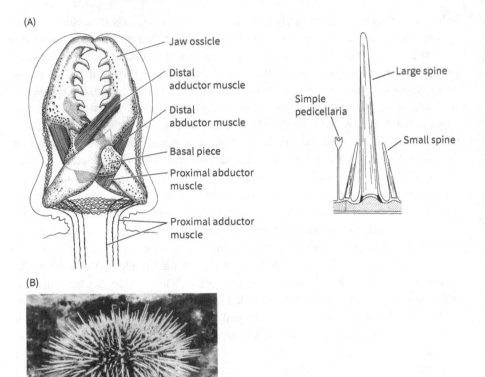

Figure 7.10 (A) Pedicellariae of echinoderms. These structures are especially common on sea stars (left) and sea urchins (right). Sea star pedicellariae are usually two-jawed, whereas urchin pedicellariae are usually three-jawed. (B) *Echinus* sp., a member of the animal phylum Echinodermata.

they afford the means of apparently sudden transitions through abortion of one of the two states of an organ.

With respect to the steps through which these curious organs have been evolved, Mr. Agassiz infers from his own research and that of Johannes Peter Müller that the pedicellariae of both sea stars and sea urchins must be looked at as modified spines. This may be inferred from their manner of embryological development, as well as from a long and perfect series of gradations from simple granules to ordinary spines in different species and genera, and then to perfect three-pronged pedicellariae. The gradations extend even to the manner in which ordinary spines and pedicellariae, with their supporting calcareous rods, are articulated to the shell. In certain sea star genera, "the very combinations needed to show that the pedicellariae are only modified branching spines" may be found. Thus we have fixed spines, with three equidistant, serrated, moveable branches articulated near their bases, and, higher up on the same spine, three other moveable branches. Now when the latter arise from the tip of a spine they form in fact a crude three-pronged pedicellaria, and such may be seen on the same spine together with the three lower branches. In this case the similarity between the fingers of the pedicellariae and the moveable branches of a spine is unmistakable. Since it is generally admitted that the ordinary spines serve for protection, there is no reason to doubt that those furnished with serrated and moveable branches are also protective; indeed they would offer particularly effective protection if the three branches began to act as a prehensile or snapping apparatus. Thus every gradation, from an ordinary fixed spine to a fixed, snapping pedicellaria, would be of use to its possessor.

In some sea star genera, the pedicellariae are not fixed or borne on an immoveable support but instead are placed at the top of a short but flexible and muscular stem; in this case they probably perform some other function in addition to defense. In sea urchins we can follow the steps by which a fixed spine becomes articulated to the shell and thus becomes moveable. I wish I had space here to give a more detailed summary of Mr. Agassiz's interesting observations on the development of pedicellariae. He notes that all possible gradations may also be found between the pedicellariae of sea stars and the hooks found on brittle stars, another group in the phylum Echinodermata, and also between the pedicellariae of sea urchins and the anchors of sea cucumbers (class Holothuroidea, another important echinoderm group).

Widely Different Organs in Members of the Same Class, Developed from One and the Same Source

One very interesting group of colonial animals, namely the Bryozoa (also called Polyzoa and Ectoprocta), are provided with curious organs called avicularia[7] (Figure 7.11A). These differ considerably in structure among the various bryozoan species. In their most perfect condition they curiously resemble the head and beak of a vulture in miniature, seated on a neck and capable of moving, as is the lower jaw as well. In one bryozoan species that I looked at, all the avicularia on the same branch often moved simultaneously backward and forward through an angle of about 90 degrees in the course of only 5 seconds, with the lower jaw wide open. When the jaws are touched with a needle, they seize it so firmly that the branch can then be shaken without the jaws losing their grip on the needle.

Mr. Mivart assumes that the avicularia of bryozoans and the pedicellariae of echinoderms are "essentially similar," so that if they were both developed through natural selection, they must have been developed independently in widely different divisions of the animal kingdom. But I see no structural similarity between three-pronged pedicellariae and avicularia. If anything, the avicularia more closely resemble the

---

[7] These are small, stalked structures that are specialized for pinching small prey or discouraging larvae of other species from attaching.

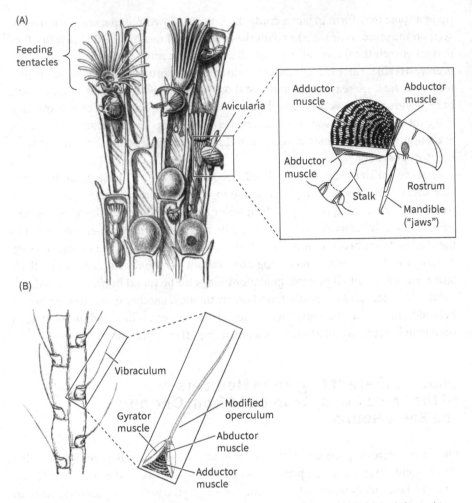

Figure 7.11 (A) A bryozoan colony (*Bugula turbinata*), with several bird-head shaped avicularia, specialized for pinching unwanted intruders. Note the ciliated tentacles on the colony's feeding zooids. (B) Vibracula, which are zooids modified for sweeping back and forth to discourage settlers and to clean debris and sediment from the colony's surface.

chelae (claws or pincers) of crabs and other crustaceans; why didn't Mr. Mivart bring up this resemblance as a special difficulty, or the resemblance between avicularia and the head and beak of a bird? Three naturalists who have carefully studied this group— Mr. George Busk, Dr. Fredrik Adam Smitt (a Swedish biologist), and Dr. Hinrich Nitsche (a German zoologist)—believe that bryozoan avicularia are homologous with the normal feeding individuals (called zooids[8]) in the colony, with the movable

---

[8] A zooid is the name for one individual in a colony of individuals, as in corals or bryozoans; all the zooids in a given colony are genetically identical, so they're really not "individuals"! Even though all

lip or lid of the often-calcareous "house" of the feeding individual corresponding to the lower and movable mandible of the avicularium. Although Mr. Busk, a Russian naval surgeon and naturalist, does not know of any gradations that now exist between a feeding individual (i.e., the zooid) and an avicularium, so that it is presently impossible to conjecture by what practical gradations the one could have been converted into the other, it by no means follows that such gradations never existed.

As the claws (chelae) of crustaceans resemble to some degree the avicularia of bryozoans, both serving as pincers, it may be worthwhile to show that a long series of serviceable gradations still exists among crustaceans. In the first and simplest stage, the terminal segment of a limb clamps down either on the square summit of the broad penultimate segment, or against one whole side; it is thus enabled to catch hold of an object. But the limb still serves as an organ of locomotion. We next find one corner of the broad penultimate segment slightly prominent, sometimes furnished with irregular teeth; the terminal segment clamps down against these. By an increase in the size of this projection, with its shape as well as that of the terminal segment slightly modified and improved, the pincers are rendered more and more perfect until we have an instrument at least as efficient as the chelae of a lobster. All of these gradations can actually be traced by looking at different existing species.

Besides the avicularia, bryozoans possess other curious organs called vibracula (Figure 7.11B). These generally consist of long bristles that are easily excited and capable of movement. The vibracula of one species I examined were slightly curved and serrated along the outer margin, and all of them on the same colony often moved simultaneously; acting like long oars, they swept a branch rapidly across the object-glass of my microscope. When a branch was placed on its face, the vibracula became entangled and made violent efforts to free themselves. They are supposed to serve as a defense and may sometimes be seen, as Mr. Busk remarks, "... to sweep slowly and carefully over the surface of the colony, removing what might be noxious to the delicate inhabitants of the 'houses' when their tentacula are protruded." The avicularia, like the vibracula, probably serve for defense, but they also catch and kill small living animals, which are probably then swept by the currents within each of the zooids' tentacles. The colonies of some species are provided with both avicularia and vibracula, some with only avicularia, and a few with only vibracula.

It is hard to imagine two objects more widely different in appearance than a vibraculum—basically a bristle—and an avicularium, which so much resembles the head of a bird, and yet they are almost certainly homologous and have been developed from the same common source, namely a single zooid with its surrounding calcareous "house." Thus we can understand how it is that in some species these organs graduate into each other, as I am told by Mr. Busk. The same is true of the avicularia

___

individuals are genetically identical, they can take different forms—some feed, some clean, some serve a protective role (as with avicularia), etc. In many bryozoan species, each feeding zooid lives in a calcium carbonate "house." Each house has a hinged lid that can be opened to allow the animal within to protrude its tentacles for feeding and gas exchange.

of several species of the bryozoan genus *Lepralia*: the moveable mandible is so much like a bristle that it is only the presence of the upper beak that serves to make its avicularian nature clear.

The vibracula may have developed directly from the lips of the houses, without having passed through the avicularian stage, but it seems more likely that they have in fact passed through this stage since, during the early stages of the transformation, the other parts of the house with the included zooid could hardly have disappeared at once. In many cases the vibracula have a grooved support at the base, which seems to represent the fixed beak; this support is completely absent in some other species. This view of the development of the vibracula, if trustworthy, is interesting: if all the species provided with avicularia had become extinct, no one with even the most vivid imagination would ever have thought that the vibracula had originally existed as part of an organ, much less one resembling a bird's head or an irregular box, or a hood. It **is interesting to see two such widely different organs developed from a common origin; and as the movable lip of the house serves to protect the zooid that lives inside, there is no difficulty in believing that all the gradations through which the lip became converted—first into the lower jaw of an avicularium and then into an elongated bristle—likewise served for protection in different ways and under different circumstances.**

With plants, Mr. Mivart only alludes to two cases: namely, the structure of the flowers of orchids and the movements of climbing plants. With respect to the orchids, he says, "… the explanation of their origin is deemed thoroughly unsatisfactory—utterly insufficient to explain the incipient, infinitesimal beginnings of structures which are of utility only when they are considerably developed." As I have fully treated this subject in another book (*The Various Contrivances by Which Orchids Are Fertilized by Insects*, 1877), I will here give only a few details on just one of the most striking peculiarities of the orchid flowers, namely their pollinia (Figure 7.12)—a discrete mass of pollen grains that is transferred to insects intact, as a single unit. A highly developed pollinium consists of a mass of pollen grains affixed to an elastic footstalk, and this to a little mass of extremely viscous material at its end. The pollinia are in this way transported by visiting insects from one flower to the stigma of another. In some orchids there is no footstalk attached to the pollen masses; the grains are merely tied together by fine threads. As these threads are not unique to orchids I won't discuss them here. I should mention, however, that at the evolutionary base of the series, in the genus *Cypripedium* (lady's slipper orchids) (Figure 7.13), we can see how the threads were probably first developed.

In other orchids the threads cohere at one end of the pollen masses, forming the very beginnings of a footstalk. That this is indeed the origin of the footstalk in orchids is shown by the aborted pollen grains that can sometimes be detected embedded within the central and solid parts.

With respect to the second chief peculiarity—namely, the little mass of viscous material attached to the end of the footstalk—a long series of gradations can be shown, each of obvious use to the plant. In most flowers belonging to other orders the stigma

Figure 7.12 Pollinia (the yellow, stalked structures) of an orchid (*Ophrys apifera*).

Figure 7.13 Lady's slipper orchid (*Crypipedium* sp.).

secretes only a little viscous material. Now in certain orchid species, similar viscous material is secreted, but only by one of the three stigmas; that stigma is sterile, perhaps because of the copious secretions it produces. When an insect visits a flower of this kind it rubs off some of the viscous material from the stigma and thus at the same time drags away some of the pollen grains. From this simple condition, which differs but little from that of many other common flowers, there are endless gradations—from species in which the pollen mass ends in a very short, free footstalk, to others in which the footstalk becomes firmly attached to the viscous material, with the sterile stigma itself much modified. In this latter case we have a pollinium in its most highly developed and perfect condition. If you carefully examine the flowers of various orchids for yourself, you will see the above series of gradations very clearly—from a mass of pollen grains merely tied together by threads with the stigma differing only slightly from that of an ordinary flower, to a highly complex pollinium that is admirably adapted for transport elsewhere by various insects. In this and in almost every other case, the inquiry may be pushed backward even further: one may ask, how did the stigma of an ordinary flower first become viscous? As we do not know the full history of any one group of organisms, it is as useless to ask such questions as it is hopeless to attempt to answer them. But at least we seem to know a good deal of the story that followed.

Let us now turn to the climbing plants. These can be arranged in a long series, from those called "twiners," whose stems simply twine around a rigid support of some sort, to those that I have called "leaf climbers," which climb using their petioles (the structures that attach the leaf to the stem), and to those provided with specialized, thread-like tendrils. In the two latter groups the stems have generally lost the power of twining, although they retain the power of revolving, which the tendrils also possess. The gradations from leaf climbers to tendril bearers are wonderfully close, and certain plants may be placed equally well in either class. But in ascending the series from simple twiners to leaf climbers, an important new quality is added: sensitivity to touch. By this means, the footstalks of the leaves or flowers, or those that have been modified and converted into tendrils, are excited to bend around and clasp any object that touches them. If you read my memoir on these plants (*On the Movements and Habits of Climbing Plants*, 1875), you will, I think, admit that all of the many gradations in function and structure between simple twiners and tendril bearers are in each case highly beneficial to the species. For instance, it is clearly a great advantage to a twining plant to become a leaf climber, and it is likely that every twiner that possessed leaves with long footstalks would have been developed into a leaf climber if the footstalks had possessed in any slight degree the requisite sensitivity to touch.

As twining is the simplest means of ascending a support and forms the basis of our series, it may naturally be asked, how did plants first acquire this twining ability, a power that could afterward be gradually improved and increased through natural selection? The power of twining depends, first of all, on the stems being extremely flexible when young, a characteristic that is common to many plants whether or not they are climbers. Second, it depends on their continually bending to all points of the compass, one after the other in succession, in the same order. By this movement the stems are inclined to all sides and are made to move round and round. As soon as the lower part of a stem strikes against any object and is stopped, the upper part still goes on bending and revolving and thus necessarily twines around and up the support. The revolving movement ceases after the early growth of each shoot.

As we see single species and single genera in many widely separated families of plants possessing the power of revolving and having thus become twiners, they must have acquired this ability independently and cannot have all inherited it from a single ancestor. Thus I predicted that some slight tendency to making movements of this kind would be found quite commonly in plants that did not climb and that this tendency had served as the basis for natural selection to work from and improve. When I made this prediction I knew of only one imperfect case, namely the young flower peduncles of a species in the genus *Maurandia* (Figure 7.14), which revolved slightly and irregularly, like the stems of twining plants, but without making any good use of this habit. Shortly afterward, Fritz Müller discovered that the young stems of plants in two other genera, *Alisma* and *Linum*—plants that do not climb and are widely separated taxonomically— revolved quite plainly, although irregularly; he further suspects that this occurs in some other plants as well. These slight movements appear to be of no use to the plants in question; certainly they are of no use in climbing, which is the point that concerns us here. Nevertheless, we can see that if the stems of these plants had been flexible, and if under the conditions to which they were exposed it had benefited them to ascend to a greater height, then the habit of slight and irregular revolving might have been increased and utilized through natural selection until, after many generations, they would become converted into well-developed twining species.

Nearly the same remarks apply to the sensitivity of the footstalks, the leaves, the flowers, and the tendrils. As a vast number of species belonging to very distinct and distantly related groups are endowed with this kind of sensitivity, it ought to be found in a nascent condition in many plants that have not become climbers—and this is indeed the case. For example, I observed that the young flower stalks of the *Maurandia* mentioned earlier curved themselves a little toward the side that was touched. The Belgian botanist Charles François Antoine Morren found that the leaves and footstalks of an *Oxalis* species (Figure 7.15) moved when they were gently and repeatedly touched, especially after exposure to a hot sun, or even when the plant was just shaken. I repeated these observations on some other *Oxalis* species with the same result: in some of them the movement was quite distinct, especially in the young leaves, while in other species it was extremely slight. Perhaps most importantly, the German botanical authority Wilhelm Friedrich Benedikt Hofmeister states that the young shoots and leaves of *all* plants move after being shaken; with climbing plants, we know that the footstalks and tendrils are sensitive only during the early stages of growth.

It is scarcely possible that the slight movements I have just described in the young and growing organs of plants, triggered by a simple touch or shake, can be of any functional importance to them. But plants do possess, in obedience to various stimuli,

Figure 7.14 Mexican viper (*Maurandya barclaiana*).

Figure 7.15 Sorrel (*Oxalis* sp.) is a perennial herb.

certain powers of movement that *are* of obvious importance to them. They move, for instance, toward light (and more rarely away from light) and away from (and more rarely in the direction of) gravity. When the nerves and muscles of animals are excited by electric currents or by the absorption of certain chemicals, the consequent movements may be called an incidental result, for the nerves and muscles have not been made specifically sensitive to those stimuli. Similarly with plants, it appears that from having the power to move in response to certain stimuli, they are also excited in an incidental manner by touch or by being shaken. Thus there is no great difficulty in admitting that, in the cases of leaf-climbing and tendril-bearing plants, it is this tendency that natural selection has taken advantage of and increased. It is, however, possible, from reasons that I discuss more fully in my memoir on this topic, that this will have occurred only with plants that have already acquired the power of revolving and had thus become twiners.

I have already tried to explain how plants first became twiners—namely, by the gradual increase of an innate tendency to slight and irregular revolving movements that were initially of no use to them; such initial movements, as well as those due to a touch or a shake, were the incidental consequences of the simple power of moving, gained for other and useful purposes.

Summary

I have now considered enough—and perhaps more than enough—cases that were carefully selected by the skillful naturalist Mr. Mivart in his attempt to prove that natural selection is incompetent to account for the beginning stages of useful structures and behaviors. And I have shown, I hope, that there is no great difficulty in answering all of his objections. He has, in fact, provided me with a good opportunity to enlarge a little on gradations of structure often associated with changed functions, an important subject that I did not treat at sufficient length in previous editions of this work. Let me now briefly review the various examples that I have presented.

With the giraffe, the continued preservation of the individuals of some now-extinct, high-reaching ruminant that had the longest necks, legs, and so forth, and could consequently browse a little above the average height of other animals, along with the continued destruction of those which could not browse so high, is all that would have been needed to eventually produce this remarkable quadruped. Similarly, in the case of the many insects that imitate leaves and various other natural objects in their surroundings, there is no reason to doubt that an accidental resemblance to some common object was in each case the foundation for the work of natural selection. That accidental advantage, subsequently perfected through the occasional preservation of any slight variations that made the resemblance at all closer, would then have been carried on as long as the insect continued to vary and as long as a more and more perfect resemblance to that object increased its chances of escape from sharp-sighted enemies. In certain species of whales there is a tendency to form irregular little points of horn on the palate; and so again it seems quite within the scope of natural selection to preserve all favorable variations until the points were eventually converted first into lamellated knobs or teeth, like those we now see on a goose's beak, and then into short lamellae, like those of our domestic ducks, and then into lamellae as perfectly formed as those of the shoveler duck, and finally into the gigantic plates of baleen that we now see in the mouths of Greenland whales. In the family of ducks, the lamellae are first used as teeth, then only partly as teeth and partly as a sifting apparatus, and finally almost exclusively for sifting food from the water.

Habit or continued use can have done little or nothing toward the development of such structures as the lamellae, horn, or whalebone discussed above, as far as we can judge. On the other hand, phenomena such as the movement of the lower eye of a flatfish to the upper side of its head and the formation of a prehensile tail in other organisms may be attributed largely to continued use, together with inheritance.[9] With respect to the mammary glands of the higher animals, the most probable conjecture is that, in the early mammalian ancestor, the cutaneous gland over the whole surface of a marsupial sack secreted a nutritious fluid and that the functioning of these glands was later improved through natural selection and then concentrated into a distinct area, thus forming a functional mammary gland. Similarly, there is no more difficulty in understanding how some of the branched spines of some ancient echinoderm, which served originally as a defense, became gradually developed through natural selection into three-pronged grasping pedicellariae than there is in understanding how crustacean pincers must have developed through slight, useful modifications in the terminal and penultimate segments of a limb that was originally used only for locomotion. With the avicularia and vibracula of bryozoans, we have organs that look very different but that must have developed from the same source in an ancient ancestor. With the vibracula in particular, we can understand how successive gradations in form might have been of use to their possessors. With the pollinia of orchids, we can see the steps by which the threads that originally served to tie the pollen grains together became fused to form stalks; we can similarly trace the steps by which viscous material, such as that secreted by the stigmas of ordinary flowers and still serving nearly the same purpose, became attached to the free ends of the stalks; all of these gradations clearly benefit the plants in question. With respect to climbing plants, I need not repeat what I have so recently described.

It has often been asked, "If natural selection is such a potent force, why has this or that structure not been gained by other species that would clearly benefit from having it?" But it is not reasonable to expect a precise answer to that question, considering our ignorance of the past history of each species and of the factors that control its present numbers and range. In most cases, only general reasons can be assigned. For

---

[9] This seemed a reasonable suggestion at the time, but of course we now know that these remarkable traits also have a genetic basis and have been achieved entirely through natural selection.

example, many coordinated modifications are almost indispensable for adapting a species to new habits of life, and it may often have happened that in many species the requisite parts did not vary in the right manner or to the necessary degree or at the right times. Many species must also have been prevented from increasing in numbers through destructive agencies that had nothing to do with whether or not certain structures were present, even though we can imagine that those structures would have been advantageous to the species and so could have been gained through natural selection. In that case, as the struggle for life did not depend on having such structures, they could not have been acquired through natural selection. In many cases, complex and long-enduring conditions, often of a peculiar nature, are necessary for a structure to develop over the many generations, and the requisite conditions may seldom have concurred. The belief that all species should have gained through natural selection every structure that we think would have benefitted them is not consistent with what we understand about how natural selection acts.

Mr. Mivart does not deny that natural selection has in fact effected some things; he simply considers natural selection as "demonstrably insufficient" to account for the phenomena that I have explained as having been brought about by its actions. His chief arguments have now been considered, and the others will be considered in the pages to come. His arguments seem to me to have little weight in comparison with those supporting the power of natural selection, aided perhaps by the other factors that I have often mentioned. I should note that some of the facts and arguments that I have used here have been advanced for the same purpose in an able article recently published in the *British and Foreign Medico-Chirurgical Review*, a London journal devoted to practical medicine and surgery.

At the present time (1872), almost all naturalists admit that evolution has occurred in some form, and all evolutionary biologists admit that species have the ability to change. Mr. Mivart believes that species change through what he calls "an internal force or tendency," about which nothing is known. But there is no need, I think, to invoke any mysterious internal force as a mechanism beyond the tendency to ordinary variability, which, through the aid of selection by man, has given rise to many well-adapted domestic races and which, through the aid of natural selection, would equally well give rise by gradual steps to natural races of species. The final result will generally have been, as already explained, an advance—or in some few cases, a retrogression—in organization.

Reasons for Disbelieving in Great and Abrupt Modifications

Mr. Mivart is further inclined to believe—and he has support from some naturalists on this—that new species appear "with suddenness and by modifications appearing at once." For instance, he supposes that the anatomical differences we see between

Figure 7.16 *Hipparion*. The members of this genus lived for about 22 million years, but went extinct about 780,000 years ago.

the modern horse and members of the extinct genus *Hipparion* (Figure 7.16) (in particular the three vestigial outer toes that it had in addition to its hoof) arose suddenly. He also finds it difficult to believe that the wing of a bird "was developed in any other way than by a comparatively sudden modification of a marked and important kind"; apparently he believes the same to be true for the wings of bats and those of the now-extinct pterodactyl. This conclusion, which implies great breaks or discontinuity in the series, appears to me to be improbable in the highest degree. Now everyone who believes in slow and gradual evolution, as I do, will of course admit that specific changes may have been as abrupt and as great as any single variation that we now occasionally see in nature, or even under domestication. But as species are more variable when domesticated or cultivated than when they are living under their natural conditions, the great and abrupt variations that occasionally arise under domestication are much less likely to have occurred in nature. Of these latter variations, several may be attributed to reversion to a previous structure; the characters that thus reappear were, it seems likely, initially gained in a gradual manner. A still greater number must be called monstrosities, such as six-fingered men, porcupine men (with their uneven skin forming ridges and spikes), Ancon sheep, and Niata cattle.[10] As these differ so greatly in character from natural species, they throw very little light on our subject. Excluding such cases of abrupt variations, the few that remain would at best constitute, if found in nature, questionable species, closely related to their parental types.

My reasons for doubting whether natural species have changed as abruptly as have, on occasion, our domestic races, and for entirely disbelieving that they have changed in the remarkable manner suggested by Mr. Mivart, are as follows. For one thing, we

---

[10] The genetic basis for such mutations was, of course, not at all understood in Darwin's time.

know that abrupt and strongly marked variations occur in our domesticated organisms only rarely and at rather long intervals of time. If such changes occurred in nature, they would be liable, as I explained earlier, to be lost by accidental causes of destruction and by subsequent interbreeding; that is certainly the case for organisms under domestication, unless abrupt variations of this kind are deliberately and carefully preserved. Thus in order for a new species to suddenly appear in nature in the manner supposed by Mr. Mivart, it is almost necessary to believe—in opposition to all analogy—that several wonderfully changed individuals appeared simultaneously within the same region. This difficulty, as in the case of unconscious selection by humans (see pp. 21–25 in Chapter 1), is avoided on the theory of gradual evolution through the preservation of a large number of individuals that varied more or less in any favorable direction and of the destruction of a large number of individuals that varied in an opposite manner or not at all.

There can be no doubt that many species have evolved in an extremely gradual manner. The species and even the genera of many large natural families are so closely related that it is often difficult to distinguish between them. On every continent as we go from north to south or from lowland to upland, and so forth, we meet with a host of closely related or representative species, as we also do on certain distinct continents that we have reason to believe were formerly connected. But in making these and the following remarks, I am compelled to mention certain topics that I will not discuss more fully until later. Look at the many outlying islands around a continent and see how many of their inhabitants can be raised only to the rank of questionable species. So it is if we look to past times and compare the species that have only recently gone extinct with those still living in the same areas, or if we compare the fossilized species embedded in the substages of the same geological formation. It is indeed clear that multitudes of species are related in the closest manner to other species that still exist, or that have recently existed; it will hardly be maintained that such species—so closely resembling each other—have been developed in an abrupt or sudden manner. Nor should it be forgotten, when we look to the special parts of allied species instead of to distinct species, that numerous and wonderfully fine gradations can be traced, connecting together widely different structures.

Many large groups of facts make sense only on the principle that species have been evolved by very small steps over long periods of time—the fact, for instance, that the species included in the larger genera (i.e., those containing a great many species) are more closely related to each other, and present a greater number of varieties, than do the species in the smaller genera. The former are also grouped in little clusters, in the same way that varieties cluster around species, and they also present other analogies with varieties, as I showed in Chapter 2 of this book. On that same principle we can understand how it is that specific characters are more variable than generic characters, and how the parts that are developed to an extraordinary degree or manner are more variable than other parts of individuals of the same species. Many analogous facts could be given, all pointing in the same direction.

Although a great many species have almost certainly been produced by steps not greater than those separating fine varieties, some may have been developed in a different and more abrupt manner. Such an admission, however, ought not to be made without strong evidence being given. The vague and in some respects false analogies—as pointed out to me by the American philosopher of science Mr. Wright, whom I mentioned earlier—that have been presented in support of this view, such as the sudden crystallization of various inorganic substances or the falling of a faceted spheroid from one facet to another, hardly deserve consideration. One class of facts, however—namely, the sudden appearance of new and distinct forms of life in our geological formations—supports at first sight the belief in abrupt change. But the value of this evidence depends entirely on the perfection of the geological record in relation to periods remote in the history of the world. If the record is as fragmentary as many geologists strenuously assert, there is nothing strange in new forms appearing as if they had been suddenly developed.

Unless we admit anatomical transformations as prodigious as those advocated by Mr. Mivart, such as the sudden development of the wings of birds and bats, or the sudden conversion of an ancient *Hipparion* into a modern horse, hardly any light is thrown by the belief in abrupt modifications on the rarity of connecting links among species in our geological formations. But against the belief in such abrupt changes, embryology enters a strong protest! It is notorious that the wings of birds and bats, and the legs of horses and other quadrupeds, are indistinguishable at early stages of embryonic development and that they become gradually differentiated by insensibly fine steps as development proceeds. Embryological resemblances of all kinds can be accounted for, as we shall later see, by the ancestors of our existing species having varied after early youth and having transmitted their newly acquired characteristics to their offspring at a corresponding age. The embryo is thus left almost unaffected and serves as a record of the past condition of the species. **Thus it is that during the early stages of their embryonic development existing species so often resemble ancient and extinct forms belonging to the same class.** On this view of the meaning of embryological resemblances, and indeed on any view, it is incredible to think that an animal should have undergone such momentous and abrupt transformations as those summarized above and yet should not bear even a trace of any such sudden modification in its embryological development; instead, every detail in its structure is seen to develop by insensibly fine steps.

He who believes that some ancient wingless form was suddenly transformed through an internal force into, for instance, one furnished with wings, will be almost compelled to assume, in opposition to all analogy, that many individuals varied in that way simultaneously. It cannot be denied that such abrupt and great changes of structure are widely different from those that most species apparently have undergone. He will further be compelled to believe that many structures beautifully adapted to all the other parts of the same creature and to the surrounding environment have been suddenly produced; of such complex and wonderful coadaptations he will not be able to assign a shadow of an explanation. He will also be forced to admit that these great and sudden transformations have left no trace of their action on the embryo. To admit all of this, it seems to me, is to enter into the realm of miracle and to leave that of science.

Key Issues to Talk and Write About

1. In *The Origin of Species*, Darwin draws on the work of a great many individuals. Find out two interesting things about one of the people that Darwin mentions in this chapter. Choose from the following:

   Louis Agassiz
   Samuel White Baker
   Pierre Paul Broca
   Heinrich Georg Bronn
   George Busk
   William Henry Flower
   Albert Günther
   Wilhelm Friedrich Benedikt Hofmeister
   Joseph Hooker
   Edward Lambert (the Porcupine Man)
   Edwin Ray Lankester
   George Henry Lewes
   Charles Lyell
   St. George Jackson Mivart
   Charles François Antoine Morren
   Johannes Peter Müller
   Jean Octave Edmond Perrier
   Antoine Laurent de Jussieu Augustin Saint-Hilaire
   Ramsay Traquair
   Carl Wilhelm von Nägeli
   Alfred Russel Wallace
   Chauncey Wright

2. In this chapter, Darwin spends considerable time talking about "bryozoans," which he refers to as "colonial animals." What are colonial animals? What properties do all colonial animals possess?

3. Carefully read the paragraph on page 192 that begins, "There is much force in the above objections." Following the instructions given in Chapter 1 (see page 28), write a one-sentence summary of that paragraph, being careful to include all the key points you think Darwin is trying to get across.

   Try the same exercise with the paragraph on page 196 that begins, "First let us consider the giraffe."

   Then try the exercise with the paragraph on page 218 that begins, "Let us now turn to the climbing plants."

4. Rewrite the following sentences from Darwin's original to make them clearer and more concise:
    a. "To this conclusion Mr. Mivart brings forward two objections."
    b. "It is also known with the higher animals, even after early youth, that the skull yields and is altered in shape, if the skin or muscles be permanently contracted through disease or some accident."
    c. "The mammary glands are common to the whole class of mammals, and are indispensable for their existence."
    d. Try simplifying this one a little bit, too, from an earlier draft of this manuscript: "Habit or continued use can have done little or nothing toward the development of such structures as the lamellae, horn, or whalebone discussed above."
5. On page 200, Sir Charles Lyell asks, why haven't seals and bats given birth on such islands to forms that can live well on land? Darwin offers a reasonable explanation, but might it still be possible for seals or bats to gradually become good terrestrial animals on these oceanic islands, far in the future? See if you can come up with a situation in some group of islands that might eventually make it possible for seals or bats to become truly terrestrial.
6. Find out more about any one of the organisms that Darwin talks about in this chapter, such as the platypus, baleen whales, the Bryozoa (also called Ectoprocta), *Hipparion*, the pterodactyl, or orchids.
7. List all of the other organisms that Darwin includes as examples in this chapter.

Bibliography

Darwin, C. 1868. *The Variation of Animals and Plants Under Domestication*. London.
Darwin, C. 1875. *On the Movements and Habits of Climbing Plants*. London.
Darwin, C. 1877. *The Various Contrivances by Which Orchids Are Fertilized by Insects*. New York.

8
Instinct

In this chapter, Darwin explains what instinctive behaviors are and how they can be explained by natural selection—by the gradual accumulation of small but beneficial variations in behavior over long periods of time. He begins by asserting that instinctive behaviors vary among individuals and shows that such behaviors—and changes in those behaviors—are passed along to offspring. He then discusses three particular examples in considerable detail: the instinct that leads the cuckoo to lay her eggs in the nests of other birds, the remarkable slave-making instinct found in certain ant species, and the complex honeycomb construction behavior of honeybees. Darwin also shows that the different instincts are developed to different degrees in different species, giving us a sense of the steps that these instincts may have gone through before reaching their remarkable state of development in some species today. He also gives us examples of the sorts of pressures that may have selected for the evolution of those instincts in the animals that reveal them today. Finally, Darwin tackles the perplexing problem of accounting for the evolution of sterile worker ants—ants that differ considerably in form and behavior from their parents and yet leave no offspring of their own.

Many instincts are so wonderful that they may well appear to overthrow my whole theory. I admit that I can say nothing about the source of these amazing mental powers, but then neither can I say anything about the origin of life itself. We are concerned here only with the diversities of instinct and other mental functions that we now find among related animals, how that diversity has come about, and how natural selection has probably led to the remarkable behaviors that we see in so many species today.

It is difficult to define instinct precisely, although it would be easy to show that several distinct mental actions are commonly included under this heading. Everyone understands what is meant when someone says, for example, that "instinct" compels the cuckoo to migrate and to lay her eggs in other birds' nests. Actions performed by an animal without prior experience or instruction, and those that are performed by many individuals in the same way without their knowing why they are performing them, are usually said to be "instinctive."

The Readable Darwin. Second Edition. Jan A. Pechenik, Oxford University Press. © Oxford University Press 2023.
DOI: 10.1093/oso/9780197575260.003.0009

Frédéric Cuvier (the younger brother of Georges Cuvier) and several other met-aphysical philosophers have compared instinct with habit. This comparison gives, I think, an accurate notion of the frame of mind under which an instinctive action is performed, but not necessarily of its origin. How unconsciously we perform many ha-bitual actions, often in direct opposition to our conscious will! Yet they may be mod-ified by will or reason. Individual habits easily become associated with other habits, with certain periods of time, and with different physiological states of the body, and, once acquired, they often remain constant throughout life. There are other interesting similarities between instincts and habits. For example, with instincts, one action fol-lows another by a sort of rhythm, as in repeating a well-known song; if a person is in-terrupted while singing a song, or in repeating anything in fact by rote, he is generally not able to simply pick up where he left off, but instead must go back a ways to recover the habitual train of thought. The Swiss entomologist Pierre Huber also found this to be the case with a certain caterpillar, one which makes a very complicated silk ham-mock to pupate in: it suspends its cocoon from the upturned edges of a leaf by means of delicate silk threads. If he took a caterpillar that had completed its hammock up to, say, the sixth step of its construction and put it into one completed up to only the third step of construction, the caterpillar simply reperformed the fourth, fifth, and sixth steps of the process. However, if he took a caterpillar from a hammock that had been constructed up to only the third stage and put it into one finished up to the sixth stage, so that much of its work was already done for it, instead of deriving any benefit from this it was much embarrassed, and, in order to complete the hammock, it had to start from the third stage where it had left off on its own hammock. The caterpillar basi-cally tried to do work that had already been finished for it.

If we suppose that habitual actions can be inherited by offspring—and it can in-deed be shown that this does sometimes happen—then the resemblance between an instinct and what was originally a habit becomes so close as to be indistinguishable. But it would be a serious error to suppose that most instincts have been acquired by habit in one generation and then transmitted by inheritance to succeeding genera-tions. If the three-year-old Mozart, instead of playing the piano with wonderfully little practice, had played a tune with no practice at all, he might truly be said to have done so instinctively. But of course that was not the case. Indeed, it can be clearly shown that the most wonderful instincts with which we are acquainted—namely those of the hive bee and of many ants—could not possibly have been acquired by habit. I will talk about these particular instincts shortly.

Instincts are clearly as important as bodily structures for the welfare of every spe-cies. As surrounding conditions change over time, it is at least possible that slight modifications of instinct might benefit a species; if it can be shown that instincts do vary among individuals, even just a small amount, then I can see no difficulty in nat-ural selection preserving and continuing to accumulate variations of instinct to any extent that was beneficial. This is, I believe, how all the most complex and wonderful instincts have in fact originated. Just as modifications of anatomical structures arise from and are increased by use or habit and are diminished or lost by disuse, so I do not doubt it has also been with instincts.[1] But I believe that the instincts we see today have mostly been shaped by natural selection acting on what may be called spontaneous variations of instincts—that is, of variations produced by the same unknown causes that produce slight variations in body structures among individuals of a species.

Indeed, complex instincts can only be produced through natural selection—by the slow and gradual accumulation of many slight, yet beneficial variations. Thus, **as in the case of anatomical structures, we ought to find in nature not the actual transitional gradations by which each complex instinct has been acquired—for we could find these only in the direct ancestors of these individuals, ancestors which no longer exist—but rather some evidence of such transitions by comparing the behaviors of related species.** At least we should try to find evidence that gradations of some kind are possible, and this we can certainly do. In fact, I have been surprised to find, even though we know nothing about the instincts of species that are now extinct, how very easy it is to find gradations in different species leading to the most complex instincts that we know of. Evolutionary changes in instinct may sometimes be facilitated by particular species having different instincts at different stages of development, or at different times of year, or when placed under different environmental conditions; in that case, any one of those instincts might be preserved by natural selection. Indeed, such instances of diversity of instinct in the same species can be shown to occur in nature.

As with anatomical structures, the instincts of any individual are good for that individual and have never, as far as we can judge, been produced for the exclusive good of others. One of the strongest instances of an animal apparently performing an action for the sole benefit of another that I am aware of is that of aphids (Figure 8.1) voluntarily yielding up their sweet excretions to ants, as first described by the Swiss entomologist Pierre Huber in 1810. The aphids give up these honeydew droplets voluntarily, as shown by the following observations. Several years ago, I removed all the ants from a group of about 12 aphids on a dock plant (a member of the genus *Rumex*) and kept them off the plant for several hours. I then felt sure that the aphids would want to excrete. So I watched them carefully with a magnifying lens for some time, but none of them excreted anything. I then tickled and stroked them with a hair in the same manner, as well as I could,[2] as the ants do with their antennae, but again not a single aphid excreted anything. Afterward I allowed one ant to visit the aphids. The ant immediately seemed, by its eager way of running about, to be well aware of what a rich flock of aphids it had discovered. It soon began to play with its antennae on the abdomen of one aphid and then that of another. Each aphid, as soon as it felt the antennae, lifted up its abdomen and excreted a limpid drop of sweet juice, which was eagerly devoured by the ant. Even the very young aphids behaved in this way, showing that the action was instinctive and not the result of experience.

---

[1] As noted in prior chapters, an introductory course in Mendelian genetics would have answered so many of Darwin's questions about the origins and inheritance of instincts. But the rest of his argument is solid.

[2] I wish I could see Darwin doing this now on a YouTube video. Would the video go viral?

Figure 8.1 Ant, farming aphids.

Huber makes it very clear that the aphids show no dislike of the ants that milk them, and, if no ants are present they eventually eject their excretion anyway. But as the excretion is extremely viscous, having it removed is clearly convenient for the aphids, although they probably don't excrete solely for the good of the ants that feed on it.[3] Although there is no evidence that any animal performs any action solely for the good of another species, even so, each tries to take advantage of the instincts of others for their own benefit, just as each takes advantage of the weaker bodily structures of other species.

Instinct must vary in nature and those variations must be passed along to offspring if natural selection is to act; therefore I really should give as many examples as possible. However, lack of space prevents me from doing so. I can only assert that instincts do indeed vary—for instance, the migratory instinct varies considerably, both in direction and extent; indeed, it has been completely lost in some species. So it is with the nests of birds, which vary in structure both in relation to the habitat chosen and with the nature and temperature of the region inhabited, but often from causes wholly unknown to us. The great naturalist and painter John James Audubon has described several remarkable cases of differences in the nests of the same species living in the northern United States and the southern United States.

---

[3] We now know that the ants protect the aphids from many predators and parasites; letting the ants obtain their sugary secretion probably promotes the association between ants and aphids, benefitting the aphids through the protection that comes with it. Thus, both partners benefit.

Now some people have asked, if instinct is variable, why has it not granted to the bee "the ability to use some other material for building its hives when wax was not available?" But what other natural material could bees use? They will work, as I myself have seen, with wax hardened with the red pigment vermilion or softened with lard, and the English horticulturist Thomas Andrew Knight has seen his bees using a cement of wax and turpentine that he provided to them rather than laboriously collecting a particular resin from tree buds; thus bees do in fact show some flexibility in their use of materials for hive construction. Similarly, it has recently been shown that bees will gladly forgo searching for pollen if they are provided instead with oatmeal.

Fear of particular enemies is certainly instinctive, as may be seen in nestling birds, though it is also strengthened by experience and by seeing fear of the same enemy expressed by other individuals. The fear of man is acquired slowly by the various animals that inhabit desert islands. We see an example of this even in England, in the greater wildness of all our large birds in comparison with our small birds, as the large birds have been the most persecuted by humans. We may safely attribute the greater wildness of our large birds to this cause, for on uninhabited islands large birds do not fear people any more than small birds do, and the magpie, which is so very wary of us in England, is tame in Norway, as is the hooded crow in Egypt. I have given other examples in my book, *The Voyage of the Beagle*.[4]

The mental qualities of animals born in nature also vary a great deal among individuals of a given species. I am also aware of occasional and strange habits in wild animals, which, if advantageous to the species, might have given rise through natural selection to new instincts. But I am well aware that these general statements, without the facts in detail to support them, will produce but a feeble effect on the reader's mind. I can only repeat my assurance that I do not speak without good evidence. I just don't have the space to elaborate further here.

Inherited Changes of Habit or Instinct in Domesticated Animals

The idea, or even the probability, that variations in instinct are indeed inherited by offspring in nature can be made more convincing by considering a few cases in domesticated animals. We will thus be able to see the part that habit and the selection of so-called spontaneous variations have played in modifying the mental qualities of these organisms. It is indeed notorious how much domestic animals vary in their mental qualities. One cat naturally takes to catching rats, for example, another one to catching mice; these tendencies are known to be inherited. One cat, according to Mr.

---

[4] *The Voyage of the Beagle*, published in 1839, describes Darwin's travels and scientific observations as he journeyed for nearly five years aboard the *H.M.S. Beagle*, captained by Robert FitzRoy. This was the voyage that first got Darwin thinking about the origin of species.

Charles St. John, always brought home game birds while another brought back only hares and rabbits, and another hunted on marshy ground and caught woodcocks or snipes almost every night.

A number of curious and authentic instances could also be given of the oddest tricks, associated with certain frames of mind or periods of time, being inherited as well. Let us look to the familiar case in breeds of dogs: it is well known that young pointers will sometimes point and even back other dogs the very first time that they are taken out, without any training and without ever having seen another dog behave in this way. I have seen a particularly striking incidence of this myself. And retrieving is certainly inherited to at least some degree by retrievers, as is a tendency to run around a flock of sheep, instead of directly at them, by shepherd dogs. These actions are performed by the young without prior experience, and in nearly the same manner by each individual, performed with eager delight by the members of each breed, and without the purpose being known—for the young pointer can no more know that he points to aid his master than the white butterfly knows why she lays her eggs on cabbage leaves. I cannot see how these actions differ essentially from true instincts. If we were to behold one kind of wolf—a young wolf, without any prior training—stand motionless like a statue as soon as it scented its prey, and then slowly crawl forward with a peculiar gait, and then another kind of wolf rushing around, instead of at, a herd of deer, and driving them to a distant point, we would surely call these actions instinctive. The instincts of domestic animals may be far less fixed than natural instincts, but they have been acted on by far less rigorous selection and have been transmitted for an incomparably shorter amount of time and under less fixed conditions of life.

To see how strongly these domestic instincts, habits, and dispositions are inherited, and how curiously they become mingled, simply consider what happens when different breeds of dogs are crossed. It is well known that a cross with a bulldog has affected the courage and obstinacy of greyhounds for many generations. Similarly, a cross with a greyhound has produced a whole family of shepherd dogs with a pronounced tendency to hunt hares. These domestic instincts, when tested in this way by crossing, resemble natural instincts in the way that they become curiously blended together and for a long time exhibit traces of the instincts of either parent. For example, the French naturalist Charles Le Roy describes a dog whose great-grandfather was a wolf; this dog showed a trace of its wild parentage in only one way: when called by its master, it never came in a straight line.

Some people have referred to domestic instincts as actions that have become inherited solely from long-continued and compulsory habit, but this is not true. No one would ever have thought of teaching the tumbler pigeon to tumble, or probably could have taught it to do so even if he had thought to try; yet I have witnessed this action performed by young birds that had never yet seen a pigeon tumble. We may believe that some particular pigeon showed a slight tendency to this strange habit many years ago, and that the long-continued selection of the best-tumbling individuals in successive generations made tumblers what they are today. I hear from the authority on pigeon breeding Mr. Bernard Peirce Brent that, near Glasgow, Scotland, there are house tumbler pigeons that can't fly 18 inches without going head over heels.

Similarly, I doubt that anyone would have thought of training a dog to point unless some one dog had naturally shown a tendency to do this sort of behavior long ago. This is in fact known to happen occasionally in some breeds; I saw it myself once in a pure terrier, a breed that does not otherwise point. The act of pointing is probably, as many have thought, only the exaggerated pause of an animal preparing to spring on its prey. When the first tendency to point was once displayed, methodical selection and selective breeding, along with the inherited effects of compulsory training in each successive generation,[5] would soon complete the work. Indeed, unconscious selection is still in progress, as each person tries to procure, without intending to improve the breed, those dogs that stand and hunt the best.

Natural instincts are eventually lost under domestication. A remarkable instance of this is seen in those breeds of fowl[6] that now rarely or never wish to sit on their eggs. Familiarity alone prevents us from seeing how largely and how permanently the minds of some of our other domestic animals have also been modified over time. It is scarcely possible to doubt, for example, that the love of man has now become instinctive in dogs. All wolves, foxes, jackals, and species of the cat genus, when recently kept tame, are still most eager to attack poultry, sheep, and pigs; this tendency is in fact incurable in dogs that have been brought home as puppies from places like Tierra del Fuego and Australia, where the aboriginals have not domesticated these animals at all. On the other hand, it is rare that our civilized dogs, even when quite young, need to be taught not to attack poultry, sheep, and pigs; they seem to be born knowing not to do this. No doubt they occasionally do make an attack, and are then beaten; if they are not thus cured, then they are destroyed. Thus, both habit and some degree of selection have probably concurred in civilizing our dogs through inheritance.

Young chickens have similarly lost, wholly by habit, the fear of dogs and cats that no doubt was originally instinctive in them. For example, the English naturalist Captain Thomas Hutton tells me that the young chickens of the parent stock, *Gallus bankiva*, when reared in India under a hen, are at first excessively wild and inattentive. So it is with young pheasants reared in England under a hen. It is not that the chickens have lost all fear, just the fear of dogs and cats, for if the hen gives the danger chuckle, they will run from under her, particularly if they are young, and will then conceal themselves in the surrounding grass or thickets. This is evidently done for the instinctive purpose of allowing their mother to fly away, as we see now in wild ground birds. This instinct retained by our chickens has, of course, now become useless under domestication, for the mother hen has now almost lost the ability to fly.

[5] Here again, Darwin is probably mistaken about the ability of learned behaviors to be transmitted to future generations, although some recent studies suggest that it may in fact be possible in certain situations.

[6] "Fowl" refers to hens and other members of the avian order Galliformes.

We may conclude then that some new instincts have been acquired under domestication, while some natural instincts have been lost, at least partly by humans having selected for peculiar mental habits and actions generation after generation—habits that at first appeared from what we must, in our ignorance, call an accident. In at least some cases, inherited mental changes have been produced through selection alone, pursued sometimes deliberately and sometimes unconsciously.

Special Instincts

We shall perhaps best understand how instincts in wild animals have become modified by selection by considering a few specific cases. I have selected only three for this discussion: (1) the instinct that leads the cuckoo to lay her eggs in other birds' nests, (2) the slave-making instinct of certain ants, and (3) the cell-making power of the honeybee. These two last-mentioned instincts have generally and justly been ranked by naturalists as the most wonderful and incredible of all known instincts. But first let's consider the cuckoo.

The Instincts of the Cuckoo. Some naturalists suppose that the more immediate cause of the European cuckoo's[7] (*Cuculus canorus*) (Figure 8.2A) instinct to lay her eggs in other birds' nests (making it a "brood parasite") is that she does not lay her eggs daily, but rather at intervals of two or three days; thus if she were to make her own nest and sit on her own eggs, those laid first would have to be left for some time uncubated, or there would be eggs and young birds of different ages in the same nest. If that were the case, the process of laying and hatching might be inconveniently long, particularly as she migrates to warmer climates very early in the fall, so that the first-hatched young would probably have to be fed by the male alone. But the American cuckoo *Coccyzus erythropthalmus* is in exactly that predicament: she makes her own nest and has eggs and young birds successively hatched, all living in the same nest at the same time. What makes this especially interesting for us is that the American cuckoo does occasionally lay her eggs in other birds' nests; I have recently heard from Dr. S. A. Merrell, of Iowa in the United States, that in Illinois he once found a young cuckoo together with a young jay in the nest of a blue jay (*Garrulus cristatus*). As both birds were nearly fully feathered, there could be no mistaking their identities. And I could give several other instances of various birds occasionally laying their eggs in other birds' nests as well. Now let us suppose that the ancient ancestor of our European cuckoo had the habits of the American cuckoo, and that she only occasionally laid an egg in another bird's nest. Suppose that the old bird profited by this occasional habit through, for example, being enabled to start migrating earlier; or perhaps the young birds were made more vigorous through the misplaced instincts of another species than they would have been had they been reared by their own mother, who

---

[7] The European cuckoo (*Cuculus canorus*) is now known as the common cuckoo.

(A)

(B)

Figure 8.2 (A) The European cuckoo (*Cuculus canorus*). (B) A male brown-headed cowbird (*Molothrus ater*).

would otherwise be encumbered by having eggs and young of different ages in her nest at the same time. In both cases, then, the old birds of the fostered young would gain an advantage.

Analogy would lead us to believe that the young thus reared would probably follow the occasional and aberrant habit of their mother through inheritance and, in their turn, would be more likely to lay their eggs in other birds' nests as well and thus be more successful in ensuring the survival of their young. I believe that it is through this process, continued over long periods of time for many generations, that the strange egg-laying instinct of our European cuckoo has in fact come about. The German naturalist Mr. Adolf Müller has recently found that the female cuckoo occasionally lays her eggs on the bare ground, sits on them, and feeds her young. This rare event is probably a case of her reverting to the long-lost ancestral instinct of nest building, showing again that there is still natural variation in the nest-building instinct among individuals.

Some have objected that I have not noticed any other related instincts and structural adaptations in the cuckoo, which are spoken of as necessarily coordinated. But in all cases, speculation on an instinct known to us only from a single species is useless, for we have until now had no further facts to guide us. Mostly we know about the instincts of only two cuckoo species: the European species and the mostly nonparasitic American cuckoo. The chief points for us are as follows: first, that our common European cuckoo, with rare exceptions, lays only one egg in a foreign nest, so that the large and voracious young bird receives ample food; second, that the eggs are remarkably small—no larger indeed than those of the skylark, a bird about one-fourth the size of the cuckoo. In contrast, the American cuckoo lays full-sized eggs, showing quite clearly that the small eggs of the European cuckoo are an adaptation. Third, it is clear that the young cuckoo, soon after hatching, has the instinct, the strength, and a back that is perfectly shaped for ejecting its foster brothers from the nest; the ejected individuals then perish from cold and hunger. This has been boldly called a beneficent arrangement, so that the young cuckoo may get sufficient food and so that its foster brothers perish before they have acquired much feeling!

The Australian ornithologist Mr. Edward Pierson Ramsay has now made observations on the habits of three Australian bird species that lay their eggs in other birds' nests. Though these birds generally lay only one egg in a nest, it is not rare to find two or even three of their eggs in some of the nests. In the bronze cuckoo, the eggs vary greatly in size, with some of the eggs being 8 to 10 times longer than some of the others. Now if it had been advantageous for this species to have laid eggs even smaller than those now laid, so as to have deceived certain foster parents, or, as is more probable, to have been hatched more quickly—for there seems to be a relationship between egg size and the duration of incubation—then there is no difficulty in believing that a race or species might have been formed that would have laid smaller and smaller eggs, for these would have been more safely hatched and reared. Mr. Ramsay remarks that two of the Australian cuckoos, when they lay their eggs in another bird's open nest, preferentially choose nests that contain eggs similar in color to their own. Females of the European species apparently show some tendency toward a similar instinct but often depart from it, as is shown by their laying their dull and pale-colored eggs in nests of the hedge warbler, a species that produces bright greenish-blue eggs. Had our cuckoo invariably displayed the above instinct, it would surely have been added to those that it is assumed must have all been acquired together. According to Mr. Ramsay, the eggs of the Australian bronze cuckoo vary in color to an extraordinary degree; in this respect, then, natural selection might have secured and fixed any advantageous variation, including but not limited to egg size.

In the case of the European cuckoo, the foster parents' offspring are commonly ejected from the nest within three days after the cuckoo has hatched. As the hatchling at this young age is in a most helpless condition, the British ornithologist Mr. John Gould previously thought that the ejection must have been performed by the foster parents themselves. But he now reports receiving a trustworthy account of a young cuckoo that was actually seen in the act of ejecting its foster brothers while still blind and not even able to hold up its own head. When the observer then returned one of the ejected individuals to its nest, it was again thrown out by the cuckoo hatchling. With respect to the means by which this strange and odious instinct was acquired, if it were of great importance for the young cuckoo to receive as much food as possible soon after birth, as is probably the case, I can see no special difficulty in its having gradually acquired, during many successive generations, the blind desire, the strength, and the structure needed for the work of ejection; those young cuckoos with such habits and the best developed structures to carry them out would be the most securely reared and most likely to reach adulthood.

The first step toward acquiring the proper instinct for ejecting other eggs might have been simply an unintentional restlessness on the part of the young bird when somewhat advanced in age and strength—the habit having been afterward improved, and finally transmitted to offspring at a slightly earlier age. I can see no more difficulty in this than in the unhatched young of other birds acquiring the instinct to break through their own shells for hatching, or than in young snakes acquiring a temporary sharp tooth in their upper jaw for cutting through the tough eggshell that surrounds them, as the British anatomist Sir Richard Owen has described. For if each part is liable to be inherited at a corresponding or earlier age— propositions that cannot be disputed—then the instincts and structures of the young could be slowly modified over many generations as surely as those of the adult. And both cases must stand or fall together with the whole theory of natural selection.

Some species in the genus *Molothrus*, a widely distinct group of American cowbirds (Figure 8.2B) allied to our starlings here in England, have parasitic habits like those of the cuckoo; of particular interest to us, they present an interesting gradation in the degree to which their instincts have been perfected. The two sexes of the species *Molothrus badius*[8] sometimes live together promiscuously in flocks and sometimes pair, according to reports by an excellent observer of bird behavior, Mr. William Henry Hudson of Buenos Aires, Argentina. They either then build a nest of their own or seize on one belonging to some other bird, occasionally throwing the stranger's nestlings out of the nest. They then either lay their eggs in the nest thus appropriated or, oddly enough, build another nest on top of it. They usually then sit on their own eggs and rear their own young, although Mr. Hudson says that they are at least occasionally parasitic[9]; he has seen the young of this species following old birds of a distinct kind and clamoring to be fed by them.

The parasitic habits of another species in the same genus, the shiny cowbird *Molothrus bonariensis*, are much more highly developed than those of *M. badius*, but are still far from perfect. This bird, it seems, invariably lays its eggs in the nests of other species; remarkably, however, several birds together sometimes commence

---

[8] This species, the bay-winged cowbird, has been moved to a different genus and is now known as *Agelaioides badius*.

[9] By "parasitic," Darwin here is referring to birds that take advantage of birds belonging to a different species.

building an irregular and untidy nest of their own, placed in singularly ill-adapted situations, such as on the leaves of a large thistle. But they never, as far as Mr. Hudson has ascertained, actually complete a nest for themselves. And yet, they often lay so many eggs—typically from 15 to 20—in the same foster nest, that few or even none can be hatched. They have, moreover, the extraordinary habit of pecking holes in all of the eggs that they find in the appropriated nests, whether of their own species or of the foster parents. They also drop many eggs on bare ground, which are thus destroyed. Continuing in our series, a third species, *M. pectoris* of North America, has acquired instincts as perfect as those of the cuckoo, for it never lays more than one egg in a foster nest; thus each young bird is reared securely.

The implications of these observations are hard to miss: Mr. Hudson is a strong disbeliever in evolution, but even he appears to have been so much struck by the imperfect instincts of *M. bonariensis* that he quotes my words and asks, "Must we consider these habits, not as especially endowed or created instincts, but as small consequences of one general law, namely, transition?"

As I have previously remarked, various birds occasionally lay their eggs in the nests of other birds. This habit is fairly common within the family Gallinaceae and throws some light on the singular instinct of the ostrich. In this family several hen birds join forces and lay first a few eggs in one nest and then in another; the males then hatch these. This instinct is probably related to the hens laying a large number of eggs at intervals of two to three days, as with the cuckoo. The instinct, however, of the American ostrich, as in the case of *M. bonariensis* described above, has not as yet been perfected: a surprising number of eggs lie strewn over the plains, so that in one day's hunting I picked up no fewer than 20 such lost and wasted eggs.

Many bees are also parasitic and regularly lay their eggs in the nests of other kinds of bees. This case is even more remarkable than that of the cuckoo, for these bees have not only had their *instincts* modified in accordance with their parasitic egg-laying habits, but also their anatomy: they do not possess the pollen-collecting apparatus that would have been indispensable had they needed to store up food for their own young. Some species of wasp-like insects belonging to the family Sphegidae are likewise parasitic—at least some of the time. For example, although the sand wasp *Tachytes nigra* generally makes its own burrow and stores it with paralyzed prey to feed its larvae, the French entomologist Mr. Jean-Henri Fabre has lately shown good reasons for believing that when this insect finds a burrow already made and stored with food by sand wasps in the genus *Sphex*, it takes advantage of the prize and becomes parasitic for the occasion. In this case, as with that of the cuckoo and with the *Molothrus* cowbird species discussed earlier, I can see no difficulty in natural selection making such an occasional habit permanent if it is advantageous to the species, and as long as the insect whose nest and stored food are feloniously appropriated is not thereby exterminated.

The Slave-Making Instinct of Ants. This remarkable instinct was first discovered in the ant species *Polyergus rufescens* by Pierre Huber, a better observer than even his celebrated father, the entomologist François Huber. This ant absolutely depends on its slaves; without their help, the species would certainly become extinct within a year, for the males and fertile females do no work of any kind, and the sterile female worker ants,[10] although they are most energetic and courageous in capturing slaves, do no other work than that: they don't help to make nests, and they don't even feed their own larvae. When the old nest is found inconvenient and the ants have to migrate, the slaves take charge of the migration and actually carry their masters in their jaws. The masters are so utterly helpless that when Huber confined 30 of them without a slave, but with plenty of the food that they liked best and with their own larvae and pupae[11] to stimulate them to work, they did nothing! They could not even feed themselves, and many died of hunger. Huber then introduced a single slave (belonging to another ant species, *Formica fusca*). The slave instantly got to work feeding the survivors, making some cells, tending the larvae, and putting all to rights. What can be more extraordinary than these well-ascertained facts? If we had not known of any other slave-making ant, it would have been pointless to speculate about how so wonderful an instinct could have been perfected. But this is *not* the only slave-making ant, and we can learn much about the likely evolution of this trait by studying how far along this instinct is developed in other species (Figure 8.3).

Another slave-making ant species, *Formica sanguinea*, was likewise first discovered by Pierre Huber. It lives in parts of southern England, and its habits have been studied by the entomologist Mr. Frederick Smith of the British Museum, to whom I am much indebted for information on this and other subjects. Although fully trusting the words of Pierre Huber and Mr. Smith, I tried to approach the subject in a skeptical frame of mind; it seems to me that one may well be excused for doubting the existence of so extraordinary an instinct as that of slave-making among ants. Hence I will give my own observations in some detail.

I opened 14 nests of *F. sanguinea*, and found a few slaves (*F. fusca*, as mentioned above) in every nest. Males and fertile females of the slave species, however, are found only in their own proper communities and have never been observed in the nests of the slave-making species I was looking at (*F. sanguinea*). The slaves are black and not more than half the size of their red-colored masters, so that the contrast in their appearances is great: they are very easy to tell apart. When the nest is slightly disturbed, the slaves occasionally come out and, like their masters, are much agitated and defend the nest. When the nest is so much disturbed that the masters' larvae and pupae are exposed, the slaves and their masters work energetically together in carrying them away to a place of safety. The slaves clearly feel quite at home.

During the months of June and July I watched several nests in Surrey and Sussex, in England, for many hours in three successive years; during this entire time I never saw a slave either leave or enter the nest. It turns out that the slaves are few in number during those months, so I thought that perhaps they would behave differently when

---

[10] All worker ants (and soldiers) are diploid females, with two sets of chromosomes, one from each parent; unfertilized eggs become haploid males (see Table 8.1). Haploid individuals have only a single set of chromosomes, because the eggs were never fertilized.

[11] The pupa is the final developmental stage from which the adult eventually emerges.

Figure 8.3 (A) Slave-making ants, such as this *Polyergus mexicanus*, steal pupae from other ant colonies and bring them back to their own colonies to raise as slaves. (B) The silver ants in this photo (*Formica argentea*) do all the work in the colony of the red slave-making ants (*Polyergus breviceps*), including raising the next generation of slave-making ants. The silver ants will not reproduce in this colony; they will be replaced, as necessary, when the slave-making ants raid more pupae from other colonies.

they are more numerous; however, Mr. Smith tells me that he has watched the nests of this species for many hours during May, June, and August, both in Surrey and in Hampshire, and has never seen the slaves either leave or enter the nest, even in August, when they are quite numerous in the nest. Thus he considers them to be strictly household slaves. The masters, on the other hand, may be constantly seen bringing in materials for the nest and food of all kinds. In July 1860, however, I came across a community with an unusually large stock of slaves and observed a few slaves mingled with their masters leaving the nest and marching along the same road 20 yards to a tall Scotch fir tree. They then ascended this tree together, probably in search of aphids or citricola scale insects (*Coccus pseudomagnoliarum*), the only other insect besides aphids known to produce honeydew secretions. According to Pierre Huber, who observed this ant species on many occasions in Switzerland, the masters and slaves routinely work together in making the nest, but the slaves alone open and close the doors in the morning and evening. Mr. Huber also notes that the principle job of the slaves is to search for aphids. This difference in the usual habits of the masters and slaves in the two countries probably depends merely on the masters capturing slaves in greater numbers in Switzerland than in England.

One day I fortunately witnessed a migration of *F. sanguinea* from one nest to another. It was a most interesting spectacle to behold the masters carefully carrying their slaves in their jaws instead of being carried *by* them, as seen in the case of the other slave-making species that I talked about earlier, *Polyergus rufescens*. Another day my attention was struck by about 20 of the slave makers haunting one particular spot, and evidently not in search of food. They approached and were then vigorously repulsed by an independent community of the slave species, *F. fusca*, with sometimes as many as three of these ants clinging to the legs of the slave-making *F. sanguinea*. However, the slave-making species then ruthlessly killed their smaller opponents, *F. fusca*, carrying the dead bodies back to their nests as food, 29 yards distant. They were unable to get any pupae to rear as slaves, though. I then dug up a small parcel of the pupae of *F. fusca* from another nest and put them down on a bare spot near the place of combat; they were eagerly seized and carried off by the tyrants, who perhaps fancied that they had in fact been victorious in their recent combat after all.

At the same time I laid a small group of the pupae of another species, *F. flava*,[12] on the same place, with a few of these little yellow ants still clinging to the fragments of their nest. I was curious to see whether *F. sanguinea* could distinguish between the pupae of *F. fusca*, which they habitually make into slaves, and those of the little and furious *F. flava*, which they rarely capture. This species is sometimes, though rarely, made into slaves, as has been described by Mr. Smith. Although having a very small body, it is a very courageous ant, and I have seen it attack other ants ferociously. In one instance I found to my surprise an independent community of *F. flava* under a stone beneath a nest of the slave-making species *F. sanguinea*. When I accidentally disturbed both nests, the little yellow ants attacked their larger neighbors with surprising courage. It was soon evident that members of *F. sanguinea* did at once distinguish between pupae of the two species: whereas they would eagerly and instantly seize the pupae of *F. fusca*, as discussed earlier, they were much terrified when they came across the pupae of *F. flava*—or even pieces of earth taken from the nest of that species—and quickly ran away. But in about 15 minutes, shortly after all the little yellow ants had crawled away, *F. sanguinea* took heart and carried off their pupae.

One evening I visited another community of *F. sanguinea* and found a number of these ants returning home and entering their nests, carrying with them the dead

---

[12] This ant is now known as *Lasius flavus*, the yellow meadow ant.

bodies of the slave species *F. fusca* (showing that it was not simply a migration) along with numerous pupae. I traced a long file of ants burdened with booty for about 40 yards back, to a very thick clump of heath, whence I saw the last individual of *F. sanguinea* emerge, carrying a pupa. Although I was not able to find the desolated nest in the thick heath, the nest must have been close at hand, for two or three individuals of *F. fusca* were rushing about in the greatest agitation, and one was perched motionless with its own pupa in its mouth on the top of a spray of heath, an image of despair over its ravaged home.

Such are the facts in regard to the wonderful instinct of slave-making ants. Note the remarkable contrast between the habits of *F. sanguinea* and those of the continental *P. rufescens*. *Polyergus rufescens* does not build its own nest, determine its own migrations, or collect food for itself or for its young, and cannot even feed itself: it absolutely depends on its numerous slaves. *Formica sanguinea*, on the other hand, possess many fewer slaves, especially in the early summer. The masters determine when and where a new nest should be formed, and, when they do migrate, the masters carry the slaves. Both in Switzerland and England, the slaves seem to have the exclusive care of the larvae, and the masters alone go on slave-making expeditions, in which they steal the larvae and pupae of the other species. In Switzerland, the slaves and masters work together in making and bringing materials for the nest, and both (although chiefly the slaves) attend and "milk" the aphids to collect food for the community. In contrast, in England only the masters generally leave the nest to collect building materials and food for themselves, their slaves, and the larvae. Thus in England the masters receive much less service from their slaves than they do in Switzerland.

By what steps might the slave-making instinct of *F. sanguinea* have originated? Of course there is no way to know with certainty. But as I have seen ants that are not slave-makers carry off the pupae of other species for food—when such pupae happen to be scattered near their nests—it is possible that some of the pupae originally kept as food might have developed through pupation within the nest of the marauding species; the foreign ants thus unintentionally reared would then follow their proper instincts and do what work they could. If their presence proved useful to the species that had seized them—e.g., if it were more advantageous to this species to capture workers of another species than to procreate workers themselves—then the original habit of collecting pupae for food might be strengthened and rendered permanent by natural selection for the very different purpose of raising slaves. Originally the instinct might be carried out to only a limited extent, as in our British *F. sanguinea*, which as we have seen is less aided by its slaves than is the same species in Switzerland. But natural selection might then increase and even modify the instinct—always supposing each modification to be of use to the species—until an ant was formed as abjectly dependent upon its slaves as are individuals of *P. rufescens*.

The Cell-Making Instinct of the Hive Bee. I will not go into minute detail on this subject, but will merely outline the conclusions I have arrived at. He must be a dull man indeed who can examine the exquisite structure of a bee's honeycomb, so beautifully adapted to its end, without enthusiastic admiration (Figure 8.4A). Mathematicians

(A)

(B)

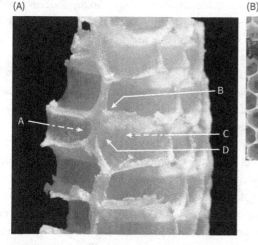

Figure 8.4 (A) This cross-section through a honeybee comb shows the double layer of cells. The two layers are staggered so that the base of each cell shares walls with the bases of three cells on the other side of the comb. In this photo, the base of cell A shares walls with the bases of cells B, C, and D (dashed lines indicate where arrows pass behind side walls). (B) Peering down into the cells in this comb, it's possible to see the junction where the bases of three cells meet on the other side of the comb.

tell us that bees have essentially solved a recondite problem—one that is extremely difficult to understand—and have made their cells in a shape that will hold the greatest possible amount of honey while using the least possible amount of precious wax in the construction process. It has been remarked that a skillful workman equipped with appropriate tools and measuring devices would find it very difficult to make cells of wax of the best form; yet a crowd of bees accomplishes this perfectly while working in a dark hive! Granting whatever instincts you please, it seems at first quite inconceivable how they can make all the necessary angles and planes, or even perceive when they are correctly made. But the difficulty is not nearly so great as it at first appears; all this beautiful work can be shown, I think, to follow from a few simple instincts.

I was led to study this subject by the entomologist Mr. George Robert Waterhouse, who has shown that the form of each hive cell stands in close relationship to the presence of adjoining cells. The following view may, perhaps, be considered only a modification of his theory. Let us look to the great principle of gradual progression through a series of stages and see whether nature does not reveal to us her method of work.

At one end of a short series we have bumblebees, which use their old cocoon to hold honey, sometimes adding short tubes of wax to them, and likewise making separate and very irregular rounded cells of wax. At the other end of the series we have the cells of the honeybee, arranged in a double layer. With honeybees, each cell, as is well known, is a hexagonal prism at the open end (Figure 8.4B), but the bottom of the cell is not flat; rather, the bottom edges of its six sides are beveled so as to form an inverted pyramid composed of three rhombuses.[13] These rhombuses are angled such that the three that form the pyramidal base of a single cell on one side of the comb form part of the bases of three adjoining cells on the opposite side.

In the stages between the simplicity of the bumblebee's honeycomb cells and the extreme perfection of those of the honeybee, we have the cells of the stingless Mexican bee *Melipona domestica*, which have been carefully described and figured by Pierre Huber. The bee itself is intermediate in structure between the honeybee and the bumblebee, but more closely related to the bumblebee. It builds a nearly regular waxen comb of cylindrical cells within which the young are hatched, along with some larger cells for storing honey. The honey-storing cells are nearly spherical and equal in size, and are aggregated into an irregular mass. But the important thing to notice is that these spherical cells are always made so close to each other that they would have intersected or broken into each other if the spheres had been completed. But this never happens: the bees build perfectly flat walls of wax between the spheres, which thus tend to intersect. Thus, each cell consists of an outer portion that is spherical in some places and of two, three, or more perfectly flat surfaces, depending on whether the cell abuts two, three, or more other cells. When one cell rests on three other cells, which, from the spheres being nearly of the same size, is very frequently and necessarily the case, the three flat surfaces are united into a pyramid. Remarkably, this pyramid is manifestly a gross imitation of the three-sided pyramidal base of the honeybee's cell. As in the cells of the honeybee, here, too, the three plane surfaces in any one cell necessarily enter into the construction of the three adjoining cells. The bees clearly save wax and, even more importantly, labor by this manner of building, for the flat walls between the adjoining cells are not doubly thick but are instead of the same thickness as the outer spherical portions. Each flat portion forms a part of two adjacent cells.

Reflecting on this case, it occurred to me that if *Melipona* had made its spheres at some uniform distance from each other, of equal sizes and arranged symmetrically in a double layer, the resulting structure would have been as perfect as the comb of the honeybee. Accordingly, I wrote to Professor William Hallowes Miller, a mathematician at Cambridge University; he has kindly read over the following statement that I had drawn up from the information he gave me and tells me that what I have to say is strictly correct. So here is the argument. Let us describe a number of equal spheres with their centers placed in two parallel layers, with the center of each sphere at the distance of the radius $\times$ 2 (= radius $\times$ 1.41421) (or at some smaller distance) from the centers of six surrounding spheres in the same layer and at the same distance from the centers of the adjoining spheres in the other and parallel layer. Then, if planes of intersection between the several spheres in both layers are formed, we will see a double layer of hexagonal prisms united together by pyramidal bases formed of three rhombuses; the rhombuses and the sides of the hexagonal prisms will have

---

[13] A rhombus is any flat shape with four sides of equal length, such as a square or a diamond.

every angle identical with the best measurements that have been made of the cells formed by honeybees.

Thus we may safely conclude that, if we could just slightly modify the instincts already possessed by *Melipona*, which are interesting but not especially wonderful, this bee could then make a structure as wonderfully perfect as that of the honeybee. We must suppose that *Melipona* has the power of forming her cells truly spherical and of equal sizes, but this would not be very surprising: she already does so to a certain extent, and we know that many other insects can make perfectly cylindrical burrows in wood, apparently just by turning around on a fixed point. We must then suppose that *Melipona* comes to arrange her cells in level layers, as she already does with her cylindrical cells. And we must further suppose—and this is the greatest difficulty—that she can somehow judge accurately at what distance to stand from her fellow laborers when several are busily making their spheres. But this is not so far-fetched, as she is already able to judge distances so well that she always creates her spheres so that they intersect to a certain extent; she then unites the points of intersection with perfectly flat surfaces. By such gradual modifications of instincts that in themselves are not especially wonderful—hardly more wonderful than those that guide a bird to make is nest—the honeybee has, I believe, acquired her inimitable architectural powers.

This theory can be tested by experiment. Following the example of Mr. William Bernhard Tegetmeier, a man well known for his work with bees (as well as with poultry and pigeons), I separated two honeycombs and put between them a long, thick, rectangular strip of wax. The bees instantly began to excavate minute circular pits in the wax, and, as they deepened these little pits, they made them wider and wider until they were converted into shallow basins, appearing to the eye as perfectly rounded portions of a sphere and of about the diameter of a single hive cell. It was most interesting to observe that wherever several bees had begun to excavate these basins near each other, they had begun their work at such a distance from each other that by the time the basins had acquired about the width of an ordinary hive cell, and were also about one-sixth of the diameter of the sphere in depth, the rims of the basins intersected or broke into each other. As soon as this occurred, the bees ceased further excavation and began to build up flat walls of wax on the lines of intersection between the basins, so that each hexagonal prism was built upon the scalloped edge of a smooth basin, instead of on the straight edges of the three-sided pyramid as in the case of ordinary cells.

Next, instead of a thick, rectangular piece of wax, I put a thick and narrow knife-edged ridge into the hive, colored with the bright red pigment vermilion. The bees instantly began excavating little basins near each other on both sides, in the same way as before. Now this time the ridge was so thin that the bottoms of the basins, if they had been excavated to the same depths as in the former experiment, would have broken into each other from the opposite sides. The bees, however, did not allow this to happen; they stopped their excavations in due time, so that the basins, as soon as they had been a little deepened, came to have flat bases. And these flat bases, formed by thin little plates of the vermilion wax left ungnawed, were situated, as far as I could judge, exactly along the planes of imaginary intersection between the basins on the opposite sides of the ridge of wax. In some parts only small portions, while in other parts large portions of a rhombic plate were thus left between the opposed basins. The bees must have worked at very nearly the same rate in circularly gnawing away and deepening the basins on both sides of the ridge of vermilion wax in order to have thus succeeded in leaving flat plates between the basins by stopping work at the planes of intersection.

Considering the great flexibility of thin wax, I do not see why the bees, while working away on the two sides of a strip of wax, should have any difficulty perceiving when they have gnawed the wax away to the proper thinness and then stopping their work. In ordinary honeycombs it has seemed to me that the bees do not always succeed in working at exactly the same rate from the opposite sides, for I have noticed half-completed rhombuses at the base of a just-started cell, which were slightly concave on one side (where I suppose the bees had excavated too quickly) and convex on the opposite side (where the bees had probably worked less quickly). On one well-marked instance, I put the comb back into the hive and allowed the bees to continue working for a short time. When I examined the cell later, I found that the rhombic plate had been completed and had become *perfectly flat*. It was absolutely impossible, from the extreme thinness of the little plate, that the bees could have done this by gnawing away the convex side. I suspect that the bees in such cases must stand on opposite sides and push and bend the warm and ductile wax into its proper intermediate plane and thus flatten it; indeed, I have tried this myself, and it is easily done.

From my experiment with the ridge of vermilion wax, we can see that if the bees were to build themselves a thin wall of wax, they could make their cells the proper shape by standing at the proper distance from each other, by excavating at the same rate, and by endeavoring to make equal spherical hollows without ever allowing the spheres to break into each other. By examining the edge of a growing comb, it is clear that bees do make a rough, circumferential wall or rim all around the comb, and that they then gnaw this away from the opposite sides, always working circularly as they deepen each cell. They do not make the whole three-sided pyramidal base of any one cell at the same time, but only that one rhombic plate that stands on the extreme growing margin, or the two plates, as the case may be. And they never complete the upper edges of the rhombic plates until the hexagonal walls are started. Some of these statements differ from those made by the justly celebrated elder Huber (Pierre's father, François), but I am convinced of their accuracy. If I had more space, I would show that they conform well with my theory.

François Huber's statement that the very first cell is excavated out of a little parallel-sided wall of wax is not, as far as I have seen, strictly correct; the cell always begins as a little hood of wax, but I will not give details here. We see how important excavation is in constructing the cells. But it would be a great error to suppose that the bees cannot build up a rough wall of wax in the proper position—that is, along the plane of intersection between two adjoining spheres. I have several specimens showing very clearly that they can do this. Even in the rude circumferential rim or wall of wax around a growing comb, flexures (curved or bent portions) may sometimes be observed, corresponding in position to the planes of the rhombic base plates of future cells. But in every case the rough wall of wax has to be finished off by being largely gnawed away on both sides.

The manner in which the bees build is curious: they always make the first rough wall from 10 to 20 times thicker than the excessively thin finished wall of the cell. We shall understand how they work by supposing masons first to pile up a broad ridge of cement and then to begin cutting it away equally on both sides near the ground until a smooth, very thin wall is left in the middle, with the masons always piling up the cut-away cement and adding fresh cement on the top of the ridge. We shall thus have a thin wall steadily growing upward but always crowned by a gigantic coping.

For all the cells—both those just begun and those completed—being thus crowned by a strong coping of wax, the bees can now cluster and crawl over the comb without injuring the delicate hexagonal walls. These walls, as Professor Miller has kindly determined for me, vary greatly in thickness, being, on average, about 72 microns (1/352 of an inch) thick, which is the average of 12 measurements made near the border of the comb. The rhomboidal plates at the base of the comb are considerably thicker, although still only about 111 microns (1/229 of an inch) thick (this time averaging 21 measurements). By this singular manner of building, the comb is continually strengthened, while using as little wax as possible.

The fact that a large number of bees all work together simultaneously would at first seem to make it harder to understand how the cells are made; one bee works for a short time on one cell and then goes to another, so that, as François Huber has remarked, a dozen individuals are already at work even in building the first cell. I saw this myself after covering the edges of the hexagonal walls of a single cell (or the extreme margin of the circumferential rim of a growing comb) with an extremely thin layer of melted vermilion wax. In every case, the color was soon most delicately diffused by the bees—as delicately as a painter could have done it with a paint brush—by atoms of the colored wax having been taken from the spot on which I had placed it and worked into the growing edges of the cells all around. The work of construction seems to be a sort of balance between many bees, all instinctively standing at the same relative distance from each other, all trying to sweep equal spheres, and then building up—or leaving ungnawed—the planes of intersection between these spheres. It was really curious to note how often the bees would pull down and rebuild the same cell in different ways in cases of difficulty, as when two pieces of comb met at an angle, sometimes returning to a shape that they had at first rejected.

When bees have a place on which they can stand in their proper position for working—on a slip of wood, for example, placed directly under the middle of a comb growing downward, so that the comb has to be built over one face of the slip—in this case the bees can lay the foundations of one wall of a new hexagon in its strictly proper place, projecting beyond the other completed cells. As long as the bees can stand at their proper relative distances from each other and from the walls of the last-completed cells, they can then strike imaginary spheres and build up a wall intermediate between two adjoining spheres. As far as I have seen, however, they never gnaw away and finish off the angles of a cell until a large part of both that cell and of the adjoining cells has been built. This capacity of bees to lay down a rough wall in its proper place between two just-commenced cells bears on a fact that at first seems to argue against the foregoing theory—namely that the cells on the extreme margins of wasp combs are sometimes strictly hexagonal. But I don't have space here to deal with this issue. Nor do I think there is any great difficulty in a single insect (e.g., a queen wasp) making hexagonal cells, if she were to work alternately on the in-side and outside of two or three cells started at the same time, always standing at the proper relative distance from the parts of the cells just begun, sweeping spheres or cylinders, and building up intermediate planes.

As natural selection acts only through the gradual accumulation of slight modifi-cations of structure or instinct, each one being profitable to the individual under its own conditions of life, one may reasonably ask how a long and graduated succession of modified architectural instincts, all leading toward the present and perfect plan of honeycomb construction, could each have profited the honeybee's ancestors. The answer is not difficult: cells constructed like those of the bee or the wasp are strong and sturdy, and they save much in labor, space, and the amount of material needed for construction. With respect to the formation of the wax used to make the combs, bees are often hard-pressed to get sufficient nectar; as Mr. Tegetmeier—a recognized authority on bees, as mentioned earlier—has informed me, experiments have shown that from 12 to 15 pounds of dry sugar are consumed by a hive of bees for the secre-tion of a single pound of wax. Thus a prodigious amount of fluid nectar must be col-lected and consumed by the bees in a hive to secrete the wax needed to construct their combs. Moreover, many bees have to remain idle for many days during the process of secretion. A large store of honey is also indispensable to support a large stock of bees during the winter, and the security of the hive is known to depend mainly on a large number of bees being supported. Thus the saving of wax by largely saving honey, and saving the time that would have been consumed in collecting that honey, must be an important element of success to any family of bees.

Of course the success of the species may sometimes depend on the number of its enemies and parasites, or on other quite distinct causes, and so be altogether independent of the quantity of honey that the bees can collect. But let us suppose that the amount of available honey determined—as, in fact, it probably has often determined—whether a bee related to our bumblebees could exist in large numbers in any country. And let us further suppose that the community lived through the winter and consequently required a store of honey to do so. In such a case, it would clearly be advantageous to our imaginary bumblebee if a slight modification in her instincts led her to make her waxen cells closer together, so as to intersect a little; for a wall in common even to two adjoining cells would save some amount of both labor and wax. Thus it would continually be more and more advantageous to our bumblebees if they were to make their cells more and more regular, nearer together, and aggregated into a mass, like the cells of *Melipona* in fact. For in this case, a large part of the exterior surface of each cell would now form part of an adjoining cell, and much labor and wax would be saved. Again, from the same cause, it would be advantageous to *Melipona* if she were to make her cells closer together and more regular in every way than at present; for then, as we have seen, the spherical surfaces would wholly disappear and be replaced by flat surfaces between cells, and *Melipona* would then make a comb as perfect as that of the honeybee. Natural selection could not lead beyond this stage of perfection in architecture, for the existing comb of the honeybee, as far as we can tell, is absolutely perfect in economizing labor and wax.

Thus it seems that the most wonderful of all known instincts—that of the honeybee—can be explained by natural selection having taken advantage of many successive slight modifications of simpler instincts, leading by slow degrees over many generations to bees sweeping equal spheres at a given distance from each other in a double layer, and to building up and excavating the wax along the planes of intersection. The bees, of course, don't know whether or not they sweep their spheres at one particular distance from the others any more than they can calculate the several angles of the hexagonal prisms they construct, or of the rhombic plates at the bottom of each cell. But the individual swarm that made the best cells with the least labor and with the least waste of costly wax will have succeeded best; they will then have transmitted through inheritance their newly acquired economical instincts to new swarms. Those swarms, in turn, will have had the best chance of succeeding in the struggle for existence.

Objections to the Theory of Natural Selection as Applied to Instincts: Neuter and Sterile Insects

With regard to my argument about the evolution of instincts, some naysayers have claimed that "the variations of structure and of instinct must have been simultaneous and accurately adjusted to each other, as a modification in the one without an immediate corresponding change in the other would have been fatal." The force of this objection rests entirely on the assumption that the changes in instincts and structures are dramatic and abrupt; but this is not the case. Consider the example of the great tit (*Parus major*) described in Chapter 6 (see Figure 6.5). This bird often holds the seeds of the yew tree between its feet on a branch, and then hammers away with its beak until it gets at the kernel within the seed. Now what special difficulty would there be in natural selection preserving all the slight individual variations in the shape of the beak that were better and better adapted to break open the seeds? Eventually a beak would be formed that was as well constructed for this purpose as that of the nuthatch. At the same time habit, or compulsion, or spontaneous variations in taste could have led the bird to become more and more of a seed-eater. In this case we can suppose that the beak will be slowly modified by natural selection, following and in accordance with slowly changing habits or taste. But let the feet of the titmouse vary and grow larger from correlation with the beak or any other unknown cause, and it is likely that such larger feet would lead the birds to climb more and more until they acquired the remarkable climbing instinct and power of the nuthatch. In this case, a gradual change of structure should slowly lead to changed instinctive habits.

Here is another case to consider. Few instincts are more remarkable than that which leads certain tropical and subtropical swifts[14] to make their nests entirely of dried and inspissated (i.e., thickened) saliva. Some birds build their nests of mud, believed to be moistened with saliva, and I have seen one of the North American swifts making its nest using sticks agglutinated with saliva, and even with flakes of this substance. Wouldn't the natural selection of individual swifts that secreted more and more saliva at last produce a species with instincts leading it to neglect other materials and to make its nest exclusively of inspissated saliva? It must be admitted, of course, that in many such instances we cannot know whether it was instinct or structure that varied first.

No doubt the origins of many instincts would be very difficult to explain and could be opposed to the theory of natural selection: cases in which we cannot see how an instinct could have originated, cases in which no intermediate gradations are known to exist, cases of instincts of such trifling importance that they could hardly have been acted on by natural selection, and cases of instincts almost identical in animals so far apart in the scale of nature that we cannot account for their similarity by inheritance from a common ancestor and consequently must believe that they were independently acquired through natural selection. Here I will confine myself to one special difficulty, one that at first appeared to me insuperable and actually fatal to my whole theory: namely, the "neuters" (i.e., sterile females) found in many insect communities. These neuters often differ markedly both in instinct and in structure from both the males and the fertile females, and yet, from being sterile, they cannot propagate their kind directly to future generations.

The subject well deserves detailed discussion, but here I will take only a single case: that of sterile worker ants. (See Box 8.1 for a brief summary of ant colony characteristics.[15]) It is difficult to know what has made the workers sterile, but no more difficult than it is to understand any similarly striking structural modification. Although we don't know the mechanism, it can, in fact, be shown that some other insects—and in fact some non-insect arthropods as well—do sometimes become sterile in nature. If such animals had been social, and it had profited the community that a number should have been annually born incapable of reproduction but capable of work, I can see no special difficulty in this having been achieved through natural selection, since the reproductively active members of the community would have benefitted from the contributions of the workers.

But I must pass over this preliminary difficulty and get to a greater one: the difficulty of explaining why the worker ants differ so widely in structure from both the

---

[14] The birds Darwin is referring to belong to the genus *Aerodramus*.
[15] Box 8.1 is my creation; it was not part of Darwin's book.

Box 8.1 Ant colonies: The cast of characters

The Queen
Each colony has one to several queens.

All queens are female.

Each queen mates for only a short time with up to 10 males, but with as few as one male.

After mating, the queen spends the rest of her life laying eggs—thousands or hundreds of thousands of them.

All fertilized eggs become either female worker ants or female soldier ants.

Unfertilized eggs become sexually active, haploid males.

Males
All males are haploid, and all have wings.

The males' only function is to fertilize the eggs of a virgin queen ant. They do no other work in the colony.

Since males are haploid, they never have a father and never leave any sons. But they can have grandsons!

Workers
All workers are female, produced from the mating of the queen with one or more males.

Most workers are sterile ("neuter") and do not lay eggs.

If the queen dies, some workers can lay eggs that eventually become haploid males, whose job it is to find a virgin female and initiate a new colony.

The workers' major roles include caring for the queen and her young, digging the nest, searching for food, feeding the males, feeding the soldiers, and defending the nest

Soldiers
All are female, produced from the mating of the queen with one or more males.

Soldier ants are sterile ("neuter") and do not lay eggs.

Soldiers are basically workers with extra large heads and especially muscular jaws (mandibles).

They are specialized for defending the nest from enemies.

---

males and the fertile females, as in the shape of the thorax, for example, and in lacking wings and sometimes even eyes, as well as differing in instinct. As far as instinct alone is concerned, the wonderful difference in this respect between the sterile female workers and the fertile females would have been even better exemplified by the honeybee, but here I will stay with ants. If a worker ant or any other neuter insect had been an ordinary animal, I should have unhesitatingly assumed that all of its characteristics had been slowly acquired through natural selection: namely by individuals having been born with slightly profitable modifications that were then inherited by the offspring and that these again varied and again were selected, and so on through many generations. But with the worker ant we have an insect differing greatly from its parents, yet absolutely sterile, so that it could never have transmitted successively acquired modifications of structure or instinct to its offspring: it has no offspring! So how is it possible to reconcile this case with the theory of natural selection?

First, remember that we have many instances, both among our domesticated animals and plants and among those in the wild, of all sorts of differences of inherited structure that are correlated with certain ages, and with either sex. We have differences correlated not only with one sex, but also with just that short period when the reproductive system is active, as in the nuptial plumage of many birds and in the hooked jaws of the male salmon. We even see slight differences in the horns of different breeds of cattle when males are castrated to produce oxen; the oxen—castrated male cattle—of certain breeds have longer horns than the oxen of other breeds, relative to the length of the horns in both the bulls and cows of these same breeds. Thus I can see no great difficulty in any character becoming correlated with the sterile condition of certain members of insect communities. The difficulty lies in understanding how such correlated modifications of structure could have been slowly accumulated by natural selection in the absence of reproduction.

Although this difficulty may appear to be insuperable, remember that selection may be applied to the family, not just to the individual, and may thus gain the desired end in that way. Cattle breeders wish the flesh and fat of their animals to be well marbled together; although an animal having those features is thus slaughtered and no longer able to reproduce, the breeder simply goes confidently back to the same stock and again succeeds. Such faith may be placed in the power of selection that a breed of cattle that always produces oxen with extraordinarily long horns could, it is probable, be formed by carefully watching which individual bulls and cows, when matched, produced oxen with the longest horns. And yet no ox could ever have propagated its kind to future generations, because oxen are castrated when young.

Here is an even better illustration: some plants produce what are called "double flowers"—flowers with extra petals, and which often contain flowers within flowers. Now according to the French botanist Bernard Verlot, some varieties of the double annual stock plant (genus *Matthiola*) (Figure 8.5) that have been carefully selected to a high degree—and for many generations—always produce a large proportion of their seedlings bearing double and quite sterile flowers along with some single-flowered and fully fertile plants. The species can of course only be propagated using the single-flowered plants; thus the single-flowered plants can be compared with fertile male and female ants, while the double sterile plants are similar to the neuter workers of the same ant community. As with the varieties of the stock plant, so with social insects: selection has been applied to the *family*, rather than to the individual, for the

Figure 8.5 (A) Single-flowered and (B) double-flowered blooms of the stock plant (*Matthiola* sp.).

sake of gaining a desired outcome. Thus we may conclude that slight modifications of structure or of instinct, correlated with the sterile condition of certain members of the community, must have been advantageous to the community as a whole: the fertile males and females have flourished, and have transmitted to their fertilized offspring a tendency to produce some sterile members with the same modifications. This process must have been repeated in social insect populations many times, eventually producing the prodigious amount of difference between the fertile and sterile females that we now see.

So we have seen that slight changes in instinct can lead to gradual changes in anatomy, and that slight changes in anatomy can gradually lead to changes in instinct. We have also seen that selection can operate not just on individuals but on entire families, so that modifications of behavior or structure of sterile organisms can still be selected for if the entire family benefits from those changes. But we have not yet touched on the acme of the difficulty: the fact that the neuters (the nonreproductive members) of several species differ not only from the fertile males and females of their species, but also from each other, and sometimes to an almost incredible degree, so that the neuters are actually divided into two or even three morphologically distinct castes. Moreover, instead of graduating smoothly into each other, the castes are instead as distinctly different from each other as are the members of any two species of the same genus, or even as the members of any two genera of the same family. Thus in the genus *Eciton* there are neuter worker ants and neuter soldier ants whose jaws and instincts are extraordinarily different. In the genus *Cryptocerus*,[16] the workers of just one of the castes carry a wonderful and distinctive sort of shield on their heads, the use of which is quite unknown. There can also be remarkable behavioral differences between the members of the different castes. In the Mexican ant genus *Myrmecocystus*, for example, the workers of one caste never leave the nest; rather, they are fed by another caste of workers found in the same nest: these workers—members of the very same species—have an enormously developed abdomen that secretes a sort of honey, which essentially plays the same role as that excreted by the aphids that are guarded and imprisoned by our European ants.

Now some readers may indeed think that I must have an overweening confidence in the principle of natural selection when I do not admit that such wonderful and well-established facts immediately annihilate the theory. But I think the facts can fit the theory perfectly well. Consider first the simpler case of neuter insects all of one caste. That these, I believe, have been rendered different from the fertile males and females through natural section, we may conclude from an analogy with ordinary variations—that the successive, slight, but profitable physical and behavioral modifications did not first arise in all the neuters in the same nest, but at first in only a few individuals. Then, by the improved survival of those communities whose females produced the greatest number of neuters possessing the advantageous modifications, all the neuters eventually came to be thus characterized. According to this view, we ought to occasionally find within the same nest some neuter insects presenting a gradation of structures; this we do in fact find—and not rarely, which is especially surprising considering the small number of European neuter insects that have so far been examined.

The British Museum entomologist Mr. Smith has shown that the neuters of several British ants differ surprisingly from each other in size and sometimes in color and that the extreme forms can be linked together by individuals taken out of the same nest; I myself have compared such perfect gradations in some species. It sometimes

---

[16] These ants are now placed in the genus *Cephalotes*.

Figure 8.6 Some driver ant workers (such as these *Dorylus* sp.) are much larger than other workers of the same species.

happens that the larger- or the smaller-sized workers are the most numerous, or that both large and small workers are equally numerous, while those of an intermediate size are relatively rare. *Formica flava*[17] has both large and small workers, as well as some few of intermediate size; moreover, as Mr. Smith has reported, the larger-sized workers have small, simple, but clearly defined ocelli (simple light receptors), while the smaller workers have only rudimentary ocelli. Having carefully dissected several specimens of these workers myself, I can confirm that the eyes are far more rudimentary in the smaller workers than can be accounted for merely by their proportionally smaller size. I fully believe that the workers of intermediate size have their ocelli in an exactly intermediate condition. Here, then, we have two bodies of sterile workers living in the same nest, differing not only in size but also in their organs of vision, and we see them connected within the same colony by a few members in an intermediate condition. Let me digress a bit and add that if the smaller workers had been the most useful to the community, and those males and females that produced more and more of the smaller workers had been continually selected until all the workers were in that condition, then we should have had a species of ant with neuters in nearly the same condition as those now seen in the genus *Myrmica*. For the *Myrmica* workers have not even rudiments of ocelli, even though the ocelli are well developed in the fertile male and female ants of this genus.

Let me consider just one more case. So confidently did I expect to occasionally find gradations of important structures between the different castes of neuter insects of the same species that I gladly accepted Mr. Smith's offer of numerous specimens from the same nest of the driver ant (*Anomma*)[18] (Figure 8.6) of West Africa. Readers will perhaps best appreciate the amount of difference between these workers by my

---

[17] This ant is now known as *Lasius flavus*, the yellow meadow ant.
[18] These ants are now placed in the genus *Dorylus*.

giving an illustration. Suppose we were to see a group of workmen building a house and that some of those workers were 5'4" tall and that many others were 16' tall. In addition, suppose that the taller workmen had heads that were four times (rather than three times) bigger than those of the shorter men and jaws that were nearly five times bigger. This is the actual level of difference that we see among these worker ants. Moreover, the jaws of the worker ants of the several body sizes also differ wonderfully in shape and in the form and number of teeth. But the important fact for us is that even though the workers can be grouped into castes of very different sizes, the castes graduate almost imperceptibly into each other, as do the differences in the structure of their jaws. I speak with confidence on this point, as my good friend the biologist Sir John Lubbock has made detailed drawings for me, using a camera lucida,[19] of the jaws that I dissected from the workers of the several sizes. Mr. Henry Walter Bates, in his interesting book *The Naturalist on the River Amazons* (1863), has described analogous cases.

With these facts before me, I believe that natural selection, by acting on the fertile ants (the parents), could create a species that should regularly produce neuters, all of large size with one form of jaw, or all of small size with widely different jaws, or even— and this is the greatest difficulty—produce one set of workers of one size and structure and simultaneously another set of workers of a different size and structure: first a graduated series is formed in a colony, as in the case of the driver ant just described, and then the extreme forms are produced in greater and greater numbers, generation after generation, through the heightened survival of the fertile parents that generated them, until none is produced with an intermediate structure.[20]

An analogous explanation has been given by Mr. Alfred Russel Wallace[21] of the equally complex case of certain Malayan butterflies regularly appearing under two or even three distinct female forms. Similarly, Fritz Müller has presented a similar explanation for certain Brazilian crustaceans that likewise appear in two very different male forms.

I have now explained how, I believe, the wonderful fact of two distinctly defined castes of sterile female workers existing in the same nest, both widely different from each other and from their parents, has originated. We can see how useful their production may have been to a community of social ants, on the same principle that a division of labor is useful in civilized human societies. Ants, however, work by inherited instincts and by inherited organs or tools, while we work through acquired knowledge and manufactured instruments. But I must confess that, even with all my faith in natural selection, I could never have appreciated that this principle could be

---

[19] A *camera lucida* is a special optical device that superimposes an image of something onto a blank piece of paper to aid in accurate drawing.

[20] We now know that different types of sterile workers can be produced within a colony simply by providing eggs with different amounts of nutrients, or through changes in temperature or other environmental conditions.

[21] As noted in earlier chapters, British naturalist Alfred Russel Wallace independently advanced the theory of natural selection; it was the impending publication of his paper on the subject that prompted Darwin to publish his own work.

efficient to such a high degree had not the case of these neuter insects led me to this conclusion. I have therefore discussed this case at some little but still insufficient length in order to show the power of natural selection, and likewise because this is by far the most serious special difficulty that my theory has encountered. The case is also very interesting as it proves that, with both animals and plants, any amount of modification may eventually be brought about by the gradual accumulation of numerous, small, spontaneous variations that are in any way beneficial, without exercise or habit having ever been brought into play: peculiar habits confined to the sterile workers, however long they might be followed, could not possibly affect anything about the males or the fertile females, which alone leave descendants.

Summary

I have tried in this chapter to show very briefly that the mental qualities of our domestic animals vary and that those variations are inherited. Still more briefly I have attempted to show that instincts also vary slightly among individuals in nature. No one will dispute that instincts are of the highest importance to each animal. Therefore there is no real difficulty, under changing conditions of life, in natural selection promoting the gradual accumulation of slight modifications of instinct that are in any way useful to the owner or to his or her descendants. I do not pretend that the facts given in this chapter strengthen my theory to any great degree; but, most importantly, none of the particularly difficult cases I describe annihilate it, to the best of my judgment. Although we know that instincts are not always absolutely perfect and are liable to mistakes, we also know that no instinct can be shown to have been produced for the good of other species, although animals certainly take advantage of the instincts of others at times. It seems equally clear that the canon of natural history "*Natura non facit saltum*" (Nature does not make jumps) is just as applicable to instinct as it is to anatomy, and is explicable through the principle of natural selection; indeed, it is otherwise inexplicable. All of these facts tend to support the theory of natural selection.

This theory is also strengthened by a few other facts regarding instincts. Consider, for example, that individuals of closely allied but distinct species, when inhabiting distant parts of the world and living under considerably different environmental conditions, commonly often retain nearly the same instincts. The principle of natural selection thus enables us to understand how it is that the thrush of tropical South America lines its nest with mud in the same peculiar manner as does our British thrush, and how it is that the hornbills of both Africa and India have the same extraordinary instinct of plastering up and imprisoning the females in a hole in a tree, with only a small hole left in the plaster through which the males feed them and also their young when hatched, and how it is that the male wrens of North America build "cock nests" to roost in, as do the males of our kitty wrens—a habit wholly unlike that of any other bird. Finally, although it may not be a logical deduction, to my imagination it is far more satisfactory to look at such remarkable instincts as the young cuckoo ejecting its foster brothers from the nest, ants making slaves, and the larvae of ichneumonid wasps feeding within the live bodies of certain caterpillars not as specially endowed or specially created instincts, but rather as small consequences of one general law leading to the advancement of all organic beings—namely, multiply, vary, and let the strongest live—and the weakest die.

Key Issues to Talk and Write About

1. In a single paragraph, try to explain what it is about the slave-making instinct of ants and the hive-making behavior of honeybees that convinces Darwin that such behaviors were molded to their present form by natural selection.
2. In what way does Darwin distinguish between habit and instinct?
3. How does Darwin explain the fact that even though neuter worker ants don't reproduce, they exist because of natural selection? In other words, how can selection possibly act on animals that don't reproduce?
4. Related birds that live thousands of miles apart in very different environmental conditions often share the same peculiar behaviors. Does this weaken or strengthen Darwin's idea that behaviors can also evolve by natural selection? Explain your reasoning.
5. In this chapter, and elsewhere, why does Darwin place so much emphasis on traits exhibited by closely related species?
6. Find out two interesting things about one of the people that Darwin mentions in this chapter. Choose from the following:
   Jean-Henri Fabre
   William Henry Hudson
   Thomas Andrew Knight
   William Hallowes Miller
   George Robert Waterhouse
7. Based on the instructions given at the end of Chapter 1 (see page 28), write a concise, stand-alone, one-sentence summary of one or two of the following six paragraphs taken from this chapter, which begin with the words:
   - "Instincts are clearly as important as bodily structures for the welfare of every species" (see page 230).
   - "To see how strongly these domestic instincts, habits, and dispositions are inherited …" (see page 234).
   - "In the case of the European cuckoo …" (see page 238).
   - "By what steps might the slave-making instinct of *F. sanguinea* have originated?" (see page 244).
   - "As natural selection acts only through the accumulation of slight gradual modifications of structure … " (see page 250).
   - "The British Museum entomologist Mr. Frederick Smith has shown that …" (see page 256).

8. Try rewriting the following sentences to make them clearer and more concise:
   a. "That the mental qualities of animals of the same kind, born in a state of nature, vary much, could be shown by many facts."
   b. "In the bronze cuckoo, the eggs vary greatly in size."
   c. "No complex instinct can possibly be produced through natural selection, except by the slow and gradual accumulation of numerous slight, yet profitable, variations."
   d. "An action, which we ourselves require experience to enable us to perform, when performed by an animal, more especially by a very young one, without experience, and when performed by many individuals in the same way, without their knowing for what purpose it is performed, is usually said to be instinctive."
   e. "It's very clear that instincts are as important as bodily structures for the welfare of every species."
9. Go through the first eight chapters in this volume and list all of the experiments that Charles Darwin himself conducted in preparing the material for this book.

Bibliography

Bates, H. W. 1863. *The Naturalist on the River Amazons*. London.
Darwin, C. 1839. *The Voyage of the Beagle*. London.

9
Hybridism

In this chapter, Darwin tightens the link between varieties and species even further. At the time, many people believed in the independent, divine creation of separate species and in the absolute reproductive isolation between those species. Here, however, Darwin shows that the members of some plant and animal species can be successfully mated with members of other species and still produce *viable offspring*, offspring that are themselves sometimes capable of reproducing. He also shows that although most varieties can be successfully crossed with other varieties of the same species, that is not always the case: some varieties produce no viable offspring when mated with members of other varieties of the same species. Indeed, there are intriguing gradations of infertility even within a variety, for plants that produce several forms of flower. This all provides strong support for Darwin's argument that species and varieties are really just points on a continuum of change and that varieties are species in the making. Species may eventually become reproductively isolated from the members of other species, but they were not created that way. Note that Darwin does not consider the role of behavioral or other "pre-zygotic" (i.e., pre-mating) isolating mechanisms in this chapter; "matings" were forced in all of the examples that he discusses. He is concerned here only with whether or not species and varieties are physiologically capable of cross-mating and producing fertile offspring.

Many naturalists believe that all species have been specially endowed with sterility[1] when crossed (i.e., mated) with individuals of a different species in order to prevent their confusion. This view certainly seems likely, for different species living in the same place could hardly have been kept distinct if they had been able to freely interbreed. The subject is an important one for us, particularly as the sterility of species when first crossed, and that of their hybrid offspring,[2] cannot, as I shall soon show, have been acquired through the preservation of successive, advantageous degrees of sterility. Sterility must instead be an incidental result of some as-yet little understood differences in the reproductive systems of the parent species.

In treating this subject, two classes of fundamentally different facts have generally been confounded: (1) the sterility of species when first cross-mated and (2) the sterility of the hybrids that are produced from successful crosses between members of different species.

---

[1] Here, "sterility" means the inability of gametes to combine successfully and produce viable offspring.

[2] *Hybrids* are offspring from a mating between the members of two different species.

The Readable Darwin. Second Edition. Jan A. Pechenik, Oxford University Press. © Oxford University Press 2023.
DOI: 10.1093/oso/9780197575260.003.0010

The members of all distinct species have, of course, perfectly functional organs of reproduction. Yet, when a member of one species is mated with a member of another species, the mating usually produces either few or no offspring. The reproductive organs of hybrids, on the other hand, are *functionally* impotent, even though the reproductive organs themselves are *structurally* perfect, as revealed by microscopic examination. Similarly, in the case of matings between the members of two distinct species, the two sexual elements (the sperm and the egg) that go to form the embryo are perfect; but in hybrids they are either not developed at all or are developed imperfectly. These distinctions are important when considering the cause of the sterility that is common to the two cases: it clearly must be caused through different mechanisms. The distinction probably has been slurred over by other authors simply because the sterility in both cases has typically been considered a special divine endowment beyond the province of our reasoning powers.

The fact that varieties—that is, different forms that are known or at least believed to be descended from common parents—as well as their mongrel[3] offspring, can be successfully crossed to obtain viable offspring is, according to my theory, also important in considering the sterility of species as it *seems* to make a broad and very clear distinction between varieties and species. I discuss this point later in this chapter.[4]

Degrees of Sterility

Here, Darwin shows that for plants, some crosses between the members of different species produce no offspring, as expected, while some other crosses are fully fertile and also that some crosses result in seed production that is reduced to various degrees. That is, there is a continuum of sterility; sterility is not absolute. Moreover, he argues that although hybrids have been said to have decreased fertility, this is probably due to procedural errors in the experiments conducted: Darwin shows that in many cases both plant and animal hybrids are perfectly fertile for many generations. Thus, Darwin argues, there is no well-defined law that species cannot interbreed. Rather, degrees of sterility between the members of different species vary with circumstance.

---

[3] *Mongrels* are offspring resulting from matings ("crosses") between different varieties of a single species.
[4] Note that Darwin ignores the role of behavioral isolating mechanisms (e.g., female choice and the timing of sexual activity) in promoting or deterring matings between different species or varieties; instead he focuses on human-induced forced matings and post-zygotic interactions. His point was that there are no well-defined and absolute reproductive barriers between species and that species are essentially just varieties that have become increasing different from each other over time.

First let us consider the sterility of species when crossed and of their hybrid off-spring. I hope to convince you that neither sterility after matings between the members of different groups or fertility provides any convincing evidence of whether the two groups represent separate species or are merely varieties of a single species. It is impossible to study the several memoirs and works of those two conscientious and admirable observers, the well-known German botanists Joseph Gottleib Kölreuter and Karl Friedrich von Gärtner, who devoted nearly their entire lives to this subject, without being deeply impressed with the high generality of at least some degree of sterility resulting from matings between members of different species. Kölreuter makes the rule of sterility universal for such matings, but then he cuts the Gordian Knot,[5] for in the 10 cases in which he found two forms that most authors have considered to be distinct species to be quite fertile when mated together, he unhesitatingly ranks them as mere varieties of the same species and not as separate species after all!

Gärtner also makes the rule of sterility universal and accordingly disputes all 10 of Kölreuter's cases of unexpected fertility. But in these and in many other cases, Gärtner is obliged to carefully count the seeds produced after the matings in order to show that there is at least some degree of sterility; for Gärtner, sterility simply means a decline in fertility, not a total failure of reproduction. He always compares the maximum number of seeds produced by two species when first crossed (and the maximum number produced by their hybrid offspring) with the average number produced by both pure parent-species in nature. He makes the same comparison between maximum seed production by the hybrid offspring and the average number of seeds produced from crosses between the pure parents. But a plant, to be reliably hybridized, must be castrated, and, even more importantly, must be isolated from other plants, in order to prevent pollen from being brought to it by insects that have visited flowers on other plants. Unfortunately, nearly all the plants experimented on by Gärtner were potted and kept in a chamber in his house. There is no doubt that such processes often decrease plant fertility. Indeed, Gärtner gives in his table of data about 20 examples of plants that he castrated and then artificially fertilized with their own pollen (i.e., they were self-fertilized), and, excluding certain groups in which that manipulation is difficult to make, half of these 20 plants had their fertility impaired to at least some degree. Moreover, as Gärtner crossed some forms repeatedly with their own pollen, such as the common red and blue pimpernels (*Anagallis arvensis* and *A. coerulea*, which the best botanists rank as mere varieties) (Figure 9.1) and found them to be absolutely sterile, we may doubt whether many species are really as sterile when inter-crossed as he believed. Clearly, Gärtner's data are not easy to interpret, and probably don't show what he thinks they show.

On the one hand, it is clear that the sterility of various species when artificially crossed is very different in degree and graduates away insensibly depending on the species studied and, on the other hand, that the fertility of pure species is very easily

---

[5] Here Darwin refers to a famous Greek legend involving Alexander the Great, in which a seemingly intractable problem is essentially solved by cheating.

Figure 9.1 The scarlet pimpernel (*Anagallis arvensis*).

affected by various circumstances; for all practical purposes, then, it is most difficult to say where perfect fertility ends and sterility begins. I think there can be no better evidence of this than that the two most experienced observers who have ever lived, namely Kölreuter and Gärtner, should have arrived at diametrically opposite conclusions about the status (Are they separate species? Or are they merely varieties of a single species?) of the exact same forms. It is similarly instructive to compare— lack of space prevents me from going into detail here—the evidence advanced by our best botanists on the question of whether certain doubtful forms should be ranked as species or varieties with the evidence from fertility adduced either by evidence from different hybridizers or from evidence provided by the same hybridizer from experiments made during different years. It can thus be shown that neither sterility nor fertility affords any clear distinction between species and varieties. The evidence from this source graduates away, so that tests of sterility are as unclear and as indistinct for defining species as are assessments made from structural and behavioral differences.

Let us consider now the sterility of hybrids in successive generations. Although Gärtner was able to rear some hybrids—carefully guarding them from a cross with either pure parent for six or seven, and in one case for 10 generations—yet he asserts positively that their fertility never increased but instead generally decreased greatly and suddenly, in keeping with his belief in divinely endowed mating barriers between species. But I am quite certain that fertility was diminished in nearly all of these cases by too close inbreeding, based on the many experiments that I myself have conducted and the many facts that I have collected showing, on the one hand, that an occasional cross with a distinct individual or variety increases offspring vigor and fertility, and on the other hand, that very close interbreeding reduces offspring vigor and fertility.

The problem is that hybrids are seldom raised by experimentalists in great numbers, and, as the parent species, or other related hybrids, are generally grown in the same garden, the visits of insects must be carefully prevented during the flowering season to avoid unintended cross pollination. Thus hybrids, if left to themselves, will generally be fertilized during each generation by pollen coming from the same flower; this self-fertilization would probably reduce their fertility, which has already been lessened by their hybrid origin. My conviction that this is the case is strengthened by a remarkable statement repeatedly made by Gärtner: namely, that if even the less fertile hybrids are *artificially* fertilized with hybrid pollen of the same kind, their fertility, notwithstanding the frequent ill effects from manipulation, sometimes increases markedly and goes on increasing. Now I know from my own experience that, in the process of artificial fertilization, pollen is as often taken by chance from the anthers of another flower as from the anthers of the flower that is to be fertilized; a cross between two flowers, though probably often on the same plant, would be thus achieved. Moreover, whenever complicated experiments are in progress, so careful an observer as Gärtner would have castrated his hybrids, and this would have ensured that each generation resulted from a cross using pollen from a different flower, either from the same plant or from another plant of the same hybrid nature. And thus, the strange fact of *increased* fertility in the successive generations of artificially fertilized hybrids, in contrast with those spontaneously self-fertilized, may be accounted for simply from too close inbreeding having been avoided.

Now let us turn to the results arrived at by another most-experienced hybridizer, namely the botanist, the Hon. and Reverend William Herbert. He is just as emphatic in his conclusion that some hybrids are perfectly fertile—as fertile, in fact, as offspring from the pure parent species—as Kölreuter and Gärtner are in their conclusion that some degree of sterility between distinct species is a universal law of nature. Yet Herbert experimented on some of the very same species that Gärtner worked with! The difference in their results may, I think, be explained at least in part by Herbert's greater horticultural skill and by his being able to conduct his studies in hot-houses. Of his many important statements I will give here only one example, namely that "every ovule in a pod of *Crinum capense* fertilized by *C. revolutum*[6] produced a plant, which I never saw to occur in a case of its natural fecundation." So here we have a perfect, or even more than commonly perfect, fertility in a first cross between two distinct plant species.

In keeping with this case of successful cross-fertilization between two species of *Crinum,* I must add that individual plants of certain species of *Lobelia, Verbascum,* and *Passiflora* can easily be fertilized using pollen from plants of a different species, but not by using pollen from the same plant, even though this pollen can be shown to be perfectly sound in that we can successfully fertilize other plants with it. In the plant genera *Hippeastrum* and *Corydalis* (as shown by Professor Hildebrand) and in various orchids (as shown by Mr. Scott and Fritz Müller), all the individuals are in this peculiar condition. So that with some species certain abnormal individuals, and in other species all the individuals, can actually be hybridized with other species much

---

[6] Both are South African herbaceous plants belonging to the genus *Crinum.*

Figure 9.2 *Hippeastrum aulicum.*

more readily than they can be fertilized by pollen from the same individual plant! To give just one example, a bulb of *Hippeastrum aulicum* (Figure 9.2)—the South American "lily of the palace" produced four flowers. Herbert fertilized three of those flowers with their own pollen, and the fourth was subsequently fertilized using pollen of a compound hybrid that was descended from three distinct species. The result was that "the ovaries of the three first flowers soon ceased to grow, and after a few days perished entirely, whereas the pod impregnated by the pollen of the hybrid made vigorous growth and rapid progress to maturity, and bore good seed, which vegetated freely." Mr. Herbert tried similar experiments over many years and always saw the same result. Thus we see on what slight and mysterious causes the lesser or greater fertility of a species sometimes depends.

The practical experiments of horticulturists, though not made with scientific precision, also deserve some notice. It is notorious in how complicated a manner species of *Pelargonium, Fuchsia, Calceolaria, Petunia, Rhododendron,* and many other plants have had to be crossed, and yet many of the resulting hybrids have seeded freely. For example, Herbert asserts that a hybrid from *Calceolaria integrifolia* and *C. plantaginea*[7]—species that are quite dissimilar in general habit—"reproduces itself as perfectly as if it had been a natural species from the mountains of Chili [Chile]." Similarly, I have taken some pains to ascertain the degree of fertility of some of the complex crosses between various species of rhododendrons, and I am assured that many of them are perfectly fertile. Mr. C. Noble, for example, tells me that he raises stocks for grafting from a hybrid between *Rododendron ponticum* and *R. catawbiense,* and that this hybrid "seeds as freely as it is possible to imagine." Indeed, had hybrids when fairly treated always gone on decreasing in fertility with each successive generation, as Gärtner believes to be the case, the fact would have been notorious

---

[7] Members of this genus are commonly referred to as "lady's purse," "slipper flower," "slipper wort," or "pocket book flower."

to nurserymen. Horticulturists raise large beds of the same hybrid and such alone are fairly treated, for by insect-mediated pollen transfer the several individuals are allowed to cross freely with each other, and the injurious influence of close inter-breeding is thus prevented. You may readily convince yourself of the efficiency of insect-mediated pollen transfer by examining the flowers of the more sterile kinds of hybrid rhododendrons, which produce no pollen, for you will find on their stigmas plenty of pollen brought there from other flowers. Clearly, species are *not* universally separated from each other reproductively.

Many fewer careful experiments have been tried with animals than with plants. If the genera of animals are as distinct from each other as are the genera of plants, then we may infer that animals more widely distinct in the scale of nature can be crossed more easily than in the case of plants. Although the hybrids themselves seem more likely to be sterile, it should be borne in mind that, owing to few animals breeding freely under confinement, few good experiments have yet been conducted. For in-stance, the canary-bird has been crossed with nine distinct finch species, but, as not one of those species breeds freely under confinement, we have no right to expect that the first crosses between them and the canary—or that their hybrids—should be per-fectly fertile. Again, with respect to the fertility in successive generations of the more fertile hybrid animals, I hardly know of even a single instance in which two families of the same hybrid have been raised at the same time from different parents to avoid the detrimental effects of close inbreeding[8]. On the contrary, brothers and sisters have usually been crossed in each successive generation, contrary to the constantly repeated admonition against such crosses by every breeder. And with such continued inbreeding it is not at all surprising to find that the inherent sterility in the hybrids should have gone on increasing from generation to generation.

Although I know of hardly any thoroughly well-authenticated cases of perfectly fer-tile hybrid animals, I have reason to believe that hybrids from the genera *Cervulus* (a genus of small Asiatic deer), *Baginalis,* and *Reevesii,* and from the pheasants *Phasianus colchicus* with *P. toraquatus,* are perfectly fertile. The zoologist M. Jean Louis Armand de Quatrefages states that hybrids obtained from matings between two moth species (*Bombyx cynthia* and *B. arrindia*) were proved in Paris to be fertile among themselves for eight generations. It has recently been asserted that two such distinct species as the hare and rabbit, when they can be made to breed together, produce offspring that are highly fertile when crossed with one of the parent species. The hybrids resulting from crosses between the common goose and Chinese geese (*A. cygnoides*; Figure 9.3), species that differ from each other so greatly that they are generally considered to be members of different genera, have often gone on to successfully breed in this country with either pure parent, and in one case they have bred successfully among themselves. This was achieved by the English naturalist Mr. Thomas Campbell Eyton, working with birds, who raised two hybrids from the same parents but from different hatches. And from these two birds he then raised no less than eight hybrids (grandchildren

---

[8] *Inbreeding* refers to matings between close relatives.

Figure 9.3 The head of a male Chinese goose.

of the pure geese) from one nest. In India, however, these cross-bred geese must be even more fertile, for I am assured by two eminently capable judges—Mr. Blyth and Captain Hutton—that whole stocks of these crossed geese are kept in various parts of the country. And as they are successfully kept for profit, where neither pure parent-species exists, they must certainly be highly, or even perfectly, fertile.

With our domesticated animals, the various races are quite fertile when crossed together; yet, in many cases the members of those races are descended from two or more wild species. From this fact we must conclude either that the original parent-species at first produced perfectly fertile hybrids, or that the hybrids subsequently became fertile when reared under domestication. This latter possibility, which was first suggested by the German botanist and zoologist Peter Simon Pallas, seems the most likely and can, in fact, hardly be doubted. Thus I have recently acquired de-cisive evidence that the offspring resulting from crosses between Indian humped cattle and common cattle are perfectly fertile among themselves. Moreover, from the observations of Rütimeyer on their important osteological[9] differences, as well as from those by Mr. Blyth on their differences in such characteristics as habits, voice, and constitution, these two forms must be regarded as good and distinct species. The same remarks may be extended to the two chief races of the pig. Thus **we have two choices: we must either give up the belief in the universal sterility of species when crossed,** or we must look at this sterility in animals not as an indelible, natural charac-teristic but rather as one capable of being removed by domestication.

[9] This refers to the branch of vertebrate anatomy dealing with bones.

Finally, considering all the known facts concerning intercrossing in plants and animals, we may conclude that although some degree of sterility is an extremely general result, both in first crosses between species and in hybrids, it cannot, as far as we can tell at present, be considered as absolutely universal.

Laws Governing the Sterility of First Crosses and Hybrids

Here, Darwin attempts to generalize from studies on both plants and animals about the rules that determine how successful crosses will be in producing offspring and in producing offspring that themselves will be capable of reproducing. He comes up with a number of general rules but shows that there are interesting exceptions to each of them, again suggesting that the reproductive boundaries between species are not absolute. Remarkably, the ability to successfully graft[10] certain plants onto other plants follows similarly complex patterns. And certainly, Darwin argues, no one has ever suggested that the ability to successfully graft one type of plant onto another type of plant is a divinely endowed characteristic.

Let us consider in a little more detail the laws governing the sterility of first crosses between members of different species and of hybrids. Our chief goal will be to see whether or not these laws indicate that species have been specially endowed with sterility to deliberately prevent them from crossing and blending together in utter confusion; in other words, to keep all species distinct. **My major point here is to show that although there do seem to be a number of general laws, there are fascinating exceptions to all of them.**

The following conclusions are drawn mainly from Gärtner's admirable work on hybridization in plants. I have taken great pains to determine how far they apply as well to animals, and, considering how limited our knowledge is in regard to hybrid animals, I have been surprised to find how generally the same rules apply to organisms in both kingdoms.

I have already shown in the previous section that the degree of fertility, both of first crosses between members of different species and of hybrids, graduates in different cases from zero in some to perfect fertility in others. It is surprising in how many curious ways this gradation can be shown. But here I can give only the barest outline of the relevant facts. For example, when pollen from a plant belonging to one family is placed on the stigma[11] of a plant from a distinctly different family, it exerts no more

---

[10] *Grafting* is a horticultural technique in which tissues from one plant are inserted into those of another, providing an opportunity for the two sets of vascular tissues to join together. In this way, many commercially important plants can be propagated in large numbers asexually.

[11] The *stigma* is the organ containing the ovary. See Figure 7.1.

influence than so much inorganic dust. In contrast to this complete sterility between such species, the pollen of different species applied to the stigma of some one species of the same genus[12] yields a perfect gradation in the number of seeds produced, up to nearly complete fertility, or even complete fertility—even, as we have seen, in certain abnormal cases, to an excess of fertility beyond that which the plant's own pollen produces.

And so it is, too, with hybrids themselves: there are some hybrids that have never produced—and probably never would produce, even with the pollen of the pure parents—a single fertile seed. But in some of these cases a first trace of fertility may be detected by the pollen of the pure parent species causing the flower of the hybrid to wither sooner than it otherwise would have done; the early withering of the flower is well known to be a sign of incipient fertilization. Again, there is a gradient: from an extreme degree of sterility resulting from a cross between species, we have self-sterilized hybrids producing more and more seeds up to perfect fertility.

The hybrids raised from two species that are very difficult to cross, and which rarely produce any offspring, are generally very sterile; that is, they produce no viable gametes. **But the difficulty in making a first cross is not always associated with the sterility of the hybrids thus produced.** There are many cases (the plant genus *Verbascum*, for example) in which even though two pure species can be crossed with unusual facility, producing numerous hybrid offspring, those hybrids are remarkably sterile. On the other hand, there are species that, although they can be successfully crossed only very rarely, or only with extreme difficulty, the hybrids, when at last produced, are themselves very fertile. Indeed we sometimes see these two opposite cases even within the same genus, as in in the genus *Dianthus*, a group of about 300 species of flowering plants belonging to the family Caryophyllaceae.[13]

The fertility of first crosses between members of different species also varies considerably among the individuals of those species; sometimes the cross succeeds brilliantly, sometimes only to a degree, and sometimes not at all, depending on the individuals involved. The same is true of crosses among plant hybrids: the degree of fertility often differs greatly even among individuals raised from seeds taken from the same capsule and exposed to the same environmental conditions.

To some extent, the fertility of first crosses, and of the hybrids thus produced, is largely governed by how closely related they are to each other—that is, to their "systematic affinity." This is clearly shown by the fact that hybrids have never been raised between species that systematists rank as belonging to different families, whereas more closely related species can usually be mated with success. **But there are many exceptions to this general truth.** I know of many cases in which we have not been able to successfully mate some closely related species, or in which we have been able to do so only with great difficulty. On the other hand, I also know of cases

---

[12] Species within a single genus are more closely related than are species in different genera within the same family.
[13] This group includes the carnation.

in which the members of very distantly related species have been interbred with ease. Within a single family of plants, for example, there may be one genus—*Dianthus*, for example—in which very many species can be crossed very easily, while in another genus, such as *Silene*,[14] the most persevering efforts have failed to produce even a single hybrid.

We sometimes see this same sort of difference among species within the same genus. For example, the many species in the tobacco genus *Nicotiana* have been more commonly crossed together than the species of almost any other genus of plants; yet Gärtner found that *N. acuminata*, which is not a particularly distinct species, obstinately refused to be fertilized by, or refused to fertilize, no less than eight other species in the same genus. Many similar cases could be given.

No one has been able to identify a kind or an amount of difference in any recognizable character that is sufficient to prevent any two species from successfully crossing. Even some plants that differ greatly in habit and general appearance, and which have strongly marked differences in every part of the flower—and even in the pollen, in the fruit, and in the cotyledons[15]—can be readily crossed. And the same is true for many annual and perennial plants, deciduous and evergreen trees, and plants living in very different habitats and exposed to extremely different climates—these can often be crossed with ease.

Let us now consider *reciprocal crosses*, by which I mean a female of one species being crossed with a male of a different species, and a male of the first species being crossed with a female of the second species—a female donkey (*Equus africanus asinus*) being first crossed with a male horse (a stallion), for example, and then a female horse (a mare) being crossed with a male donkey. Those two species can then be said to have been "reciprocally crossed." **Sometimes it is very easy to make such reciprocal crosses, and sometimes it is extremely difficult—or impossible. Such cases are highly important for us, for they prove that the capacity in any two species to be crossed often has nothing to do with how closely related they are.** Kölreuter has observed a great diversity of results from reciprocal crosses between the same two species in a great number of experiments conducted over many years. For example, he has shown that *Mirabilis jalapa*, the "four o'clock flower" of Peru, can easily be fertilized by the pollen of *M. longiflora* and that the resulting hybrids are reasonably fertile. The reverse, however, is not true: *M. longiflora* cannot be fertilized by pollen taken from *M. jalapa*; indeed he tried that experiment more than 200 times over 8 years, and failed utterly. The French botanist Gustave Thuret has observed the same results in experiments with certain seaweed species, while Gärtner has observed something similar between even more closely related forms such as *Matthiola*[16] *annua* and *M. gilabra*, which most botanists rank as only varieties of a single species. It is also a

---

[14] This is a very large genus of plants (about 700 species); both *Dianthus* and *Silene* are members of the family Caryophyllaceae.

[15] *Cotyledons* are the embryonic first leaves that develop within plants seeds.

[16] *Matthiola* is a genus of flowering plants in the mustard family, which contains about 50 species.

remarkable and well-known fact that the hybrids raised from successful reciprocal crosses, although of course compounded of the very same two species—the one species having first served as the father and then as the mother—generally differ at least somewhat in fertility and sometimes to a very high degree, even though they rarely differ in external appearance.

Gärtner has also shown that some plant species can be easily crossed with other species and that some different species within a given genus have a remarkable power of impressing their likeness on their hybrid offspring. But these two powers do not necessarily go together. For example, certain hybrids, instead of having characteristics intermediate between those of their two parents, always closely resemble just one of them; such hybrids, though resembling so closely just one of their pure parent species, are almost always extremely sterile. Similarly, among hybrids that are structurally intermediate between their parents, exceptional and abnormal individuals are sometimes born that closely resemble just one of their pure parents. Those hybrids are almost always utterly sterile, even when the other hybrids raised from seeds obtained from the same capsule show a considerable degree of fertility. These facts show very clearly how the fertility of a hybrid may have little to do with how much it resembles either of its pure parents.

In summary, we see (1) that when forms that must be considered perfectly good and distinct species are crossed, their fertility graduates from zero to perfect fertility, or even to an excess of fertility under certain conditions, depending on the species involved; (2) that fertility varies considerably *among individuals* within a species; (3) that the degree of fertility is by no means always the same in the first cross and in the hybrids produced from that cross; (4) that the fertility of hybrids is not related to the degree to which they resemble either of their parents; and (5) that the ease of making a first cross between individuals of any two species is not necessarily determined by how closely related those species are, or by the degree to which they resemble each other. This latter point (5) is clearly proven by the differences obtained in reciprocal crosses between individuals of the same two species: depending on which species is used as the father and which is used as the mother, there is generally some difference, and occasionally the widest possible difference, in the likelihood of obtaining offspring. Moreover, the hybrids produced from reciprocal crosses between species often differ in fertility, depending on which species served as the mother and which as the father.

Do these complex and singular rules support the contention that species have been deliberately endowed with sterility to prevent their becoming confounded in nature? I think not. For why should the degree of sterility differ so greatly when different species are crossed, when it should be equally important to keep any two species from blending together? And why should the degree of sterility differ among individuals within a species? And why should some species be easy to cross successfully with some other species and yet produce hybrids that are sterile, while other species are extremely difficult to cross successfully and yet produce hybrids that are quite fertile? Why should there often be so great a difference in the result of a reciprocal cross between members of same two species, depending on which species serves as the mother and which as the father? Why, it may even be asked, has the production of hybrids been permitted at all? To grant to some species the special power of producing hybrids and then to stop their further propagation by causing different degrees of sterility in those hybrids—degrees that are not related to how easy it was to obtain a successful mating between the parents in the first place—seems a very strange arrangement indeed.

It seems quite clear to me from the forgoing facts that the degree of sterility both of first crosses between species and of the hybrids produced from such crosses must be caused by some unknown differences in their reproductive systems, not by any "design" to keep the species separated. Whatever the causes are, they are clearly of a peculiar and limited nature such that, in reciprocal crosses between two species, the male sexual element of the one species will often successfully interact with the female element of the other species, while the male element of that second species cannot successfully fertilize the eggs of the first species.

Let me explain through another example—something called "grafting"[17]—what I mean by sterility being caused by "unknown differences" in the reproductive systems and not by any specially endowed quality. Our ability to successfully graft one plant onto another has no role in determining their subsequent welfare in nature; thus I presume that no one will suppose that this capacity is "specially endowed," but will rather admit that it has something to do with differences in the laws of growth in the two plants. We can sometimes see why one type of tree cannot be grafted onto another from differences in their rates of growth, or in the hardness of their wood, or in the period of the flow or the nature of their sap, and so forth. But in a multitude of cases we can assign no reason whatsoever. Great diversity in the sizes of two plants, or one plant being woody and the other herbaceous, or one being evergreen and the other deciduous,[18] or adaptation to very different climates—none of those things will always prevent the two from being successfully grafted together. As with hybridization, so it is with grafting: the capacity for successful grafting depends on degrees of relatedness, e.g., "systematic affinity." For no one has been able to graft together trees belonging to quite distinct families, whereas closely related species and varieties of the same species can usually be grafted with ease.

But this capacity for grafting, as in hybridization, is by no means absolutely determined by systematic affinity. Although members of many distinct genera within a single family have been successfully grafted together, there are also some species within a single genus that will not take each other on. Indeed, the pear can be grafted far more readily onto the quince, which is ranked as belonging to a different genus,

---

[17] As noted earlier, grafting is a horticultural technique in which tissues from one plant are inserted into those of another, providing an opportunity for the two sets of vascular tissues to join together. In this way, many commercially important plants can be propagated in large numbers asexually.

[18] *Deciduous* plants are trees and shrubs shed their leaves once each year.

than onto the apple, which belongs to the same genus.[19] Even different varieties of the pear take on the quince with different degrees of facility, and the same is true with different varieties of the apricot and peach on certain varieties of the plum.

Just as Gärtner has found that there was sometimes an innate difference in the ability of different *individuals* within the same two species to cross successfully, so the French botanist Sageret believes this to be the case with different individuals of the same two species in being successfully grafted together. Just as the ease of achieving a union is often very far from equal in reciprocal crosses between species, so it sometimes is in grafting: the common gooseberry, for instance, *cannot* be grafted on to the currant, whereas the currant *can* be grafted onto the gooseberry, though admittedly only with difficulty.

We have seen that the sterility of hybrids, which have their reproductive organs in an imperfect condition, is a quite different issue from the difficulty of uniting two pure species, which of course have their reproductive organs perfectly well developed and fully functional. And yet these two distinct classes of cases run to a large extent parallel. Something analogous occurs with grafting. The French botanist André Thouin, for example, found that although three species in the deciduous plant genus *Robinia* that seeded freely when growing attached to their own roots could be grafted easily onto a fourth *Robinia* species, those grafts were unable to produce seed. On the other hand, certain species of trees and shrubs in the genus *Sorbus* (which belong to the rose family Rosaceae, along with the pears, quinces, and apples mentioned earlier) produced twice as much fruit when grafted onto other species as when they were left attached to their own roots. This latter result reminds us of the extraordinary cases of *Hippeastrum, Passiflora* and some other plants, which seeded much more freely when fertilized with the pollen of a different species than when fertilized with pollen from the same plant (Figure 9.2).

Thus we see that there is a certain degree of parallelism in the results from grafting and of the crossing of distinct plant species. And as we must look at the curious and complex laws governing the ease with which trees can be successfully grafted onto each other as depending on some unknown differences in their *vegetative* systems, so I believe that the still more complex laws that must govern the ability to make successful first crosses between the members of different species must also depend on unknown differences in their *reproductive* systems. To some extent these differences are related to degrees of systematic affinity. But the facts that we have here discussed by no means seem to support the idea that the greater or lesser difficulty of either grafting or crossing various species has been a "special endowment" of divine creation.

---

[19] Actually, the quince and the apple are now placed in separate genera; the apple belongs to the genus *Mallus*, while the quince is the sole member of the genus *Cydonia*. Both are members of the rose family, Rosaceae, which also contains pears, cherries, plums, and a variety of other fruits. The family contains at least 3,000 species.

The Origin and Causes of the Sterility of First Crosses and Hybrids

> If sterility is not universal and divinely endowed, then how do we explain it? Here Darwin uses the limited evidence available at the time to try to figure out what causes sterility, how it might have originated, and whether it may have originated through natural selection or in some other way. But the causes ultimately remain, for Darwin, unfathomable since he knew nothing about chromosomes or genes.[20]

Many crosses between different species fail to produce offspring. And of those that do produce offspring, the hybrids are often completely sterile. At one time it seemed likely to me, as it has to others, that this sterility of first crosses and of hybrids might have been something that was slowly acquired through the natural selection of slightly lessened degrees of fertility in each generation, when such variations appeared in certain individuals of one variety when crossed with those of another variety. It would clearly be advantageous to two varieties if they could be kept from blending, for the same reason that when breeders are selecting for two varieties of plant at the same time, they must keep the plants well separated to prevent unwanted crossing.

After mature reflection, however, it now seems to me that in most cases, sterility could not have been achieved through natural selection. For example, take the case of any two species which, when crossed, produced few and sterile offspring. What could favor the survival of those individuals that happened to be endowed in a slightly higher degree with mutual infertility, and which thus approached toward absolute sterility by this one small step? We have already seen, in Chapter 8, that modifications in the structure and fertility of what are now sterile neuter insects probably did in fact come about slowly, over many generations, through natural selection, from an advantage having been thus indirectly gained by the community to which those individuals belonged over other communities of the same species. But an individual animal not belonging to a social community, if rendered slightly sterile when crossed with some other variety, would not thereby gain any advantage for itself and would not indirectly give any advantage to other individuals of the same variety; so nothing would lead to their preservation. The sterility seen among so many crossed species

---

[20] I have omitted much of this section, as Darwin's reasoning was severely limited by the fact that he knew nothing about the mechanisms of inheritance, something that must have been terribly frustrating for him.

must have a basis in something other than natural selection, but we cannot at present say what that basis is.[21]

Let us now look more closely at the factors that likely account for the sterility seen in so many first crosses and in hybrids. In the case of first crosses between the members of different species, the degree of difficulty in achieving a union of gametes and in obtaining offspring apparently depends on several distinct causes. Among plants, sometimes there is something that just physically prevents the male gamete (pollen) from reaching the ovule.[22] Such would be the case, for example, with a plant having a pistil that is too long for the pollen tubes from another plant to reach the ovarium.[23] It has also been observed that when the pollen of one species is deliberately placed on the stigma of a distantly related species, although the pollen tubes protrude, they do not—for some reason—succeed in penetrating the surface of the stigma.

Alternatively, the male gamete may reach the female gamete but for unknown reasons be incapable of causing an embryo to be developed, as seems to have been the case in some of Gustave Thuret's experiments on seaweeds. Lastly, an embryo may start to develop after fertilization but then perish shortly afterward. This situation has not been sufficiently studied, but I believe, based on observations communicated to me by Mr. Edward Hewitt, someone who has had a great deal of experience in hybridizing both pheasants and fowls, that the early death of the embryo is a very frequent cause of sterility in first crosses. Indeed, Mr. Salter has recently reported the results of his studies on about 500 eggs produced from various crosses between three species of birds in the genus *Gallus*[24] and their hybrids. Most of the eggs he examined had been fertilized. However, in many of those eggs the embryos had developed only partially before perishing, and, in many other egg, although the embryos had matured nicely, the young chickens had then been unable to break through the egg shell and had died without hatching. Of the few chickens that successfully hatched, more than 80% died within the first few days or in later weeks, "without any obvious cause, apparently from mere inability to live." Thus of the 500 eggs that he worked with, only 12 chickens could be reared.

And so it is with plants, with hybridized embryos probably often perishing in a similar manner. Certainly we know that even when fertilization is successful, hybrids

---

[21] We now know that the sterility of hybrids resulting from crosses between different species (crosses between horses and donkeys, for example) can be caused by the two species having different numbers of chromosomes. In Darwin's day, of course, chromosomes had not yet been discovered, nor had DNA. The term "chromosome" was first used to describe these cellular components in 1888, by the German anatomist Henrich von Waldyer-Hartz, and were finally recognized as the vectors of heredity in 1902, through the experiments of Theodor Boveri.

[22] The ovule is the part of the female plant that contains the egg and eventually becomes the seed, after the egg is fertilized (see Figure 7.1).

[23] When pollen lands on a stigma, it must "tunnel" into the stigma in order to reach the ovary; if these "pollen tubes" do not form properly the pollen never reaches the eggs.

[24] The genus *Gallus* includes our domesticated chickens, which are primarily descended from the wild red junglefowl (*Gallus gallus*) of India.

raised from very distinct species are sometimes weak and dwarfed and perish at an early age. Indeed, the German botanist Max Ernest Wichura has recently given some striking cases with hybrid willows. It may also be worth noting here that in some cases of parthenogenesis,[25] the embryos of silk moths from eggs that had not been fertilized pass through their early development but then perish, just like the embryos produced from a cross between distinct species. The cause is as yet unknown.

Indeed, there are many facts about the sterility of hybrids that are still beyond our understanding. For instance, how can we explain the unequal fertility of hybrids produced from reciprocal crosses, or the increased sterility in those hybrids that occasionally and exceptionally resemble closely either pure parent? All of these facts seem connected by some common but unknown bond, something that must be related to some great principle of life that remains to be uncovered.

Reciprocal Dimorphism and Trimorphism

Some plant species produce two or three distinctly different forms of flower within a single variety. Remarkably, fertility varies with the form of flower used to provide the pollen or egg, showing, in fact, just the sort of variability in fertility that we see among varieties and species. Note that this section was not included in the first edition of *The Origin*, or even in the third edition; it appears for the first time in the fourth edition, which appeared June 1866.

This topic will throw some light on the degree of sterility brought on by hybridism. Several plants belonging to distinctly different orders present two distinct forms of flower that exist in about equal numbers and that differ only in their reproductive organs: one form of flower has a long pistil with short stamens, while the other has a short pistil and long stamens (see Figure 7.1); the two also differ in the size of the pollen grains produced. With "trimorphic" species, there are three flower forms that likewise differ in the lengths of the pistils and stamens, in the size and color of the pollen grains, and in some other respects as well; as there are two sets of stamens in each of the three forms, the three forms possess altogether six sets of stamens and three kinds of pistils (Figure 9.4). These organs are proportioned in length relative to each other such that half of the stamens in two of the forms stand at the same level as the stigma of the third form. Now I have shown—and my results have been since confirmed by other observers—that in order to obtain full fertility with these plants, one must be sure that the stigma of the one form should be fertilized by

---

[25] Parthenogenesis is a form of asexual reproduction, occurring in the absence of fertilization.

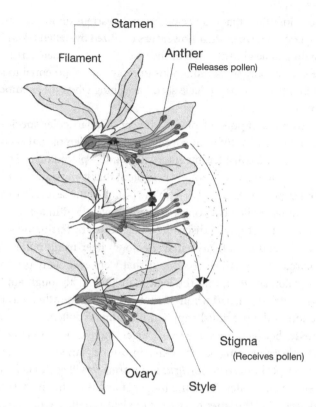

Figure 9.4 Trimorphic plants. Each species has three different forms that differ in the lengths of their pistils and stamens and in some other characteristics as well, including the size and color of their pollen grains. The stamens come in three different lengths (short, medium, and tall), but each flower has stamens of only two of those lengths: the flower at the top of the figure has tall and medium length stamens, the middle figure has tall and short stamens, and the bottom figure shows a flower with short and medium-length stamens. The stigma, which receives the pollen from other flowers, also comes in three lengths. As shown in the diagram, the stigma of each flower must be fertilized with pollen derived from a stigma of similar length from another flower. Drawing by Ardea Thurston-Shane.

pollen taken from the stamens of corresponding height in another form; i.e., the pollen from a long stamen must contact a long pistil on another flower, and the pollen from a short stamen must contact a short pistil on another flower. Thus with dimorphic plant species, two unions (which I will call "legitimate" unions) are fully fertile, and two other unions (which I will call "illegitimate" unions) are infertile to some degree. With trimorphic species, six unions will be legitimate, or fully fertile, and 12 will be illegitimate, or more or less infertile.

The degree of infertility that may be observed in various dimorphic and trimorphic plants of a single species when their flowers are fertilized by pollen taken from stamens not corresponding in height with that of the pistil (i.e., illegitimate fertilization) differs much in extent, from only small reductions in reproductive potential in some cases to utter sterility in others, just as we have seen happening when the members of distinct species are crossed.

It is well known that if pollen from one flower of a particular species is placed on the stigma of a flower of a different species, and the pollen of that second species is afterward—even after a considerable interval of time—placed on the same stigma of the same flower, its action is so strongly prepotent that it generally annihilates the effect of the foreign pollen. And so it is with the pollen of the several forms of the same species, for legitimate pollen is strongly prepotent over illegitimate pollen when both are placed on the same stigma. I discovered this myself by fertilizing several flowers, first illegitimately, and then legitimately 24 hours later using pollen taken from a peculiarly colored variety. All the seedlings that were eventually produced had that same peculiar color, showing that the legitimate pollen, although applied 24 hours after the illegitimate pollen, had wholly destroyed or in some other way prevented the successful action of the previously applied illegitimate pollen. Again, as in making reciprocal crosses between the same two species, there is occasionally a great difference in results. The same thing occurs with trimorphic plants[26]: for instance, the mid-styled form of purple loosestrife (*Lythrum salicaria*) was illegitimately fertilized with the greatest ease using pollen from the longer stamens of the short-styled form and yielded many seeds; but the latter form did not yield a single seed when fertilized with pollen from the longer stamens of the mid-style form.

In all these respects, the different forms of the same undoubted species of dimorphic or trimorphic plant when illegitimately united behave in exactly the same manner as do two distinct species when crossed. This led me to carefully observe, for four years, many seedlings that I raised from several illegitimate unions. The chief result is that these "illegitimate" plants were never fully fertile. It is possible to raise both long-styled and short-styled illegitimate plants from dimorphic species and to raise all three illegitimate forms from trimorphic plants. These can then be properly united in a legitimate manner (i.e., the pollen from a long stamen is used on a plant with a long pistil, and the pollen from a short stamen is used on a plant with a short pistil). When this is done, there is no apparent reason why they should not yield as many seeds as did their parents when legitimately fertilized. But such is not the case! In my experiments, they were all infertile to various degrees, some being so utterly and incurably sterile that they did not yield a single seed or even a seed-capsule[27] during my four seasons of study.

---

[26] With trimorphic plants, the reproductive systems come in three different forms, including three differences in the lengths of their pistils (the female part of the flower) and two differences in the lengths of their stamens (the male, pollen-producing part of the flower; see Figure 9.4).

[27] Seed capsules are a type of simple, dry fruit produced by many species of flowering plants.

The sterility of these illegitimate plants of the same species when united with each other in a legitimate manner is exactly comparable to the sterility of hybrids when crossed among themselves. If, on the other hand, a hybrid is crossed with either pure parent species, the degree of sterility is usually reduced considerably. The same holds true when an illegitimate plant is fertilized using pollen from a legitimate plant. Similarly, in the same way that the degree of sterility of hybrids does not always correspond with the degree of difficulty in making the first cross between the two parent species, so it is that the sterility of certain illegitimate plants was unusually great while the sterility of the union from which they were derived was not great. Moreover, the degree of sterility varies from seed to seed with hybrids raised from the same seed capsule, and the same is markedly true with illegitimate plants. Last, many hybrids are profuse and persistent flowerers, while other, more sterile hybrids, produce few flowers and are weak, miserable dwarfs; exactly similar cases occur with the illegitimate offspring of various dimorphic and trimorphic plants.

Taken together, there is a marked parallel between the character and behavior of illegitimate plants and the hybrids resulting from crosses between members of different species. It is hardly an exaggeration to maintain that illegitimate plants are essentially hybrids, produced within the limits of the same species by the improper union of certain flower forms, while ordinary hybrids are produced from an improper union between so-called distinct species.

We have also seen previously that there is the closest similarity in all respects between first illegitimate unions and first crosses between the members of distinct species. Allow me to illustrate. Suppose that a botanist found two well-marked varieties of the long-styled form of the trimorphic plant *Lythrum salicaria* (purple loosestrife, as mentioned earlier)—and such varieties do in fact exist—and that he wanted to determine by crossing whether they were actually members of different species. But to make the case sure, he would then raise the plants grown from his supposedly hybridized seed, and he would find that the seedlings were miserably dwarfed and utterly sterile and that they behaved in all other respects like ordinary hybrids. He might then maintain that he had actually proven that his two varieties were as good and as distinct species as any in the world. But of course he would be completely mistaken in taking this view.

These facts about degrees of sterility in dimorphic and trimorphic plants are important because, for one thing, they show us that the physiological test of lessened fertility, both in first crosses and in hybrids, is no safe criterion for defining species. They also suggest that there is some unknown bond that connects the infertility of illegitimate unions with that of their illegitimate offspring, leading us to extend the same view to first crosses and hybrids. These facts also show that there can be two or three forms of the same species that are sterile when united in certain ways, even though they don't differ at all in either structure or constitution. Remember that it is the union of the sexual elements of individuals of the same form—two long-styled forms, for example—that results in sterility, while it is the union of the sexual elements proper to two distinct forms that is fertile. Hence the case appears at first sight exactly the reverse of what occurs in the ordinary unions of the individuals of the same species and with crosses between distinct species. It is, however, doubtful whether this is really so; but I will not enlarge on this obscure subject here. We may infer, however, from our consideration of dimorphic and trimorphic plants, that the sterility of distinct species when crossed (and of their hybrid offspring) depends exclusively on the nature of their sexual elements[28] and not on any difference in their structure or general constitution. We are led to the same conclusion by considering reciprocal crosses, in which the male of one species cannot be united, or can be united only with great difficulty, with the female of a second species, while the converse cross can be achieved with ease. That excellent observer Gärtner similarly concluded that species when crossed are sterile owing to differences confined to aspects of their reproductive systems. We have yet to understand just what those aspects are.

The Fertility of Varieties When Crossed, and of Their Mongrel Offspring, Is Not Universal

> Although varieties of any particular species can usually be successfully mated with each other, resulting in viable offspring that are themselves capable of eventually mating, that turns out to not always be the case. Here Darwin provides several such exceptions, all from botanical experiments. In a number of cases, for example, it has proven surprisingly difficult to obtain viable offspring from matings between different varieties of the same species. And different varieties of a given species can also differ in the degree to which they can be successfully mated with members of a different species.

It may be strongly argued that there *must* be some essential distinction between species and varieties, inasmuch as varieties, however much they may differ from each other in external appearance, may usually be crossed with ease and yield perfectly fertile offspring. I fully admit that this is generally the case. But the evidence often involves circular reasoning: looking at varieties produced in nature, if two forms previously said to be varieties of a species are found to be sterile with each other to any degree, they are at once ranked by most naturalists as distinct species! For example, the blue and red pimpernel (flowers belonging to the genus *Anagallis*; Figure 9.4), which most botanists consider to be separate varieties, are said by Gärtner to be quite sterile when crossed; thus he now ranks them with conviction as undoubted separate

---

[28] Darwin is pretty much on target here; without really knowing it, he is really talking about genetic differences.

species of that genus.[29] With that sort of circular reasoning, the fertility of all varieties produced in nature will assuredly have to be granted.

Let us turn now to varieties that we believe were produced under domestication. The perfect fertility of so many domestic races when crossed—as with the various breeds of pigeons (see Chapter 1), or dogs, or cabbages, for example—is a remarkable fact, considering how much they differ in physical appearance. This is especially remarkable considering how many species we know of that are utterly sterile when crossed even when they resemble each other quite closely. The main issue here is not, I think, why domestic varieties have not become mutually infertile when crossed, but why this has so generally occurred with *natural* varieties, as soon as they have been permanently modified in a sufficient degree to take rank as separate species. We are far from understanding the cause, something that should not be particularly surprising, seeing how profoundly ignorant we are in regard to the normal and abnormal action of the reproductive system.

Although varieties of a given species are usually fertile when intercrossed, it is impossible to resist the evidence of a certain amount of sterility occurring in at least a few cases, which I will briefly summarize. The evidence for sterility here is at least as good as that from which we believe in the sterility of crosses between a multitude of species. It is also worth noting that the evidence I am about to present is derived from hostile witnesses, who in all other cases consider fertility and sterility to be safe criteria for defining species.

First, for several years, Carl Friedrich von Gärtner kept in his garden, growing near to each other, two varieties of corn (also known as maize): a dwarf kind of corn with yellow seeds and a tall variety with red seeds. Both have separated sexes, and they never naturally crossed. He then attempted to fertilize 13 flowers of the one kind with pollen taken from the other; only a single head produced any seeds, and this one head produced only 5 grains. (Note that the manipulation involved in obtaining the pollen and effecting fertilization could not have been injurious here, as the plants have separated sexes.) No one, I believe, has ever thought these varieties of maize to be distinct species, and it is important to note that the hybrid plants thus raised were themselves perfectly fertile. Thus even Gärtner did not consider the two varieties to actually be distinct species, despite the massive failure of 12 of the 13 crosses and the limited success of the thirteenth. Clearly, varieties of a given species are not always interfertile.

Similarly, Girou de Buzareingues, a well-known French expert in plant physiology and agronomy, crossed three varieties of gourd, which, like the corn plant, has separate sexes. He asserts, in his 1833 paper, that the degree of their mutual fertilization was less as the differences between them were greater. How far we may trust these experiments I know not; but the forms he experimented on are ranked by the French botanical authority Augustin Sagaret (who mainly bases his classification on the test

---

[29] Indeed, the blue pimpernel plant is now placed in a different genus, the genus *Lysimachia*.

of infertility) as varieties, and the French botanist Charles Victor Naudin has come to the same conclusion. So why does the degree of fertility among acknowledged varieties vary with the degree of dissimilarity in form?

The following case is far more remarkable and seems at first incredible. But it is the result of an astonishing number of experiments made during many years on nine figwort species (genus *Verbascum*) by so good an observer, and so hostile a witness, as Gärtner[30]. He found that the yellow and white varieties when crossed produced less seed than when the similarly colored varieties of the same species were crossed. Moreover, he asserts that when yellow and white varieties of one species were crossed with yellow and white varieties of a different species, more seed was produced by the crosses between the similarly colored flowers than between those which were differently colored. Mr. Scott has also experimented with the various species and varieties of *Verbascum*; although he was unable to confirm Gärtner's results on the crossing of the distinct species, he found that the differently colored varieties of the same species did indeed yield fewer seeds when crossed than did crosses between the similarly colored varieties, in the proportion of 86 to 100. And yet these varieties differ only in the color of their flowers. Indeed, one variety can sometimes be raised from the seed of another.

The German botanist Joseph Kölreuter, whose accuracy has been confirmed by every subsequent observer, has experimented on five forms of tobacco plant (genus *Nicotiana*) that are commonly reputed to be varieties and which he tested by the severest trial—namely by reciprocal crosses. He found their mongrel offspring to be perfectly fertile. But one of these five varieties, when used either as the father or as the mother in being crossed with a different species in the same genus (*Nicotiana glutinosa*), always yielded hybrids that were less sterile than those produced when the four other varieties were crossed with *N. glutinosa*. Thus the reproductive system of this one variety must have differed in some manner from that of the others. In any event, he has proven the remarkable fact that one particular variety of common tobacco was more fertile than four other varieties when crossed with a widely distinct species of the same genus!

From these facts one can clearly no longer assert that varieties of any given species are always fertile when crossed. We have, of course, great difficulty in ascertaining the infertility of varieties in nature, for a supposed variety, if proved to be infertile to any degree, would almost universally be ranked as a distinct species. It does seem, however, that fertility does not constitute a fundamental distinction between varieties and species. The general sterility of crossed species may safely be looked at, not as a special acquirement or endowment, but rather as being incidental and caused by changes of an unknown nature in their sexual elements.

---

[30] Gärtner crossed more than 1,000 flowers in his experiments and documented the results over a period of 18 years, counting an unimaginable number of seeds in the process!

Hybrids and Mongrels Compared, Independently of Their Fertility

As defined here, "hybrids" result from successful crosses between individuals of different species, while "mongrels" result from crosses between different varieties of a single species. In this, the final subsection of Chapter 9, Darwin presents evidence that there are no consistently clear distinctions between the performance of hybrids and mongrels. The evidence adds additional support to the idea that different species originated as varieties within an ancestral species and that species are basically just especially well-defined varieties.

Independently of the question of fertility, the offspring of species and of varieties when crossed may be compared in several other respects. Gärtner, who so strongly wished to be able to draw a distinct line between species and varieties, could find very few, and, as it seems to me, quite unimportant differences between the so-called hybrid offspring of species and the so-called mongrel offspring of varieties. Indeed, those differences agree most closely in many important respects.

I shall here discuss this subject with extreme brevity. The most important distinction is that, in the first generation, the offspring of crossed varieties—so-called mongrels—vary more than the offspring of crossed species (hybrids); but Gärtner admits that hybrids derived from species that have been cultivated over many generations are often variable in the first generation, and I have myself seen striking instances of this fact. Gärtner further admits that hybrids between very closely allied species are more variable than those from very distinct species, and this shows that the difference in the degree of variability graduates away. When mongrels and the more fertile hybrids are propagated for several generations, an extreme amount of variability in the offspring in both cases is notorious; but in some few instances that have been reported, both hybrids and mongrels long retain a uniform character. The variability, however, in the successive generations of mongrels is, perhaps, greater than it is in hybrids.

Continuing our comparison of mongrels and hybrids, Gärtner states that mongrels are more liable than hybrids to revert to either parent form; but this, if it be true, is certainly only a difference in degree. Moreover, Gärtner expressly states that the hybrids from long-cultivated plants are more subject to reversion than hybrids from species in their natural state. This probably explains the singular difference in the results arrived at by different observers. Thus the German botanist Max Ernest Wichura, who experimented on uncultivated species of willows, doubts whether hybrids ever revert to their parent forms in successive generations, while the French botanist Charles Victor Naudin, who, on the other hand, experimented chiefly on cultivated plants, insists in the strongest terms on the almost universal tendency to reversion in hybrids. Gärtner further states that when any two species, even when most closely allied to each other, are crossed with a third species, the hybrids are widely different from each other; but if two very distinct varieties of one species are crossed with another species, the hybrids do not differ much. But this conclusion, as far as I can make out, is founded on a single experiment and seems directly opposed to the results of several experiments made some years ago by Kölreuter.

These trivial differences are the *only* differences that Gärtner has been able to point out between hybrid and mongrel plants. On the other hand, the degrees and kinds of resemblance in mongrels and in hybrids to their respective parents—more especially in hybrids produced from nearly related species—follow, according to Gärtner, the same laws. When two species are crossed, one of the species sometimes has a prepotent power of impressing its likeness on the hybrid. So I believe it to be with varieties of plants, and, with animals, one variety certainly often has this prepotent power over another variety. Hybrid plants produced from a reciprocal cross generally resemble each other closely, and so it is with mongrel plants from a reciprocal cross. Both hybrids and mongrels can eventually be reduced to either pure parent form by repeated crosses in successive generations with either parent.[31]

These several remarks apparently also apply to animals, but the subject is here much more complicated, partly owing to the existence of secondary sexual characters, but more especially owing to a prepotency in transmitting likeness running more strongly in one sex than in the other, both when one species is crossed with another species and when one variety is crossed with another variety. For instance, I think those authors are correct when they insist that the donkey (the "ass," *Equus africanus asinus*) has a prepotent power over the horse, so that both the mule and the hinny[32] resemble the donkey more closely than the horse. But this prepotency runs more strongly in the male donkey than in the female donkey, so that the mule, which results from mating a male donkey with a mare, is more like a donkey than is the hinny, which results from crossing a female donkey with a stallion (Figure 9.5).

Much stress has been laid by some authors on the supposed fact that it is only with mongrels that the offspring are not intermediate in character, but instead closely resemble one of their parents; but this also occurs sometimes with hybrids, although I admit that this occurs much less frequently than with mongrels. Looking to the cases that I have collected of cross-bred animals closely resembling one of the parents, the

---

[31] Here Darwin is talking about what we now call "dominance."

[32] A mule results from a cross between a female horse (a mare) and a male donkey, while a hinny results from a cross between a male horse (a stallion) and a female donkey. Both mules and hinnys are generally sterile because their parents have different numbers of chromosomes. Donkeys, by the way, are basically domesticated horses; both belong to the same family, *Equus*.

Figure 9.5 (A) Donkey. (B) Horse. (C) Mule. (D) Hinny.

resemblances seem chiefly confined to characters almost monstrous in their nature and which have suddenly appeared—such as albinism, melanism, deficiency of tail or horns, or additional fingers and toes; they do not relate to characters that have been slowly acquired through selection. A tendency to sudden reversions to the perfect character of either parent would, also, be much more likely to occur with mongrels, which are descended from varieties often suddenly produced and semi-monstrous in character, than with hybrids, which are descended from species slowly and are naturally produced. On the whole, I entirely agree with Dr. Prosper Lucas, an expert in studies of heredity, who, after arranging an enormous body of facts with respect to animals, comes to the conclusion that the laws of resemblance of the child to its parents are the same whether the two parents differ little or much from each other—namely, in the union of individuals of the same variety, or of different varieties, or of distinct species.

Independently of the question of fertility and sterility, in all other respects there seems to be a general and close similarity in the offspring of crossed species, and of crossed varieties. If we look at species as having been specially created, and at varieties as having been produced by secondary laws, this similarity would be an astonishing fact. But it harmonizes perfectly with the view that there is no essential distinction between species and varieties.

Summary

First crosses between forms that are sufficiently distinct to be ranked as species, and their hybrids, are very generally, but not universally, sterile. Sterility varies widely in degree, and is often so slight that the most careful experimentalists have arrived by this test at diametrically opposite conclusions in ranking forms as being either separate species or varieties of a single species. The degree of sterility is innately variable among individuals of the same species and is eminently susceptible to the action of favorable and unfavorable conditions. The degree of sterility does not strictly follow systematic affinity but is governed by several curious and complex laws. It is generally different—and sometimes widely different—in reciprocal crosses between the same two species. It is not always equal in degree in a first cross and in the hybrids produced from this cross.

In the same manner as in grafting trees, where the ability of one species or variety to take on another depends on differences, generally of an unknown nature, in their vegetative systems, so in crossing, the greater or less facility of one species to unite with another depends on unknown differences in their reproductive systems. There is no more reason to think that species have been specially endowed with various degrees of sterility to prevent their crossing and blending in nature than to think that trees have been specially endowed with various and somewhat analogous degrees of difficulty in being grafted together in order to prevent their inarching[33] in our forests.

The sterility of first crosses and of their hybrid progeny has not been acquired through natural selection. In the case of first crosses it seems to depend on several circumstances, in some instances chiefly on the early death of the embryo. In the case of hybrids, it apparently depends on their whole organization having been disturbed by being compounded from two distinct forms, the sterility being closely allied to that which so frequently affects pure species when exposed to new and unnatural conditions of life. He who will explain these latter cases will be able to explain the sterility of hybrids.

The facts given on the sterility of the illegitimate unions of dimorphic and trimorphic plants and of their illegitimate progeny perhaps render it probable that some unknown bond in all cases connects the degree of fertility of first unions with that of their offspring. The consideration of these facts on dimorphism, as well as of the results of reciprocal crosses, clearly leads to the conclusion that the primary cause of the sterility of crossed species is confined to differences in their sexual elements. But why, in the case of distinct species, the sexual elements should so generally have become more or less modified, leading to their mutual infertility, we do not know; but it seems to stand in some close relation to species having been exposed for long periods of time to nearly uniform conditions of life.

---

[33] "Inarching" is a type of grafting in which two plants are grafted together while remaining on their own roots; one plant is later severed from its roots after the graft has been successful.

It is not surprising that the difficulty in crossing any two species, and the sterility of their hybrid offspring, should in most cases correspond, even if due to distinct causes, for both depend on the amount of difference between the species that are crossed. Nor is it surprising that the facility of effecting a first cross and the fertility of the hybrids thus produced, and the capacity of being grafted together—though this latter capacity evidently depends on widely different circumstances—should all run, to a certain extent, in parallel with the systematic affinity of the forms subjected to experiment; for systematic affinity includes resemblances of all kinds. First crosses between forms known to be varieties, or sufficiently alike to be considered as varieties, and their mongrel offspring, are very generally but not, as is so often stated, invariably fertile. Nor is this almost universal and perfect fertility surprising when it is remembered how liable we are to argue in a circle with respect to varieties in a state of nature and when we remember that the greater number of varieties have been produced under domestication by the selection of mere external differences.

Independently of the question of fertility, in all other respects there is the closest general resemblance between hybrids and mongrels: in their variability, in their power of absorbing each other by repeated crosses, and in their inheritance of characters from both parent forms. Finally, then, although we are as ignorant of the precise cause of the sterility of first crosses and of hybrids as we are about why animals and plants removed from their natural conditions become sterile, yet the facts given in this chapter seem fully consistent with the belief that species originally existed as varieties.

Key Issues to Talk and Write About

1. What are hybrids? What are the differences between a species, a variety, a hybrid, and a "mongrel"?
2. As Darwin points out, many people believed that the members of different species were specially created to not be able to mate with each other and produce offspring. And yet the males of some species can be mated with the females of other species and still produce viable offspring. Summarize one example that Darwin gives of this finding for plants that you find especially interesting. What are some of the difficulties that Darwin brings up in interpreting the results of such studies?
3. What do we now understand about the genetic mechanisms that cause sterility, things that Darwin knew nothing about? What do we now understand about why some plant grafts produce viable seeds and fruits, whereas other grafts do not?
4. Which of the experiments described in this chapter were actually conducted by Darwin?
5. The names Kölreuter and Gärtner appear many times in this chapter. Find out three interesting things about the lives of one of these botanists.

10

On the Imperfection of the Geological Record

Throughout *The Origin of Species*, Darwin has been arguing that all modern species of animals and plants are descended from other species, many of which looked very different from the descendants that are with us today. If he is correct, then why don't we see all the intermediate stages preserved as fossils? Does the geological evidence support his ideas about evolution? This is the issue that Darwin deals with in this and the following chapter. In previous chapters, we have seen that Darwin was extremely knowledgeable about the natural history, distribution, and behavior of both plants and animals. Here we see that he was also expert in geology. While a student at Cambridge, he had accompanied one of the founding fathers of geology, Professor Adam Sedgwick, on a geological tour of northern Wales, and he later became an avid reader of Charles Lyell's geological works while aboard the Beagle.

In this chapter, he makes the case that we can't expect to see many intermediate stages in the fossil record because that record is very incomplete. Indeed, he argues that fossils are expected to form—and persist—only under very unusual circumstances that occur sporadically. Note that the lack of fossils that were intermediate between those of whales and certain land mammals had long been used as evidence against evolution; however, such intermediate stages were finally discovered in the early 1980s. Note also that, in Darwin's day, there was no method for determining how old a fossil was. Geologists knew which sedimentary layers the fossil was found in and tried to estimate the time taken to form such thick surrounding layers of sediment. So they had some idea of the relative ages of the different sedimentary layers and thus of the fossils within them, but not the actual age.[1]

In Chapter 6, I listed the chief objections that might be justly brought against the views that I have maintained in this book. Most of them have now been discussed. One—namely, the physical distinctness of different species and their not being blended together by innumerable transitional links—is a very obvious difficulty. I explained why such transitional links are not now commonly found, even though present circumstances— with extensive and continuous areas of land with gradually changing physical conditions—might be expected to favor such transitions. In

---

[1] Modern dating techniques based on decay rates of radioactive elements only started being developed in the 1930s.

The Readable Darwin. Second Edition. Jan A. Pechenik, Oxford University Press. © Oxford University Press 2023.
DOI: 10.1093/oso/9780197575260.003.0011

Figure 10.1 A cliff showing distinct geological strata: layers of sedimentary or igneous rock that have distinct, defining characteristics.

earlier chapters I tried to show that the lives of each species depend more on their interactions with other life forms than on climate, and therefore that the factors which really govern the suitability of different regions as habitats do not in fact graduate away quite so insensibly as heat or moisture do. I also explained that intermediate varieties, because they are rarer than the forms that they connect, will generally be overwhelmed and exterminated during the course of further modification and improvement.

The main reason, however, that innumerable intermediate links are not now encountered everywhere throughout nature relates to the very process of natural selection, through which new varieties continually supplant their parent forms over long periods of time. But in the same proportion as this process of extermination of various forms has acted on an enormous scale, so must the number of intermediate varieties that have formerly existed be truly enormous. Why, then, do we not see such intermediate links appearing as fossils in every geological formation and every stratum[2] (Figure 10.1)? Geology assuredly does not reveal any such finely graduated organic chain of links between now existing species; this is perhaps the most obvious and serious objection that can be brought against my theory of evolution by natural selection.

---

[2] A geological formation is made of distinctive layers, called *strata*, that share a variety of distinctive physical properties, such as color, texture, grain size, and composition. The conspicuousness of different formations is what allows geologists to correlate strata in distant locations.

Figure 10.2 A tapir. Tapirs are herbivorous mammals. The adults weigh 500–800 lbs. Note the prehensile trunk, which they use to grab branches for ingesting leaves or fruit. Tapirs are mostly found in the forests and grasslands of Southeast Asia and Central and South America, although one species lives high in the South American Andes Mountains.

The explanation lies, I think, in the extreme imperfection of the geological record. Let me explain. First of all, we must always remember what those intermediate forms must have looked like, if my theory is correct. I have found it difficult when looking at any two species to avoid picturing to myself forms that are directly intermediate between them. But this is a wholly false way of thinking: rather, **we should be looking for forms that are intermediate between each species and a common but unknown ancestor.** This ancestor will generally have differed in some respects from all of its modified descendants.

Consider our various pigeon breeds, for example, as described in Chapter 1. The fantail and pouter pigeons are both descended from the rock pigeon, *Columba livia*. If we possessed all the intermediate varieties that have ever existed between these two derived breeds, we should have an extremely close series between them and the ancestral rock pigeon. But we should not expect to see any varieties that are directly intermediate between the fantail and the pouter—none, for instance, combining a tail somewhat expanded (as in today's fantail pigeon) with a crop that is somewhat enlarged (as in today's pouter; see Figures 1.2 and 1.3C). In fact, these two breeds have become so greatly modified over many hundreds of generations that if we had no evidence regarding their origins it would not have been possible, by simply comparing their structures with that of the rock pigeon, to have determined whether they had descended from the rock pigeon or from some related form.

And so it is with natural species. If we consider forms that are very distinct—horses and tapirs[3] (Figure 10.2) for example—we have no reason to suppose that there have

---

[3] Tapirs are large herbivorous mammals that live in the jungles and forests of South America, Central America, and Southeast Asia.

ever been any links directly intermediate between them; rather, there must have been forms between each and some unknown common parent. That common parent will have had many characteristics of both modern horses and tapirs, but in some points of structure they may have differed considerably from both, even perhaps more than horses and tapirs now differ from each other. Thus, in all such cases, we should not be able to recognize the parent form of any two or more species even if we carefully compared the structure of the parent with that of its modified descendants unless, at the same time, we had a nearly perfect chain of all the intermediate links between the two present forms.

On the other hand, my theory does not rule out the possibility that one of two living forms might indeed have descended from the other. A horse, for example, might have descended directly from a tapir; if so, then direct intermediate links must have existed between them. But such a case would imply that one form had remained completely unaltered for an extremely long time while its descendants had undergone a vast amount of change during the same period; the principle of competition between organism and organism, and even between child and parent, will make this a very rare and unlikely event: in all cases the new and improved forms of life, with greater fitness, tend to supplant the old and unimproved forms.[4]

According to the theory of natural selection, all living species within any genus have been connected with the parent species of that genus by differences not greater than those we see today between the natural and domesticated varieties of the same species. And those parent species, which are now generally extinct, were, in their turn, similarly connected, long, long ago, with even more ancient forms—and so on backward through time, always converging to the common ancestor of each great class. Thus the number of intermediate and transitional links between all living and extinct species must have been inconceivably great. But surely then, if my theory is correct, such organisms must indeed have lived at one time upon the Earth.

On the Passage of Time, as Inferred from the Rate of Deposition and Extent of Denudation of Sediments

Aside from our not finding fossil remains of such infinitely numerous connecting links between species, between genera, between families, and so forth, one might also object that there cannot possibly have been enough time for so great an amount of organic change to have occurred, as such changes happen only slowly. But unless you have had some practical experience as a geologist, your mind will have great difficulty comprehending the vast amount of time over which fossils have been accumulating.

---

[4] In what sense are they improved? Because natural selection favors forms with greater fitness, current understanding suggests that some forms may have indeed remained stable for long periods of time because their divergent offspring may have had no distinct advantage over them.

If you can read Sir Charles Lyell's grand book, *Principles of Geology*[5] (which future historians will recognize as having produced a revolution in natural science), and yet not admit how vast have been the past periods of time, you might as well just close this book right now. Not that it suffices to simply study the *Principles of Geology* or to read special treatises by different observers on separate geological formations and to note how each author tries to give an inadequate idea of the duration of each formation, or even of each layer (stratum).

We can best gain some idea of how much time has passed by knowing the agencies that have created these formations and by learning exactly how deeply the surface of the land has been denuded and exactly how much sediment has been deposited. As Lyell has well remarked, the extent and thickness of our sedimentary formations are the result and measure of the amount of erosion ("denudation") that the Earth's crust has undergone elsewhere over the millennia: sediment, after all, results from the wearing away of rock. Examine for yourself, then, the great piles of superimposed strata (Figure 10.2), and watch small rivers bringing down mud and the waves slowly wearing away the sea cliffs, and you will begin to comprehend something about the tremendous duration of past time, the monuments of which we see all around us.

It is good to wander along a coastal area formed of moderately hard rocks and admire the process of degradation. The tides in most cases reach the cliffs for only a few hours twice each day, and the waves eat into them only when they are charged with sand or pebbles; there is good evidence that pure water alone does not wear away rock to any measurable degree. Eventually, the base of the cliff becomes undermined and huge fragments fall down, and these, remaining fixed in one place, then have to be gradually worn away atom by atom, until, after being reduced sufficiently in size they can be rolled about by the waves and then more quickly be ground into pebbles, sand, or mud. But so often we see rounded boulders that are completely encrusted by various marine organisms, which shows how little they are abraded, and how seldom they are rolled about! Moreover, if we follow for a few miles any line of rocky cliff that is undergoing degradation along the bases of retreating cliffs, we find that it is only here and there along a short length—or around a promontory—that the cliffs are now suffering. The appearance of the surface and the vegetation elsewhere shows that years have elapsed since any waters have washed their base.

We have, however, recently learned from the observations of the Scottish geologist Andrew Ramsay, in the van of many excellent observers—of Jukes, Geikie, Cross, and others—that subaerial degradation[6] is a far more important agency than the coast action described above, or the power of the waves. The whole surface of the land is exposed to the chemical action of the air and of the rainwater, with its highly erosive dissolved carbonic acid, and in colder areas to frost as well; the disintegrated matter is

---

[5] Lyell published his book over the period 1830–1833, in three volumes; Darwin read the first two volumes during his travels aboard the *HMS Beagle*.

[6] The term "subaerial" refers to land that is currently above sea level, so that it is exposed to weathering by rain and air.

then carried down even gentle slopes during heavy rains and, to a greater extent than one might suppose, by the wind, especially in arid regions. It is then transported by the streams and rivers, which when moving rapidly deepen their channels and grind or crush the fragments to a fine powder. On a rainy day, even in a gently undulating country, we see the effects of subaerial degradation in the small muddy streams that flow down every slope. Andrew Ramsay and the British geologist William Whitaker have shown—and the observation is a most striking one—that the great lines of escarpment in the Wealden of the UK[7] and those ranging across England, which formerly were thought to be ancient seacoasts, cannot have been formed in that way, as each line is composed of one and the same formation. In contrast, our sea cliffs are everywhere formed by the intersection of various formations. This being the case, we are compelled to admit that these escarpments owe their origins mainly to the fact that the rocks of which they are composed have resisted subaerial denudation (i.e., erosion) better than the surrounding surface and that this surrounding surface has consequently been gradually lowered, while the lines of harder rock have been left projecting. Nothing impresses the mind with vast duration of time more forcibly than the conviction thus gained that subaerial agencies, which would seem to have so little power and to work so slowly, have produced such truly impressive results.

When thus impressed with the extremely slow rate at which the land is worn away through subaerial action and wave action, it is good—in order to appreciate the vast amount of time that has been required to form these impressive formations—to consider, on the one hand, the masses of rock that have been removed over many extensive areas and, on the other hand, the remarkable thickness of our sedimentary formations. I remember having been much amazed by seeing volcanic islands whose edges have been worn by the waves and pared all around into perpendicular cliffs that are now one or two thousand feet high; for the gentle slope of the lava streams, due to their formerly liquid state, showed at a glance how far the hard, rocky beds had once extended into the open ocean.

The same story is told still more plainly by geological faults[8]—those great cracks in the Earth's crust along which the strata[9] have been upheaved on one side or thrown down on the other, for thousands of feet. For since the Earth's crust cracked, whether suddenly or, as most geologists now believe, only slowly and sporadically, the surface of the land has now been so completely planed down that no trace of these vast dislocations is externally visible. The Craven fault in England,[10] for instance, extends upward for more than 30 miles, and, along this line, the vertical displacement of the strata varies from 600 to 3,000 feet. Professor Ramsay has published an account of a

---

[7] The Wealden is a district in East Sussex, England: its name comes from "the Weald," the area of high land that occupies the center of its area.

[8] A geological fault is a large crack in rock, often resulting from the action of plate tectonic forces. Rocks have slid past each other on either side of the fault, sometimes for hundreds of miles.

[9] Strata are layers of sedimentary rock that are distinctively different from the layers that are above or below.

[10] See https://en.wikipedia.org/wiki/Craven_Fault_System

downthrow in Anglesea of 2,300 feet, and he tells me that he fully believes that there is another one in Merionethshire, one of the counties in Wales, of 12,000 feet! Yet, in these cases, there is nothing on the surface of the land to show such prodigious movements: the pile of rocks on either side of the crack has by now been smoothly swept away.

On the other hand, in all parts of the world the piles of sedimentary strata are wonderfully thick. In the Cordillera, I estimated one mass of conglomerate to be 10,000 feet tall. And although conglomerates[11] have probably been accumulated more quickly than finer sediments, yet from being formed of worn and rounded pebbles, each of which bears the stamp of passing time, they show quite nicely how slowly the mass must have been heaped together. Professor Ramsay has given me the maximum thickness, from actual measurements in most cases, of the successive formations in different parts of Great Britain, and these are the results:

	Thickness
Paleozoic Strata (not including igneous beds)	57,154 feet
Secondary (Mesozoic) strata	13,190 feet
Tertiary (Cenozoic) strata	2,240 feet

Together, these add to 72,585 feet of sedimentary material, which is very nearly 13.75 British miles of depth! And some of the formations that are represented only by thin beds in England are thousands of feet thick on the European continent. Moreover, between each successive formation we have, in the opinion of most geologists, blank periods of time of enormous duration. So it is that even the lofty pile of sedimentary rocks in Britain gives an inadequate, and misleading, idea of the vast amount of time that has elapsed during their accumulation.

The consideration of these various facts impresses the mind in almost the same way, and to the same extent, as does the vain endeavor to grapple with the idea of eternity. Nevertheless, this impression is partly false. The Scottish scientist Mr. James Croll, in an interesting paper, remarks that we do not err "in forming too great a conception of the length of geological periods," but in estimating them by years. When geologists look at large and complicated phenomena, and then at the figures representing several millions of years, the two produce a totally different effect on the mind, and the figures for elapsed years are at once pronounced too small. In regard to subaerial denudation (i.e., erosion), Mr. Croll shows, by calculating the known amount of sediment annually brought down by certain rivers relative to their areas of drainage, that 1,000 feet of solid rock, as it became gradually disintegrated, would thus be removed from the mean level of the whole area in the course of 6 million years. This seems an

---

[11] Conglomerates are a coarse-grained sedimentary rock composed of rounded fragments (>2 mm diameter) lying within a matrix of finer-grained material such as sandstone.

astonishing result, and some considerations lead to the suspicion that it may be too large; but even if halved or quartered it is still very surprising.[12]

Few of us, however, truly understand what a million years really means: Mr. Croll gives an excellent illustration, suggesting that we take a narrow strip of paper, 83 feet 4 inches long, and stretch it along the wall of a large hall. Then we are to mark off at one end the tenth of an inch. This tenth of an inch will represent 100 years, and the entire strip a million years. But let it be borne in mind, in relation to the subject of this work, what 100 years implies, represented as it is by a measure utterly insignificant in a hall of the above dimensions. Several eminent breeders have so greatly modified some of the higher animals—which propagate their kind much more slowly than most of the lower animals—that within a single lifetime they have formed what well deserves to be called a new sub-breed. Few breeders have attended with due care to any one strain for more than 50 years, so that 100 years represents the work of two breeders in succession. It is not to be supposed that species in nature ever change as quickly through natural selection as domestic animals do through methodical selection. The comparison would be in every way fairer by considering only the effects that follow from our unconscious selection: that is, from the preservation of the most useful or beautiful animals, with no intention of modifying the breed. But, even by this process of unconscious selection, various breeds have been noticeably changed in the course of only two or three centuries.

Species, however, probably change much more slowly, and within the same country only a few change at any one time. This slowness follows from all the inhabitants of the same area being already so well adapted to each other that new niches open up in any particular area only after long intervals, due to physical changes of some kind or through the immigration of new forms into the area. Moreover, variations or individual differences of the right nature, by which some of the inhabitants might be better fitted than others to their new places under the altered circumstances, would not always occur immediately. Unfortunately, we have no means of determining how many years it takes to modify a species; but to the subject of time we will return later.[13]

On the Poorness of Paleontological Collections

Now let us turn to our richest geological museums. What a paltry display we behold! Everyone admits that our collections are imperfect. Indeed, the remark of that admirable palaeontologist, Edward Forbes, should never be forgotten: very many of our fossil species are known and named from single and often broken specimens, or from a few specimens collected from one place. Why are our paleontological collections so

---

[12] Actually, rates of erosion in drainage basins are now known to be far higher than these estimates.

[13] It wasn't until about 1920 that scientists came up with a way of dating fossils, based on the rates at which certain elements (e.g., Carbon-14, Potassium-40, and Uranium-235) decayed.

imperfect? For one thing, only a small portion of the Earth's surface has so far been geologically explored, and no part with sufficient care, as proven by the important discoveries being made every year in Europe. Moreover, no organisms that are entirely soft-bodied can be preserved,[14] and even shells and bones decay and disappear when left on the bottom of the sea where sediment is not accumulating to bury them.

We probably take a quite erroneous view when we assume that sediment is being deposited over nearly the whole bed of the sea, and quickly enough to embed and preserve fossil remains. Throughout an enormously large proportion of the ocean, the clearness and bright blue tint of the water bespeaks its purity and lack of sediment. The many cases on record of a geological formation conformably covered, after an immense interval of time, by another and later formation without the underlying bed having suffered any wear and tear in the interval, seem explicable only on the view of the bottom of the sea typically lying for ages and ages in an unaltered condition. The animal and plant remains that do become embedded, if in sand or gravel, will, when the beds are eventually upraised, generally be dissolved by the percolation of rainwater full of carbonic acid. Moreover, there is no guarantee that materials will ultimately be preserved as fossils just because they are common. Some of the many kinds of marine animals that live on the beach between the high- and low-tide water mark[15] seem to be preserved as fossils only rarely. For instance, members of the several species in the barnacle subfamily Chthamalinae presently coat intertidal rocks all over the world in infinite numbers that are too great to count. These animals are found only intertidally, with the exception of a single Mediterranean species that inhabits deep water; fossils of that one Mediterranean species have been found in Sicily, whereas not one other barnacle species has yet been found fossilized in any tertiary formation. And yet it is known that members of the barnacle genus *Chthamalus* existed during the Cretaceous ("Chalk") period.[16]

In addition, many great deposits, requiring a vast length of time for their accumulation, are entirely destitute of any organic remains without our being able to assign any reason for that. One of the most striking instances is that of the Flysch formation, which consists of layers of limestone shale and sandstone several thousand feet thick—up to 6,000 feet thick in some places—and extending for a least 300 miles from Vienna to Switzerland. Although this great mass has been most carefully searched, no fossils, except for a few vegetable remains, have so far been found.

With respect to the terrestrial productions that lived during the Secondary[17] and the even older Paleozoic periods, our fossilized evidence is extremely fragmentary.[18]

---

[14] More recently, people have indeed discovered preserved soft-bodied organisms, notably in the Burgess Shale of Canada; these fossils, discovered by palaeontologist Charles Walcott in 1909, are more than 500 million years old.

[15] Tides are caused by the Earth's rotation and the gravitational attraction of the sun and moon. Typically, there are two high tides and two low tides each day. In some parts of the world—in the Bay of Fundy, Canada, for example—the difference between water levels at low tide and high tide can be more than 52 feet

[16] The Cretaceaous period, known as the *Chalk period* in Darwin's day, extends from 66 million to 145 million years ago.

[17] This is now known as the Mesozoic Era (see Table 10.1).

[18] Dead organisms become fossilized only under particular conditions. Indeed, it has been estimated that fewer than 1% of all species that have ever lived on Earth have been found as fossils.

For instance, until recently not a single land snail shell was known from either of these vast periods, with the exception of one species discovered by Sir Charles Lyell and Dr. Dawson in the carboniferous strata of North America; but land snail shells from these periods have now been found in the lias formation of Great Britain.[19]

With regard to the fossilized remains of mammals, a glance at the historical table published in Charles Lyell's *Manual* will bring home the truth—far better than pages of detail—regarding how accidental and rare is their preservation. Nor is their rarity surprising when we remember that most known fossil remains of tertiary mammals have been discovered either in caves or in lacustrine deposits[20] and that not a cave or true lacustrine bed is known belonging to the age of our Mesozoic or Paleozoic formations.

But the imperfection of the geological record largely results from another and more important cause than the rare preservation of remains: namely, from the fact that the several distinct geological formations that we see today are separated from each other by great intervals of time. This doctrine has been emphatically admitted even by many geologists and paleontologists, who, like the geologist and naturalist Edward Forbes, entirely disbelieve in species changing over time. When we see the formations tabulated in written works, or when we study them in the field, it is difficult to avoid believing that they are closely consecutive in time. But they are not. We know, for instance, from the Scottish geologist Sir Roderick Murchison's great work in Russia what wide gaps there are in that country between the superimposed geological formations; and so it is in North America and in many other parts of the world.[21] The most skillful geologists, if their attention had been confined exclusively to these large territories, would never have suspected that during the periods that were blank and barren in their own country, great piles of sediment, charged with new and peculiar life forms had been accumulated elsewhere. And if, in each separate territory, hardly an idea can be formed of the length of time that has elapsed between consecutive formations, we may infer that this could not be determined anywhere. The frequent and great changes in the mineralogical composition of consecutive formations generally imply great changes in the geography of the surrounding lands from which the sediment was derived; this accords well with the belief that vast intervals of time have elapsed between the creation of each formation.

We can, I think, see why the geological formations of each region are almost invariably intermittent; that is, why they have not followed each other closely in time. During my nearly five-year voyage aboard the *HMS Beagle* (1831–1836), I examined many hundreds of miles of the South American coastline that have been upraised several hundred feet within the recent geological period. Scarcely any fact struck me

---

[19] The lias formation is one of the most geologically important geological formations in Britain, containing deposits from 180 to 205 million years ago.

[20] Lacustrine deposits are sedimentary deposits formed at the bottoms of ancient lakes.

[21] The idea here is that there have been many long periods of time during which there are no preserved sedimentary deposits in some places. Thus an adjacent layer of the geological record that we see today could have begun millions of years after the previous layer was deposited.

more when examining this coastline than the absence of any recent deposits sufficiently extensive to have lasted for even a short geological period. Along the entire
west coast, which is inhabited by a very distinctive marine fauna, tertiary beds[22] are
so poorly developed that no record of several successive and peculiar marine faunas
is likely to be preserved to a distant, future age. A little reflection will explain why no
extensive formations with recent or tertiary remains can be found anywhere along the
rising coast of the western side of South America, even though the supply of sediment
must for ages have been great from the enormous degradation of the coastal rocks
and from muddy streams entering the sea. The explanation, no doubt, is that the littoral and sublittoral deposits are continually being worn away by the grinding action
of the coast waves as the land slowly and gradually rises above the water.

We may, I think, conclude that if we are eventually to find any fossil remains, sediment must be accumulated in extremely thick, solid, or extensive masses in order to
withstand the incessant actions of the waves when the sediments are first upraised,
and during successive lowerings and raisings, as well as the effects of subsequent subaerial degradation. Such necessary thick and extensive accumulations of sediment
may be formed in only two ways: (1) either in the profound depths of the sea, in which
case the bottom will not be inhabited by the many and highly varied forms of life that
we see in the more shallow seas; when upraised, the mass will then give an imperfect
record of the organisms that existed in the neighborhood during the period of its accumulation. Or (2), sediment may be deposited to a great thickness and extent over
a shallow bottom as long as that bottom continues to slowly subside (i.e., to slowly
sink). In this latter case, as long as the rate of subsidence[23] and the rate at which sediment is being supplied nearly balance each other, the sea will remain shallow and
favorable for many and varied forms for a very long time, and thus a rich fossiliferous
formation, thick enough to resist a large amount of erosion when eventually upraised,
may be gradually formed.

I am convinced that nearly all of our ancient formations that are throughout the
greater part of their thickness rich in fossils have thus been formed in shallow water
during periods of slow subsidence. Indeed, since publishing my views on this subject in 1845, I have watched the progress of Geology and have been surprised to note
how author after author, in describing this or that great geological formation, has also
come to the conclusion that the sediments comprising it were accumulated during
periods of subsidence. Let me add that the only ancient tertiary formation on the west
coast of South America that has been bulky enough to resist such degradation as it has
as yet suffered, but which will hardly last to a distant geological age, must certainly
have been deposited during a downward oscillation of level and thus gained considerable thickness.

---

[22] Tertiary beds were formed as recently as about 2.6 million years ago, and as long ago as about 66 million years ago. In geological terms, tertiary beds are relatively recent.

[23] To "subside" in this case means that the land is sinking to greater depths. The word "subsidence" is
pronounced "sub-SIGH'-dence." If that region later becomes elevated above the water level, erosion can
remove much of its fossil history.

All geological facts tell us plainly that each area has undergone slow oscillations of level and that these oscillations have apparently affected wide spaces. Consequently, formations rich in fossils and sufficiently thick and extensive enough to resist subsequent degradation when raised above sea level will have been formed over wide spaces during periods of subsidence but only where the supply of sediment was sufficient to keep the sea shallow and to embed and preserve the remains before they had time to decay. In contrast, when the sea bed has remained stationary and has not subsided, thick deposits will not have been able to accumulate in the shallow parts, which are parts most favorable to life. Still less can this have happened during the alternate periods of elevation; or, to speak more accurately, the beds that were then accumulated will generally have been destroyed by being upraised and exposed to wave action and erosion.

These remarks apply mostly to littoral and sublittoral deposits.[24] In the case of an extensive and shallow sea, such as that within a large part of the Malay Archipelago,[25] where the depth varies from 30 or 40 to 60 fathoms,[26] a widely extended formation might be formed during a period of elevation and yet not suffer excessively from denudation during its slow upheaval. But the thickness of the formation could not be great for, owing to the elevating movements, it would be less than the depth in which it was formed; nor would the deposit be much consolidated, nor be capped by overlying formations, so that it would run a good chance of later being worn away by atmospheric degradation and by the action of the sea during subsequent oscillations of level. The English geologist and mathematician Mr. William Hopkins, however, has suggested that if one part of the area subsided after rising but before being denuded, the deposit formed during the rising movement, though not thick, might afterward become protected by fresh accumulations and thus be preserved for a long period.

Mr. Hopkins also believes that sedimentary beds of considerable horizontal extent have rarely been completely destroyed. But all geologists—except for those few who believe that our present metamorphic schists and plutonic rocks once formed the primordial nucleus of the globe—will admit that these latter rocks have been stripped of their coverings to an enormous extent. Indeed, it is scarcely possible that such rocks could have been solidified and crystallized while uncovered, but if such metamorphic action occurred at profound depths of the ocean, the former protecting mantle of rock may not have been very thick.

Admitting, then, that igneous and metamorphic rocks[27] such as gneiss, micaschist, granite, and diorite were once necessarily covered up, how can we account for

---

[24] *Littoral* means "along the shore"; for marine habitats, it is the area between high tide and low tide, areas of land that are exposed to air periodically at low tide. Thus *sublittoral* is the zone that begins just below the lower edge of the littoral zone; it is always covered by seawater.

[25] This refers to the chain of islands between the mainland of Southeast Asia and Australia, including Indonesia and the Philippines. It is also where Alfred Russel Wallace spent much of his time collecting specimens while coming up with his own ideas about the origin of species.

[26] A *fathom* is a nautical term, referring to a depth of 6 feet (or 1.8 meters).

[27] Igneous rocks (e.g., granite and basalt) are formed by the solidification of molten materials; metamorphic rocks (e.g., marble, slate, and schist) result when preexisting rocks are modified by exposure to extreme heat (150–200°C) and pressure (1,500 bars). The resulting rocks are much denser and less susceptible to erosion.

the naked and extensive areas of such rocks that we now see in many parts of the world, except on the belief that they have subsequently been completed denuded of all overlying strata? There is no doubt that such extensive areas do indeed exist: the granitic region of Parime in South America is described by Friedrich Alexander von Humboldt as being at least 19 times larger than Switzerland! South of the Amazon River in South America, the Austrian geologist Ami Boue colors an area composed of rocks of this nature as equal to that of Spain, France, Italy, part of Germany, and the British Islands combined. This region has not been carefully explored, but, from the concurrent testimony of travelers, the granitic area is very large: indeed, the German geologist Ludwig von Eschwege gives a detailed section of these rocks stretching from Rio de Janeiro in Brazil for 260 geographic miles inland in a straight line; I traveled for 150 miles in another direction and saw nothing but granitic rocks. I collected many specimens along the whole coast from near Rio de Janeiro to the mouth of the Rio de la Plata, a distance of 1,100 miles, and found that they all were granitic. Inland, along the whole northern bank of the Plata I saw, besides modern tertiary beds, only one small patch of slightly metamorphosed rock, which alone could have formed a part of the original capping of the granitic series.

Turning to a well-known region, namely to the United States and Canada, as shown in the American geologist Professor H. D. Rogers's beautiful map, I have estimated the areas by cutting out and weighing the paper, and I find that the metamorphic (excluding "the semi-metamorphic") and granitic rocks exceed—in the proportion of 19 to 12.5—the whole of the newer Paleozoic formations. In many regions the metamorphic and granitic rocks would be found much more widely extended than they appear to be if all the sedimentary beds that rest unconformably on them were removed and which could not have formed part of the original mantle under which they were crystallized. Thus it is probable that whole formations have been completely denuded in some parts of the world, with not a wreck left behind to tell the tale.

One remark is here worth a passing notice. During periods of elevation, the area of the land and of the adjoining shallow parts of the sea will be increased, and new habitats will often be created; as previously explained, these circumstances will promote the eventual formation of new varieties and new species. But, during such periods, there will generally be a blank space in the geological record, leaving us with no direct evidence of what took place after the land and adjoining sea rose. On the other hand, during times of subsidence, when the land is sinking, the inhabited terrestrial areas and the number of inhabitants in those areas will decrease, either through death or migration to other areas; this will be true everywhere except on the shores of a continent when first broken up into an archipelago. Consequently, during periods of subsidence, although there will be much extinction, few new varieties or new species will be formed since smaller populations generate fewer varieties than large populations, as discussed earlier. And it is during these very periods of subsidence that the deposits which are richest in fossils have been accumulated.

On the Absence of Numerous Intermediate Varieties in Any Single Formation

From these several considerations, it cannot be doubted that the geological record, viewed as a whole, is extremely imperfect. But if we confine our attention to any one existing formation, it becomes much more difficult to understand why we do not typically find within that formation closely graduated varieties that are intermediate in structure between those related species that lived at its commencement and those that lived at the close of its creation. There are, however, several cases on record of just that: the same species presenting clearly different varieties in the upper and lower parts of the same formation (i.e., at the start and close of its creation). Thus the German geologist Adolfovich Trautschold gives a number of such instances with ammonites (Figure 10.3), while Franz Hilgendorf has described a most curious case of 10 graduated forms of the air-breathing land snail *Planorbis multiformis* in the successive beds of a freshwater formation in Switzerland. However, it must be admitted that such formations are rare. I can give several reasons why each individual geological formation should not commonly include a

Figure 10.3 Ammonites. Ammonites are related to squid, octopus, and other cephalopods, but, like the chambered *Nautilus*, they lived inside a shell that was divided internally into many different compartments by calcareous walls known as *septa*. Many thousands of ammonite species are found as fossils, with different species or genera found in geological strata formed in different specific time periods; thus, ammonites are commonly used as index fossils. Ammonites became extinct at the end of the Cretaceous period, about 66 million years ago, although the fossil record extends back to at least 240 million years ago.

graduated series of links between the species that lived at its commencement and at its close, even though each formation has indisputably required a vast number of years for its deposition. However, I can't say which of the following considerations have played the most important roles.

Although each geological formation may have required a very long period of time to form, that amount of time is probably short compared with the amount of time needed for one species to change into another. I am aware that two paleontologists, Heinrich Bronn and Henry Woodward, whose opinions are worthy of much deference, have concluded that the average duration of each formation is two or three times as long as the average life span of a species. But it seems to me that insuperable difficulties prevent us from knowing whether or not this is truly the case. When we see a species first appearing in the middle of any geological formation, it would be rash in the extreme to infer that it had not previously existed anywhere else. Similarly, when we find a species disappearing in a formation before the last layers have been deposited, it would be equally rash to suppose that all members of that species then became immediately extinct. For one thing, we forget how small the area of Europe is compared with the rest of the world; nor have the several stages of the same formation throughout Europe been correlated yet with perfect accuracy.

Moreover, we may safely infer that marine animals of all kinds have undergone a large amount of migration, driven by climatic and other changes; thus, when we see a species first appearing in any particular geological formation, the probability is not that it had first originated there at that time, but rather that it only then first immigrated into that area from somewhere else. It is well known, for instance, that several species appear somewhat earlier in the Paleozoic beds of North America than in those of Europe, time having apparently been required for their migration from the American to the European seas.[28] In examining the most recent deposits in various quarters of the world, it has everywhere been noted that some few still existing species that are common in the deposit have become extinct in the immediately surrounding sea; or, conversely, that some of the species are now abundant in the neighboring sea but are rare or even absent in this particular deposit. It is an excellent lesson to reflect on the extensive and well-documented amount of migration of the inhabitants of Europe during the last glacial epoch,[29] which forms only a part of one whole geological period. Likewise, we should reflect on the changes of level, on the extreme changes of climate, and on the great lapse of time all included within this single glacial period.

Yet it may be doubted whether, in any quarter of the world, sedimentary deposits—including fossil remains—went on accumulating within the same area throughout that entire period. For instance, it seems unlikely that sediment was deposited near

---

[28] We now know that these land masses were actually continuous by the late Paleozoic; there were no separate European and American seas acting as barriers to migration.

[29] The most recent glacial period on Earth took place in the Pleistocene epoch, which ended about 12,000 years ago and began about 2.6 million years earlier.

the mouth of the Mississippi River during the entire glacial period within that limit of depth at which marine animals can best flourish; indeed, we know that great geographical changes occurred in other parts of America during this time.[30] When such beds as were deposited in shallow water near the mouth of the Mississippi during some part of the glacial period shall have been upraised, organic remains will probably first appear and then disappear at different levels, owing to the migration of species and to geographical changes. And in the distant future, a geologist, examining those beds, would be tempted to conclude that the average duration of life of the embedded fossils had been less than that of the glacial period, instead of its having really been far greater (i.e., extending from before the glacial epoch to the present day).

Conditions allowing for a truly continuous record of morphological change for any organisms living in any particular area must be extremely rare. In order to get a perfect gradation between two related forms found in the upper and lower parts of the same geological formation, the deposit would have to have gone on accumulating continuously long enough to allow for the slow process of biological variation—a very long period of time indeed. Thus the deposit must be very thick, and the species undergoing change must have lived in the same area for the entire time. But we have seen that a thick formation, filled with fossils throughout its entire thickness, can accumulate only during a period of subsidence. Moreover, if the same marine species are to live in the same area during that time, the water depth must also stay approximately the same, something that can happen only if the rate of sedimentation very nearly counterbalances the rate of subsidence. However, this same movement of subsidence will tend to submerge the area that is the source of the sediment and thus diminish the supply while the surface continues to move downward. In fact, a nearly exact balancing between the supply of sediment and the amount of subsidence is probably very rare; indeed, more than one paleontologist has observed that very thick deposits are usually devoid of organic remains except near their upper or lower limits.

It would seem that each separate formation, like the whole pile of formations in any one country, has generally been accumulated only intermittently. When we see, as is so often the case, a formation composed of beds[31] of widely different mineralogical composition, we may reasonably suspect that the process of deposition has been more or less interrupted. Nor will the closest inspection of a formation give us any idea of how long it took for its deposition.

Many instances could be given of beds that are only a few feet thick in one area but that represent formations that are elsewhere thousands of feet thick, and which therefore must have required an enormous period of time for their accumulation; yet no one ignorant of this fact would have ever suspected the vast lapse of time represented by the thinner formation. Many cases could also be given of the lower beds of a

---

[30] And we now know that there were also substantial changes in sea level during that period.

[31] A "bed" is the smallest recognizable division of a geological formation or stratigraphic rock series. A geological bed can vary from as little as 1 centimeter to as much as several meters in thickness.

formation having been upraised, denuded, submerged, and then covered again by the upper beds of the same formation—facts showing what wide, yet easily overlooked, intervals have occurred in its accumulation. In other cases we have the plainest evidence in great fossilized trees, still standing upright as they grew, of many long intervals of time and changes of level during the process of deposition, things that would never have been suspected if the trees had not been preserved. Thus Sir Charles Lyell and Dr. John Dawson found beds in Nova Scotia from the Carboniferous Period[32] that were 1,400 feet thick, with ancient root-bearing strata, one above the other, at no less than 68 different levels. Therefore, when we find the same species appearing at the bottom, middle, and top of a formation, the probability is that it has *not* lived on the same spot during the whole period of deposition, but rather that it has disappeared and reappeared—perhaps many times—during the same geological period. Consequently, if it were to have undergone a considerable amount of modification during the deposition of any one geological formation, any one section would not include all the fine intermediate gradations in form that must, according to my theory, have existed, but instead would show abrupt—though perhaps slight—changes of form.

It is all-important to remember that naturalists have no golden rule by which to distinguish between species and varieties of a species; they allow for some small amount of variability within each species, but, when they meet with a somewhat greater amount of difference between any two forms, they rank both as separate species unless they are able to connect them together by a series of very close intermediate gradations. From the reasons just assigned, we can seldom hope to see this in any one geological section. Suppose that two distinct fossilized species—let's call them species B and species C—are found in one bed and that a third species, species A, is found in an underlying, older bed; even if species A were strictly intermediate in form between species B and C, it would simply be ranked as a third and distinct species, unless at the same time it could be closely connected by intermediate varieties with either one or both forms. Nor should it be forgotten, as I have explained before, that even if species A was the actual ancestor of both species B and C, it would not necessarily be strictly intermediate between them in all respects. Thus we might well obtain the parent species and its several modified descendants from the lower and upper beds of the same formation, respectively, but unless we obtained numerous transitional gradations, we should not be able to recognize their blood relationship and should consequently rank them as distinct species.

It is notorious on what excessively slight differences many paleontologists have established their species; they are especially likely to do this if the specimens come from different substages of the same geological formation. Some experts in the study of snail, bivalve, and other mollusc shells ("conchologists") are, however, now sinking many of the very fine species described by the French naturalist Alcide D'Orbigny

---

[32] We now know that these beds were formed about 300–360 million years ago.

and others into the rank of varieties; thus, on this view, we do in fact find evidence of the kinds of change that, on my theory, we ought to find. Similarly, look at the more recent tertiary deposits from about 26 million years ago and which include many shells that most naturalists believe to be identical with those of existing species; but some excellent naturalists, including Louis Agassiz and François Jules Pictet, maintain that all of these tertiary species are in fact distinctly different species, although the distinction is admitted to be very slight. **So here, unless we believe that these eminent naturalists have been misled by their imaginations and that these late tertiary species really present no difference whatever from their now-living representatives, or unless we admit, in opposition to the judgment of most naturalists, that these tertiary species are all truly distinct from the recent ones, we have clear evidence of the frequent occurrence of slight modifications of the kind required for the theory of evolution by natural selection.** Similarly, if we look to rather wider intervals of time, namely, to distinct but consecutive stages within the same great geological formation, we find that the embedded fossils, though universally ranked as distinctly different species, yet are far more closely related to each other than are the species found in more widely separated geological formations. Here again we have undoubted evidence of change in the direction required by my theory. I shall return to this topic in Chapter 11.

With animals and plants that propagate rapidly and do not wander much, there is reason to suspect, as we have previously seen, that their varieties are generally at first confined to small, local areas and that such local varieties do not spread widely and supplant their parent form until they have become modified and perfected to some considerable degree. According to this view, there is a very small chance of discovering, in any one geological formation in any one country, all the early stages of transition between any two forms—for the successive changes are supposed to have been local or confined to some one spot.

Most marine animals have a wide geographical range. With plants we have seen that those having the widest ranges most often present different varieties; thus, with shells and other marine animals, it is probably those species that had the widest geographical range, far exceeding the limits of the known geological formations in Europe, that have most often given rise first to local varieties and, ultimately, to new species. And again, this would greatly lessen the chance of our being able to trace the stages of transition in any one geological formation.

Dr. Falconer insists that an even more important consideration leads to the same result: namely, that the period of time during which each species underwent modification, though long as measured by years, was probably short in comparison with that during which it remained without undergoing any change. It should not be forgotten that, at the present day, with perfect live specimens for examination, two contemporary forms can seldom be connected by intermediate varieties—and thus proved to be the same species—until many specimens are collected from many places; but with fossilized species this can rarely be done. We shall, perhaps, best perceive the improbability of our being able to connect species by numerous, fine, intermediate fossil links by asking ourselves whether, for instance, geologists in some distant future period will be able to prove that our different breeds of cattle, sheep, horses, and dogs are descended from *a single* ancestral stock, as we know they are, or from *several* ancestral stocks; or, again, whether certain snail and bivalve species inhabiting the shores of North America—species that are ranked by some conchologists as specifically distinct from their European counterparts but by other conchologists as only varieties—are really varieties or are actually separate species. This could be accomplished by future geologists only if they discover in a fossil state numerous intermediate gradations; such success is improbable in the highest degree.

Writers who believe in the immutability of species have asserted over and over again that geology yields no linking forms. This assertion, as we shall see in the next chapter, is certainly erroneous. As Sir J. Lubbock has remarked, "Every species is a link between other allied forms." If we take a genus having 20 species, recent and extinct, and destroy four-fifths of them, no one doubts that the remaining species will stand much more distinct from each other. If the most extremely different forms in the genus happen to have been thus destroyed, the genus itself will stand more distinct from other allied genera. Granted that geological research has not yet revealed the former existence of infinitely numerous gradations, as fine as existing varieties, connecting together nearly all existing and extinct species, but this ought not to be expected; and yet this has been repeatedly advanced as a most serious objection against my views.

It may be worthwhile to sum up the foregoing remarks on the causes of the imperfection of the geological record using an imaginary illustration. The Malay Archipelago, a cluster of islands found between mainland Southeast Asia and Australia, is about the size of Europe from the North Cape to the Mediterranean, and from Britain to Russia, and therefore equals all the geological formations that have been examined with any accuracy, excepting those of the United States of America. I fully agree with Mr. Godwin-Austen that the present condition of the Malay Archipelago, with its numerous large islands separated by wide and shallow seas, probably represents the former state of Europe during the period that most of our formations were accumulating. The Malay Archipelago is now one of the biologically richest regions in the world; and yet, even if we could collect all the species that have ever lived there, how imperfectly would they represent the natural history of the world!

But we have every reason to believe that the terrestrial productions of the archipelago would be preserved in an extremely imperfect manner in the formations that we suppose are accumulating there now. Not many of the strictly littoral animals, or of those that lived on naked submarine rocks, would be embedded. Moreover, those embedded in gravel or sand would not endure to a distant time. Wherever sediment did not accumulate on the bed of the sea, or where it did not accumulate at a sufficient rate to protect organic bodies from decay, no remains could be preserved for future examination.

Formations rich in fossils of many kinds—and sufficiently thick to last to an age as distant in futurity as the secondary formations lie in the past—would generally be formed in the archipelago only during periods of subsidence. These periods of subsidence would be separated from each other by immense intervals of time, during which the area would be either stationary or rising; while rising, the fossiliferous formations on the steeper shores would be destroyed almost as soon as they were accumulated by incessant erosion such as we now see on the shores of South America. Even throughout the extensive and shallow seas within the archipelago, sedimentary beds of great thickness could hardly be accumulated during the periods of elevation (or become capped and protected by subsequent deposits) so as to have a good chance of enduring to a very distant future. During the periods of subsidence, there would probably be much extinction of life; during the periods of elevation, there would be much variation, but the geological record would then be less perfect.

It may be doubted whether the duration of any one great period of subsidence over the whole or part of the archipelago, together with a contemporaneous accumulation of sediment, would exceed the average duration of the same specific life forms; yet without these contingencies, the transitional gradations between any two or more species would not be preserved. If such gradations were not all fully preserved, what were actually transitional varieties would merely appear as so many new—though closely allied—species. It is also probable that each great period of subsidence would be interrupted by oscillations of level and that slight climatical changes would intervene during such lengthy periods; in such cases, the inhabitants of the archipelago would migrate and no closely consecutive record of their modifications could be preserved in any one geological formation.

Very many of the marine inhabitants of the Maylay Archipelago now range thousands of miles beyond its confines; analogy plainly leads to the belief that it would be chiefly these far-ranging species, though only some of them, that would most often produce new varieties. These varieties would at first be local or confined to one place, but, if they possessed any decided advantage or became further modified and improved, they would slowly spread and supplant their parent forms. When such varieties returned to their ancient homes, as they would differ from their former state in a nearly uniform, though perhaps extremely slight degree, and as they would be found embedded in slightly different substages of the same geological formation, they would, according to the principles followed by many paleontologists, be ranked as new and distinct species.

If there is some degree of truth in these remarks, then surely we have no right to expect to find, in our geological formations, an infinite number of those fine transitional forms, which, on our theory, have connected all the past and present species of the same group into one long and branching chain of life. We ought only to look for a few links, and such assuredly we do find—some more distantly related to each other, some more closely related to each other—and these links, let them be ever so similar and closely related, if found in different stages of the same formation, would, by many paleontologists, be ranked as distinct species. But I do not pretend that I should ever have suspected how poor the fossil record was, even in the best preserved geological sections, had not the absence of innumerable transitional links between the species that lived at the commencement and close of each formation argued so strongly against my theory.

On the Sudden Appearance of Whole Groups of Allied Species

The abrupt manner in which whole groups of species suddenly appear in certain formations has been emphasized by several paleontologists (e.g., Agassiz, Pictet, and Sedgwick) as a fatal objection to the belief in the transmutation of species. If numerous species belonging to the same genera or families have really all started into life at the same time in the distant past, the fact would indeed be fatal to my theory of evolution through natural selection. For the development of a group of forms, all of which are descended from some one ancestor by natural selection, must have been an extremely slow process, and the ancestors must have lived long before their modified descendants appeared.

But we continually overrate the perfection of the geological record and falsely infer that because we have found no representatives of certain genera or families beyond a certain age, they did not exist before that age. In all cases, positive paleontological evidence may be implicitly trusted; negative evidence, on the other hand, is worthless, as experience has so often shown. We continually forget how large the world is compared with the area over which our geological formations have so far been carefully examined, and we forget that groups of species may elsewhere have long existed and have slowly multiplied before they invaded the ancient archipelagoes of Europe and the United States to be preserved as fossils. Moreover, we do not make due allowance for the enormous intervals of time that have elapsed between our consecutive geological formations, longer perhaps in many cases than the time required for the accumulation of each formation. These long intervals will have given time for the multiplication of species from some one parent form, and, in the succeeding formation, such groups or species will appear as if they were suddenly created.

Let me here recall a remark formerly made: namely, that it might require a long succession of ages to adapt an organism to some new and peculiar line of life (for instance, to fly through the air) and consequently that the transitional forms would often long remain confined to a single region—but that, once this adaptation had taken place and a few species had thus acquired a great advantage over other organisms, a comparatively short time would be sufficient to produce many divergent forms, which would then spread rapidly and widely throughout the world. Professor François Pictet, in commenting on early transitional forms in his excellent review of this work, and taking birds as an illustration, cannot see how the successive modifications of the anterior limbs of a supposed pre-bird prototype could possibly have been of any advantage. But look at the penguins of the Southern Ocean: Do not these birds have their front limbs in this precise intermediate state of "neither true arms nor true wings?" Yet these birds hold their place victoriously in the battle for life, for they exist in infinite numbers and of many kinds. I do not suppose that we here see the real transitional grades through which the wings of birds have passed, but what special difficulty is there in believing that it might profit the modified descendants of the penguin first to become enabled to flap along the surface of the sea like the logger-headed duck and, ultimately, to rise from its surface and glide through the air?[33]

I will now give a few examples to illustrate the foregoing remarks and to show how liable we are to error in supposing that whole groups of species have suddenly been produced. Even in so short an interval as that between the first and second editions of Pictet's great work on Paleontology, published in 1844–1846 (first edition) and in 1853–1857 (second edition), the author modified his conclusions about the first appearance and disappearance of several groups of animals considerably; a third edition would require still further changes. I may recall the well-known fact that, in geological treatises published not many years ago, mammals were always spoken of as having abruptly come in at the commencement of the tertiary period.[34] And now one of the richest known accumulations of fossil mammals belongs to the middle of the secondary (Mesozoic) era[35]; indeed, true mammals have been discovered in the new red sandstone at nearly the commencement of this great era. Cuvier used to emphasize that no monkey occurred in any stratum from the Tertiary period, but extinct species have now been discovered in India, South America, and in Europe, as far back as the Miocene epoch. And had it not been for the rare accident of the preservation of footsteps in the new red sandstone of the United States, who would have ventured to suppose that at least 30 different bird-like animals, some of gigantic size, existed during that period? Not a fragment of bone has been discovered in those beds. Not long ago, paleontologists maintained that the whole class of birds came suddenly into existence during the Eocene epoch; but now we know, on the authority of Professor Owen, that a bird certainly lived during the deposition of the upper greensand. Even more recently, that strange bird, the *Archeopteryx* (Figure 10.4), with a long lizard-like tail and bearing a pair of feathers on each joint, and with its wings furnished with two free claws, has been discovered in the oolitic slates of Solenhofen.[36] Hardly any recent discovery shows more forcibly than this how little we as yet know of the former inhabitants of the world.[37]

I may give another instance, which, from having passed under my own eyes, has much struck me. In my monograph on fossil sessile cirripedes,[38] I stated that, from the

---

[33] Apparently Darwin never knew that penguins have actually descended from winged ancestors. Instead, he is assuming that penguins are similar to the early ancestors of flying birds and are therefore examples of the sorts of species that eventually evolved wings. As we now know, however, they actually illustrate the evolution of wing loss from previously winged, flying ancestors as an adaptation to living a marine lifestyle.

[34] As we know now, this would be about 66 million years ago.

[35] The middle of the Mesozoic era would be about 150 to 200 million years ago.

[36] Solenhofen is a German city famous for its exquisitely detailed fossils, which are about 155 million years old. These fossils appear suddenly in the "Solenhofen Limestone."

[37] And now we have the rich fossil finds from China, including feathered dinosaurs, leading to the conclusion that birds in fact evolved from dinosaur ancestors!

[38] Cirripedes are a group of crustaceans more commonly known as barnacles. Darwin published four monumental monographs on these animals: two on presently living cirripedes in 1851, and two on fossil cirripedes—one in 1854, and one in 1855.

(A)

(B)

Archéoptéryx.

Figure 10.4 Archeopteryx. This is the oldest known fossilized bird, with fossils preserved from about 150 million years ago. These animals had feathers, wings, and hollow bones, like modern birds, but they also had teeth, a bony tail, and legs, as found in traditional dinosaur fossils.

large number of existing and extinct tertiary species, from the extraordinary abundance of the individuals of many species all over the world, from the Arctic regions to the Equator, inhabiting various zones of depths, from the upper tidal limits to 50 fathoms, from the perfect manner in which specimens are preserved in the oldest tertiary beds, from the ease with which even a fragment of a valve can be recognized—from all these circumstances, I inferred that, had sessile cirripedes existed during the secondary periods, they would certainly have been preserved and discovered. But as not even one species had then been discovered in beds of this age, I concluded that this great group must have been suddenly developed at the commencement of the tertiary series. This was a sore trouble to me, adding, as I then thought, one more instance of the abrupt appearance of a great group of related species. But my work had hardly been published when a skillful paleontologist, M. Bosquet, sent me a drawing of a perfect specimen of an unmistakable sessile cirripede that he himself had extracted from the chalk of Belgium.[39] And, as if to make the case as striking as possible, this cirripede was a member of the genus *Chthamalus*, a very common, large, and ubiquitous genus of which not one fossilized species has as yet been found even in any tertiary stratum. Still more recently, a representative of the genus *Pyrgoma*, a member of a distinct subfamily of sessile cirripedes, has been discovered by Mr. Woodward in

---

[39] What was formerly known as the "Chalk period" is now known as the Cretaceous period. As noted at the start of this chapter, the Cretaceous period extends from 145 to 66 million years ago; see Table 10.1 (page 318). The Latin word *creta* means "chalk." Chalk rocks are sedimentary limestone, composed of the mineral calcite. As it turns out, this barnacle fossil was actually collected from a much more recent deposit; other barnacle fossils from the Cretaceous were, however, subsequently found.

the upper chalk; so we now have abundant evidence that this group of animals did indeed exist during the secondary (Mesozoic) era.

The case most frequently insisted on by paleontologists of the apparently sudden appearance of a whole group of species is that of the teleost fishes, low down—according to Agassiz—in the late "Chalk" period. This group includes most of the presently existing bony fish species. But certain Jurassic and Triassic forms are now commonly also admitted to be teleosts, and even some more ancient, Paleozoic forms have thus been classified by one high authority. If the teleost fishes had really appeared suddenly in the northern hemisphere at the start of the chalk formation, that fact would have been highly remarkable; but it would not have formed an insuperable difficulty unless it could likewise have been shown that, at the same period, the species were suddenly and simultaneously also developed in other quarters of the world. It is almost superfluous to remark that hardly any fish fossils are known from south of the Equator; and, by running through François Pictet's treatise on paleontology, it will be seen that very few fish species are known from several geological formations in Europe. Some few families of fish now have a confined range; the teleostean fishes might formerly have had a similarly confined range, spreading widely only after they had been largely developed in some one sea. Nor have we any right to suppose that the seas of the world have always been as freely open from south to north as they are at present. Even at this day, if the Malay Archipelago were to become converted into land, the tropical parts of the Indian Ocean would form a large and perfectly enclosed basin in which any great group of marine animals might be multiplied; they would remain confined there until some of the species became adapted to a cooler climate and were then enabled to double around the southern capes of Africa or Australia and thus reach other and distant seas.

From these considerations, and from our ignorance of the geology of other countries beyond the confines of Europe and the United States, and from the revolution in our paleontological knowledge brought about by the remarkable discoveries of the last dozen years, it seems to me to be about as rash to dogmatize on the succession of organic forms throughout the world as it would be for a naturalist to land for five minutes on a barren point in Australia and then to discuss the number and range of its inhabitants.

On the Sudden Appearance of Groups of Allied Species in the Lowest Known Fossiliferous Strata

Here is another and related difficulty, one that is much more serious: the manner in which species belonging to several of the main divisions of the Animal Kingdom suddenly appear in the oldest known fossiliferous rocks.[40] Most of the arguments that

---

[40] This was a major argument in favor of special creation: the sudden appearance of complex animals in the fossil record.

Figure 10.5 A trilobite. Trilobites are one of the earliest known arthropods, with an extensive fossil record that extends from about 520 million years in the past. The smallest trilobites were only about 1 mm long, but the largest species grew as long as 72 cm. The group underwent a long decline, with the last trilobites finally going extinct about 250 million years ago.

have convinced me that all the existing species of the same group are descended from a single ancestor apply with equal force to the earliest known species. For instance, it cannot be doubted that all the Cambrian and Silurian trilobites[41] (Figure 10.5) are descended from some one crustacean ancestor, which must have lived long before the Cambrian age and which probably differed greatly from any known animal. Some of the most ancient animals (e.g., members of the shelled cephalopod genus *Nautilus* and the brachiopod genus *Lingula*) do not differ much from living species; and it cannot on our theory be supposed that these ancient species were the ancestors of all the species belonging to the same groups that have subsequently appeared, for they are not in any degree intermediate in character. Consequently, if the theory be true, it is indisputable that long periods must have elapsed before the oldest Cambrian stratum was deposited, as long as—or probably far longer than—the whole interval from the Cambrian age to the present day and that, during these vast periods, the world swarmed with living creatures.

Here we encounter a formidable objection for it seems doubtful whether the Earth has been for a long enough time in a state suitable for the habitation of living creatures. Sir William Thompson concludes that the consolidation of the crust can hardly

---

[41] Trilobites are an ancient group of crab relatives (crustaceans) that were extremely abundant in the world's oceans for almost 400 million years. The entire group went extinct at the end of the Permian period, about 252 million years ago. They are now found only as fossils.

have occurred less than 20 million or more than 400 million years ago, but probably not less than 98 million or more than 200 million years. These very wide limits show how doubtful the data are, and other elements may have hereafter to be introduced into the problem. Mr. Croll estimates that only about 60 million years[42] have elapsed since the Cambrian period, but this, judging from the small amount of organic change since the commencement of the glacial epoch, appears a very short time for the many and great mutations of life that have certainly occurred since the Cambrian formation, and the previous 140 million years can hardly be considered as sufficient for the development of the varied forms of life that were already in existence during the Cambrian period. It is, however, probable, as Sir William Thompson insists, that the world at a very early period was subjected to more rapid and violent changes in its physical conditions than those now occurring; such changes would have tended to induce changes at a correspondingly rapid rate in the organisms that then existed.

To the question of why we do not find rich fossiliferous deposits belonging to these assumed earliest periods prior to the Cambrian period, I can give no satisfactory answer. Several eminent geologists, with Sir R. Murchison at their head, were until recently convinced that we beheld in the organic remains of the lowest Silurian stratum the first dawn of life. Other highly competent judges, such as Lyell and Forbes, have disputed this conclusion. We should not forget that only a small portion of the world has so far been explored with accuracy. Not very long ago M. Joachim Barrande added another and lower stage, abounding with new and peculiar species, beneath the then known Silurian system, and now, still lower down in the Lower Cambrian formation, Mr. Hicks has found in South Wales beds rich in trilobites and also containing various molluscs and annelids. The presence of phosphatic nodules and bituminous matter, even in some of the lowest azotic rocks, probably indicates life at these periods, and the existence of the *Eozoon*[43] in the Laurentian formation of Canada is generally admitted. There are three great series of strata beneath the Silurian system in Canada, in the lowest of which the *Eozoon* is found. Sir W. Logan states that their "united thickness may possibly far surpass that of all the succeeding rocks, from the base of the Paleozoic series to the present time. We are thus carried back to a period so remote, that the appearance of the so-called primordial fauna (of Barrande) may by some be considered as a comparatively modern event." The *Eozoon* belongs to the most lowly organized of all classes of animals, but it is highly organized for its class; it existed in countless numbers, and, as Dr. Dawson has remarked, certainly preyed on other minute organic beings, which must have lived in great numbers. Thus the words that I wrote in 1859, about the existence of living beings long before the Cambrian period, and which are almost the same with those since used by Sir W. Logan, have proved true.

---

[42] Actually we now know this to be more like 540 million years!

[43] The *Eozoon* (first reported in 1865 by John William Dawson) is a structure found in the Archæan limestones of Canada and other regions. They were initially thought to be the oldest known fossils, an ancient species of a gigantic, single-celled foraminiferan, but it eventually came to be seen as a "pseudo-fossil," formed by chemical and physical processes caused by high temperatures.

Nevertheless, the difficulty of assigning any good reason for the absence of vast piles of strata rich in fossils beneath the Cambrian system is very great. It does not seem probable that the more ancient beds have been completely worn away by denudation or that their fossils have been wholly obliterated by metamorphic action, for if this had been the case we should have found only small remnants of the formations next succeeding them in age, and these would always have existed in a partially metamorphosed condition. But the descriptions which we possess of the Silurian deposits over immense territories in Russia and in North America do not support the view that the older a formation is the more invariably it has suffered extreme denudation and metamorphism.

The case at present must remain inexplicable and may be truly urged as a valid argument against the views here entertained. To show that it may hereafter receive some explanation, I will give the following hypothesis. From the nature of the organic remains that do not appear to have inhabited profound depths in the several geological formations of Europe and of the United States, and from the amount of sediment, miles in thickness, of which the formations are composed, we may infer that, from first to last, large islands or tracts of land from which the sediment was derived occurred in the neighborhood of the now existing continents of Europe and North America. This same view has since been maintained by Agassiz and others. But we do not know what the state of things was in the intervals between the several successive geological formations. Did Europe and the United States during these intervals exist as dry land, or as an underwater surface near land on which sediment was not deposited, or as the bed of an open and unfathomable sea? We don't know.

Looking to the existing oceans, which are three times as extensive in area as the land, we see them studded with many islands; but hardly one truly oceanic island (with the exception of New Zealand, if this can be called a truly oceanic island) is as yet known to afford even a remnant of any Paleozoic or secondary geological formation. Hence, we may perhaps infer that, during the Paleozoic and secondary eras, neither continents nor continental islands existed where our oceans now extend; for, had they existed, paleozoic and secondary formations would in all probability have been accumulated from sediment derived from their wear and tear and would have been at least partially upheaved by the oscillations of level, which must have intervened during these enormously long periods.

If, then, we may infer anything from these facts, we may infer that, where our oceans now extend, oceans have extended from the remotest period of which we have any record. And, on the other hand, that where continents now exist, large tracts of land have existed—subjected, no doubt, to great oscillations of level—since the Cambrian period. The colored map appended to my volume on Coral Reefs[44] led me

---

[44] Darwin's very first monograph, *The Structure and Distribution of Coral Reefs*, published in 1842, included his very original (and, as it turns out, correct) thinking about how coral reefs and atolls were formed.

to conclude that the great oceans are still mainly areas of subsidence, the great archipelagoes still areas of oscillations of level, and the continents areas of elevation. But we have no reason to assume that things have thus remained from the beginning of the world. Our continents seem to have been formed by a preponderance, during many oscillations of level, of the force of elevation.

But may not the areas of preponderant movement have changed in the lapse of ages? At a period long before the Cambrian epoch, continents may have existed where oceans are now spread out, and clear and open oceans may have existed where our continents now stand. Nor should we be justified in assuming that if, for instance, the bed of the Pacific Ocean were now converted into a continent, we should find sedimentary formations there, in recognizable condition, that are older than the Cambrian strata, supposing such to have been formerly deposited. For it might well happen that strata that had subsided some miles nearer to the center of the Earth, and which had been pressed on by an enormous weight of superincumbent water, might have undergone far more metamorphic action than strata that have always remained nearer to the surface. The immense areas in some parts of the world—for instance in South America—of naked metamorphic rocks that must have been heated under great pressure have always seemed to me to require some special explanation, and we may perhaps believe that we see in these large areas the many formations long anterior to the Cambrian epoch in a completely metamorphosed and denuded condition.

The several difficulties here discussed—namely, (1) that though we find in our geological formations many links between the species that now exist and that formerly existed, we do not find infinitely numerous fine transitional forms closely joining them all together; (2) the sudden manner in which several groups of species first appear in our European formations; and (3) the almost entire absence, as at present known, of formations rich in fossils beneath the Cambrian strata—are all undoubtedly of the most serious nature. We see this in the fact that the most eminent paleontologists, namely, Cuvier, Agassiz, Barrande, Pictet, Falconer, E. Forbes, etc., and all our greatest geologists, including Lyell, Murchison, and Sedgwick, have unanimously, often vehemently, maintained the immutability of species. But Sir Charles Lyell now gives the support of his high authority to the opposite side, and most geologists and paleontologists are much shaken in their former belief. Those who believe that the geological record is in any degree perfect will undoubtedly at once reject my theory. **For my part, following out Lyell's metaphor, I look at the geological record as a history of the world imperfectly kept and written in a changing dialect.** Of this history we possess the last volume alone, relating only to two or three countries. Of this volume, only here and there has a short chapter been preserved and, of each page, only here and there a few lines. Each word of the slowly changing language, more or less different in the successive chapters, may represent the forms of life that are entombed in our consecutive geological formations and which falsely appear to have been abruptly introduced. On this view, the difficulties that I have discussed earlier are greatly diminished or even disappear completely.

Table 10.1 Geological terms and ages, from the Precambrian Era, starting more than 4 billion years ago, up to the start of the Paleozoic Era

Paleozoic Era	541 mya to 252 mya
Cambrian period	541 to 485 mya
Ordovician period	485 to 444 mya
Silurian period	444 to 419 mya
Devonian period	419 to 359 mya
Carboniferous period	359 to 299 mya
Permian period	299 to 252 mya
"Secondary" or Mesozoic Era	252 to 66 mya
Triassic period	251 to 201 mya
Jurassic period	201 to 145 mya
Cretaceous period	145 to 66 mya
"Tertiary" or Cenozoic Era	66 to the present
Tertiary period	66 to 2.6 mya
Paleocene epoch	66 to 56 mya
Eocene epoch	56 to 34 mya
Oligocene epoch	34 to 23 mya
Miocene epoch	23 mya to 5.3 mya
Pliocene epoch	5.3 to 2.6 mya
Quaternary period	2.6 mya to the present
Pleistocene epoch	2.6 mya to 12,000 years ago
Holocene epoch	12,000 years ago to the present

Mya, million years ago.

Key Issues to Talk and Write About

1. Briefly summarize the main arguments that Darwin gives, hoping that they will explain why there were so few fossils showing clear intermediate links between ancestral species and species living today.
2. How does Darwin explain why museum fossil collections in his time included such a small number of individuals?
3. For either the paragraph on page 299, beginning with the words "We can, I think, see why the geological formations ..." or the paragraph on page 305, beginning with the words "Conditions allowing for a truly continuous record of ... ," make a list of the key points in that paragraph. Then try to work those

key points into a single sentence that summarizes that paragraph's content and does so in a way that would be perfectly clear to someone who has never read the original paragraph.

4. What is "subsidence"? Briefly explain the essential role that Darwin believes subsidence played in the accumulation of fossils over time.

5. List the arguments that Darwin is quite open about, that, if true, would devastate his argument for the gradual evolution of species by natural selection.

11

On the Geological Succession of Organic Beings

In the previous chapter, Darwin explained why he thinks the fossil record is so incomplete. In this chapter, he considers the evidence for and against the idea that the geological record that we have, incomplete as it is, does in fact support the idea of slow evolutionary changes through natural selection, rather than the idea that species were created suddenly and have not changed over time. For example, he notes that the fossil record clearly documents changes within different groups and shows the members of various groups changing at different rates. Indeed, some groups have changed substantially over time, while some other have changed very little, and, within any group of organisms, the number of species in that group has tended to increase gradually over time. Also, the geological record clearly shows that while some species disappear from the fossil record—usually very gradually—we also see the appearance of completely new species over time. Moreover, once a species disappears from the fossil record, it never reappears. Perhaps two of his most convincing arguments are that all extinct species fall within the same general taxonomic groupings that exist today and that distinctive types of mammals (e.g., marsupials) that are now found only in particular parts of the world are clearly related to fossilized mammals found only in those same places. All of these facts argue against special creation, but can be fully explained on the theory of gradual descent with modification through natural selection.

Let us now see whether the various facts and laws that we presently have relating to the geological succession of animal and plant species agree more with the common view that species are unchangeable or with the view that species have become gradually modified over time through the processes of variation and natural selection.

Our leading geologist, Charles Lyell, has shown very convincingly that new species have appeared on Earth very slowly, one after another, both on the land and in the sea: indeed, new material collected every year tends to fill in the blanks between the morphological stages and to make the stages between the extinct and existing forms of life more gradual and more apparent. In some of the most recently formed geological beds, only one or two species are now extinct, and only one or two of the fossilized species have appeared there for the first time, either locally, or, as far as we

The Readable Darwin. Second Edition. Jan A. Pechenik, Oxford University Press. © Oxford University Press 2023.
DOI: 10.1093/oso/9780197575260.003.0012

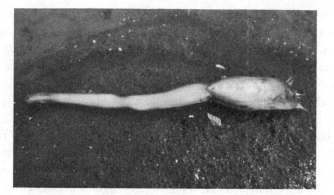

Figure 11.1 *Lingula*, brachiopod. Like bivalved molluscs, brachiopods have two calcified shells. However, unlike the bivalves, whose shells are on the left and right sides of the body, the shells of brachiopods are on the upper and lower surfaces of the body. Although only about 300 brachiopod species exist at present, more than 12,000 fossilized species have been discovered so far. *Lingula* is often referred to as a "living fossil," as extremely similar looking individuals appear in the fossil record from more than 440 million years ago.

can know, for the first time on the face of the Earth. The older Mesozoic Era[1] formations are more disjointed; however, as the geologist and paleontologist Heinrich Bronn has remarked, neither the appearance nor the disappearance of the many species embedded in each formation is ever seen to have happened simultaneously.

Moreover, species belonging to different genera and different classes have clearly not changed at the same rate or to the same extent. In the older tertiary beds[2] of the Cenozoic era, a few shells representing still-living species may still be found, surrounded by a multitude of forms that are now extinct. Similarly, the Silurian brachiopod *Lingula* (Figure 11.1) differs but little from species in this genus that are still with us, whereas most of the other Silurian shelled animals and all of the crustaceans have changed a great deal.

Organisms living on land seem to have changed faster than those in the sea. There is some indication that higher organisms, including vertebrates, have changed more quickly than less complex organisms have, although there are certainly exceptions to this rule. As the Swiss entomologist and paleontologist François Jules Pictet has noted, the amount of organic change is not the same in each successive so-called geological formation. Yet if we compare any but the most closely related formations, all species in those formations will be seen to have undergone some small degree of change. And when a species has disappeared from the face of the Earth, we have no

[1] Darwin referred to this as the "Secondary" period, but it is now known as the Mesozoic era. I have changed "Secondary" to "Mesozoic" throughout the chapter. The Mesozoic era ended about 66 million years ago (see Table 10.1).

[2] The Tertiary period began with the mass extinction of dinosaurs (except for the ancestors of modern birds) about 66 million years ago, at the end of the Mesozoic, and it extends to the start of the most recent period of glaciation, which was about 2.6 million years ago.

reason to believe that the identical form ever reappears. The strongest apparent exception to this generalization is that of the so-called colonies of Monsieur Joachim Barrande, which intrude for a period in the midst of an older formation and then allow the preexisting fauna to reappear; but, as Lyell suggests, this is probably a simple case of temporary migration from a distinct geographical province.

These various facts agree well with our theory of evolution, which includes no fixed law of development that would cause all the inhabitants of any area to change abruptly, or simultaneously, or to an equal degree. Rather, the process of modification must be slow. Moreover, it will generally affect only a few species at any one time since each species varies independently of all others. The degree to which any variations that arise will be accumulated in a population through natural selection—thus causing permanent modification—will depend on many complex contingencies, such as the degree to which the variations are beneficial; on the likeliness of intercrossing; the degree to which physical conditions are changing in the area and how quickly they are changing; the amount of immigration into the area from elsewhere; and the lifestyles of the other inhabitants with which the varying species are forced to compete. Thus it is by no means surprising that one species should retain the same form for a much longer time than others; or, if changing, that it should change to a lesser degree over the same amount of time.

We find similar relations between the existing inhabitants of distinct countries. For example, the land snails and coleopterous insects[3] of Madeira have come to differ considerably from their closest relatives in Europe, whereas the marine snails and birds from these two areas have remained unaltered. We can perhaps understand the apparently quicker rate of change in terrestrial and in more complex organisms compared with marine and less complex organisms through the more complex relations of the higher beings to their organic and inorganic conditions of life. When many of the inhabitants of any area have become modified and improved, we can also understand, on the principle of competition and from the all-important interactions between organisms in the struggle for life, that any form which did not become in some degree modified and improved over time to keep up with the changes in those around them would be liable to extermination. Thus we can see why all the species in the same region do eventually—if we look over sufficiently long intervals of time—become modified: if they hadn't become modified, they would have become extinct.

For members of the same class, the average amount of change during long and equal periods of time, may, perhaps, be nearly the same. But as the accumulation of enduring geological formations, rich in fossils, depends on great masses of sediment being deposited in areas that are subsiding, our fossil formations have been almost necessarily accumulated at wide and irregularly intermittent intervals of time; consequently, the amount of organic change exhibited by the fossils embedded in

---

[3] These are members of the insect order Coleoptera, known as "beetles." The front pair of coleopteran wings are hardened into conspicuous wingcases, called *elytra*.

consecutive layers is not equal. On this view, each formation does not represent a new act of independent creation, but rather an occasional picture: a photograph taken more or less at random, in a continuous but slowly changing drama.

We can clearly understand why a species should never reappear once it has gone extinct, even if the very same conditions of organic and inorganic life should recur. For though the offspring of one species might be adapted to fill the place of another species in the economy of nature (and no doubt this has occurred in innumerable instances) and thus supplant it, yet the old form and the new form would not be identical: both would almost certainly inherit different characters from their distinct ancestors, and organisms that already differed would vary in a different manner. For instance, suppose that all of our fantail pigeons were somehow destroyed. Pigeon fanciers might then make a new breed hardly distinguishable from the present breed. But suppose that the parent rock pigeons were likewise destroyed; indeed, under nature, we have every reason to believe that parent forms are in fact generally supplanted and eventually exterminated by their improved offspring. In that case, it is incredible to think that a fantail, identical in all respects with the existing breed, could possibly be raised from any other species of pigeon or even from any other well-established race of the domestic pigeon; the successive variations would almost certainly be in some degree different, and the newly formed variety would probably inherit from its ancestor some characteristic differences.

Taxonomic genera and families follow the same general rules in their appearance and disappearance as do individual species, changing more or less quickly and changing in a greater or lesser degree over time. Once a group has disappeared from the geological record, it never reappears; that is, its existence, as long as it lasts, is continuous. I am aware that there are some apparent exceptions to this rule, but the exceptions are surprisingly few, so few that E. Forbes, Pictet, and Woodward—though all three are strongly opposed to such views as I maintain—admit its truth, and the rule strictly accords with the theory. For all the species of the same group, however long that group may have lasted, are the modified descendants one from the other and all from a common ancestor. In the brachiopod genus *Lingula*, for instance, the species that have successively appeared at all ages must have been connected to each by an unbroken series of generations, from the lowest Silurian stratum formed nearly 444 million years ago to the present day.

We saw in Chapter 10 that whole groups of species sometimes falsely appear to have been abruptly developed, and I have attempted to explain this fact which, if true, would be fatal to my views. But such cases are certainly exceptional: as a general rule, we see a gradual increase in the number of species in a group until the group reaches its maximum number, and then, sooner or later, a gradual decrease. If the number of species included within a genus—or the number of genera included within a family—be represented by a vertical line of varying thickness ascending through the successive geological formations in which the species are found, the line will sometimes falsely appear to begin at its lower end, not in a sharp point, but abruptly; it then gradually thickens upward, often keeping of equal thickness for a time, and ultimately thins out in the upper beds, marking the decrease and final extinction of the species. This gradual increase in the number of species within a group is strictly conformable with the theory; for the species of the same genus, and the genera of the same family, can increase only slowly and progressively, the process of modification and the production of a number of allied forms necessarily being a slow and gradual process: one species first gives rise to two or three varieties, and these then become slowly converted into new distinct species, which in their turn produce by equally slow steps other varieties that in turn eventually give rise to new species, and so on, like the branching of a great tree from a single stem, until the group eventually becomes large.

On Extinction

We have as yet only spoken incidentally of the disappearance of species and of groups of species from the fossil record. On the theory of natural selection, the extinction of old forms and the production of new and improved forms are intimately connected together. The old notion of all the inhabitants of the Earth having been swept away by catastrophes at successive periods has generally been abandoned by geologists, even by such as the French geologist Jean-Baptiste Elie de Beaumont and the Scottish geologists Roderick Murchison, and M. Barrande, whose general views would naturally lead them to this conclusion. On the contrary, we have every reason to believe, from the study of the 2.6 to 66 million-year-old Tertiary formations, that species and groups of species gradually disappear one after another, first from one spot, then from another, and finally from the entire world. In some few cases, however, as by the breaking of an isthmus and the consequent bursting in of a multitude of new inhabitants into an adjoining sea, or by the final sinking (through subsidence) of an island into the sea, the process of extinction may have been rapid. Both single species and whole groups of species last for very unequal periods; some groups, as we have seen, have endured from the earliest known dawn of life to the present day, while some others have disappeared before the close of the Paleozoic period, which ended some 245 million years ago. *No fixed law seems to determine the length of time during which any single species or any single genus endures.* There is reason to believe that it takes longer for an entire group of species to go extinct than it takes for their production: if their appearance and disappearance be represented, as before, by a vertical line of varying thickness, the line will taper more gradually at its upper end—which marks the progress of extermination—than at its lower end, which marks the first appearance and the early increase in number of the species. In some cases, however, the extermination of whole groups of organisms, such as the ammonites[4] some 66 million years ago,[5] has been wonderfully sudden.

---

[4] Ammonites are an extinct group of shelled molluscs related to the modern octopus, squid, and cuttlefish (see Figure 10.3).

[5] This massive Cretaceous–Tertiary extinction, commonly known as the K-T extinction, saw the demise of about 75% of all animal and plant species on Earth, likely caused by the consequences of a collision between the Earth and a massive comet or asteroid.

(A)

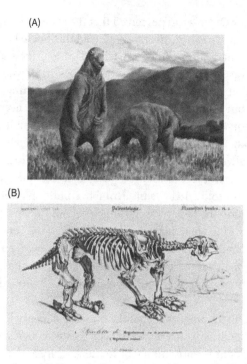

(B)

Figure 11.2 Megatherium. (A) *Megatherium* was a 13-foot-tall giant sloth, endemic to South America. The oldest known fossils are from about 5.4 million years ago. The entire genus went extinct about 12,000 years ago, at the end of the Pleistocene, very likely due to intensive hunting by humans. (B) The skeleton of *Megatherium*.

The extinction of species has been involved in the most gratuitous mystery. Some authors have even supposed that, as the individual has a definite length of life, so must species have a predetermined, definite duration of existence. No one can have marveled more than I have done at the extinction of species. When I found in La Plata the tooth of a horse embedded with the remains of extinct giant quadrupeds such as *Megatherium* (Figure 11.2), *Toxodon*, the mastodons, and other extinct monsters that all co-existed at a very late geological period with a variety of shelled invertebrates that are still with us, I was filled with astonishment; for, seeing that the modern horse, since its introduction by the Spaniards into South America, has run wild over the whole country and has increased in numbers at an unparalleled rate, I asked myself what could previously have exterminated the former horse from South America under conditions of life that seem to be so favorable?[6] But my astonishment was

---

[6] Thousands, or perhaps even millions of years ago, it turns out, horses were quite common in South America but then went extinct. They were later reintroduced there by Europeans and have continued to do well since then.

groundless. Professor Owen soon perceived that the tooth, though so like that of the existing horse, belonged to an extinct species. Had this horse been still living, but in some degree rare, no naturalist would have felt the least surprise at its rarity, for rarity is the attribute of a vast number of species of all classes, in all countries. If we ask ourselves why this or that species is rare, we answer that something is unfavorable in its conditions of life—but what that something is, we can hardly ever tell. On the supposition of the fossil horse still existing as a rare species, we might have felt certain, from the analogy of all other mammals, even of the slow-breeding elephant, and from the history of the naturalization of the domestic horse in South America, that under more favorable conditions it would in a very few years have occupied the whole continent. But we could not have told what the unfavorable conditions were that prevented its increase, whether some one or several contingencies, and at what period of the horse's life they had acted, and to what degree each had acted. If the conditions had gone on, however slowly, becoming less and less favorable over time, we assuredly should not have perceived the fact; yet the fossil horse would certainly have become rarer and rarer and finally extinct—its place would then have been assumed by some more successful competitor.

It is always difficult to remember that the increase of every living creature is constantly being checked by various unperceived hostile agencies and that these same unperceived agencies are amply sufficient to cause rarity and, finally, extinction. So little is this subject understood that many people have expressed considerable surprise that such great monsters as the Mastodon and the more ancient Dinosaurians have become extinct, as if mere bodily strength gave victory in the battle of life. On the contrary, a larger size would in some cases result in quicker extermination from the greater amount of food required to stay alive, as noted by Richard Owen. Before man inhabited India or Africa, some other cause must have checked the continued increase of the existing elephant. A highly capable judge, Dr. Falconer, believes that it is chiefly insects which, from incessantly harassing and weakening the elephant in India, check its increase; this was also Bruce's conclusion with respect to the African elephant in Abyssinia. It is certain that insects and blood-sucking bats control the existence of the larger naturalized quadrupeds in several parts of South America.

In the more recent tertiary geological formation,[7] we see in many cases that organisms became rare before they become extinct, and we know that this has also been the progress of events with those animals that have been exterminated in more recent times, either locally or entirely, through human actions. I may repeat what I published in 1845: namely, that to admit that species generally become rare before they become extinct—and to feel no surprise at the rarity of a species and yet to marvel greatly when the species ceases to exist—is much the same as admitting that sickness in an individual can be the forerunner of death: to feel no surprise at sickness, but, when

---

[7] This was about 66 million to 2.5 million years ago.

the sick man dies, to wonder and to suspect that he died by some deed of violence rather than by natural causes.

The theory of natural selection is grounded on the belief that each new variety—and ultimately each new species—is produced and maintained by having some advantage over those other species with which it comes into competition; the consequent extinction of less-advantaged forms almost inevitably follows. It is the same with our domestic productions: when a new and slightly improved variety of some animal or plant has been raised, it at first supplants the less-improved varieties in the same neighborhood; when much improved, it is transported far and near, as with our short-horn cattle, and takes the place of other breeds in other countries. Thus the appearance of new forms and the disappearance of old forms, both those naturally produced and artificially produced, are bound together. In flourishing groups, the number of new specific forms that have been produced within a given time has at some periods probably been greater than the number of the old specific forms that have been exterminated; but we know that the number of species has not gone on indefinitely increasing, at least during the later geological epochs, so that, looking to later times, we may believe that the production of new forms has caused the extinction of about the same number of old forms.

The competition will generally be most severe—as I have formerly explained and illustrated by examples—between forms that are most like each other in all respects. Thus the improved and modified descendants of a species will generally cause the extermination of the parent species due to their substantial similarity; and, if many new forms have been developed from any one species, the nearest relatives of that species (i.e., the species of the same genus) will be the most vulnerable to extermination. **This is how, I believe, a number of new species descended from one species—creating a new genus—comes to supplant an old genus within the same family.** But it must also have often happened that a new species belonging to some one group has seized on the place occupied by a species belonging to a distinctly different group and thus has caused its extermination. If many allied forms become developed from the successful intruder, many will have to yield their places, and it will generally be the allied forms that will suffer from some inherited inferiority in common. But whether it be species belonging to the same or to a different class that have yielded their places to other modified and improved species, a few of the sufferers may often persevere for a long time, from being particularly well-fitted to some peculiar line of life or from inhabiting some distant and isolated station in which they will have escaped severe competition. For instance, some species of *Trigonia*, a great, mostly extinct group of saltwater clams, still survive in the Australian seas and nowhere else, as far as we know. Similarly, a few members of the great and almost extinct group of Ganoid fishes still inhabit our fresh waters. Therefore, the utter extinction of a group of organisms is generally, as we have seen, a slower process than its production.

With respect to the apparently sudden extermination of whole families or orders—as of trilobites at the close of the Paleozoic period[8] and of ammonites at the close of the Mesozoic era[9]—we must remember what has already been said on the probable wide intervals of time between our consecutive geological formations, and, in those missing intervals, there may have been much slow extermination. Moreover, when many species of a new group have taken possession of an area, whether by sudden immigration or by unusually rapid development, many of the older species will have been exterminated in a correspondingly rapid manner and the forms that thus yield their places will commonly be related, for they will partake of the same inferiority in common.

It seems to me, then, that the manner in which single species and whole groups of species have become extinct accords well with the theory of natural selection. We need not marvel at extinction; if we must marvel, let it be at our presumption in imagining for a moment that we understand the many complex contingencies on which the existence of each species depends. If we forget for an instant that each species tends to increase over time to an excessive degree and that some check on its growth is always in action (although seldom perceived by us), the whole economy of nature will be utterly obscured. Only when we can precisely say why this species has more individuals than that species, or why this species and not another can be naturalized in a given country, then, and not until then, we may justly feel surprise at why we cannot yet account for the extinction of any particular species or group of species.

On the Forms of Life Changing Almost Simultaneously Throughout the World

Scarcely any paleontological discovery is more striking than the fact that the fossilized forms of life change almost simultaneously in geological deposits throughout the world. Thus our European Chalk formations[10] of late Cretaceous limestone can be recognized in many distant regions, and under the most different climates, where not a fragment of the mineral chalk itself can be found; namely, in North America, in equatorial South America, in Tierra del Fuego at the southern tip of South America, at the Cape of Good Hope, and in the Indian peninsula. For at these distant points, the organic remains in certain beds present an unmistakable resemblance to those of the Chalk. It is not that the same species are met with, for in some cases not one species is identically the same; yet they all belong to the same families, genera, and sections of genera, and sometimes are similarly characterized in such trifling points as mere superficial sculpture. Moreover, other forms, which are not found in the Chalk

---

[8] The Paleozoic ended about 252 million years ago.
[9] As mentioned earlier, the Mesozoic era ended about 66 million years ago.
[10] The "Chalk" consists mostly of calcium carbonate left from the tiny shells of single-celled organisms (foraminiferans and coccoliths).

of Europe but that occur in the geological formations either above or below, occur in the same order at these distant points of the world. In the several successive Paleozoic formations of Russia, Western Europe, and North America, a similar parallelism in the forms of life has been observed by several authors; and so it is, according to Lyell, with the European and North American tertiary deposits. Even if the few fossil species that are common to the Old and New Worlds were kept wholly out of view, the general parallelism in the successive forms of life in the Paleozoic and Tertiary stages would still be very clear, and the several formations could be easily correlated.

These observations, however, relate to the marine inhabitants of the world: we do not yet have sufficient data to judge whether the productions of the land and of freshwater at distant points have changed in the same manner. We may doubt whether they have thus changed: if the *Megatherium*, *Mylodon*, *Macrauchenia*, and *Toxodon* had been brought to Europe from La Plata without any information in regard to their geological position, no one would have suspected that they had co-existed with still-living shelled marine species; but as these anomalous monsters co-existed with the mastodon and horse, it might at least have been inferred that they had lived during one of the later Tertiary stages.

When the marine forms of life are spoken of as having changed simultaneously throughout the world, it must not be supposed that this expression relates to things happening in the same year, or even to the same century, or even that it has a very strict geological sense; for if all the marine animals now living in Europe, and all those that lived in Europe during the Pleistocene period (a very remote period as measured by years, including the whole glacial epoch)[11] were compared with those that now exist in South America or in Australia, the most skillful naturalist would hardly be able to say whether the present inhabitants of Europe or the Pleistocene inhabitants of Europe most closely resembled those of the southern hemisphere. So, again, several highly competent observers maintain that the existing species of the United States are more closely related to those that lived in Europe during certain late tertiary stages than they are to the present inhabitants of Europe; and, if this is so, it is evident that fossiliferous beds now deposited on the shores of North America would hereafter be liable to be classed with somewhat older European beds. Nevertheless, looking far into the future, there can be little doubt that all the more modern marine formations—namely, the upper Pliocene, the Pleistocene[10] and strictly modern beds of Europe, North and South America, and Australia—would be correctly ranked as simultaneous in a geological sense, in that they contain fossil remains that are related to some degree and do not include those forms that are found only in the older underlying deposits. The fact of various forms of life changing simultaneously (in the above large, geological sense) at distant parts of the world has greatly struck those admirable observers, Messrs. Edouard de Verneuil and Adolphe d'Archiac. After referring to the parallelism of the Paleozoic forms of life in various parts of Europe, they add,

---

[11] The Pleistocene is now commonly referred to as the "Ice Age."

"If struck by this strange sequence, we turn our attention to North America, and there discover a series of analogous phenomena, it will appear certain that all these modifications of species, their extinction, and the introduction of new ones, cannot be owing to mere changes in marine currents or other causes more or less local and temporary, but must depend on general laws which govern the whole animal kingdom." M. Barrande has made similarly forcible remarks to precisely the same effect. It is, indeed, quite futile to look to changes of currents, climate, or other physical conditions as the cause of these great changes in the forms of life throughout the world, under the most different climates. We must, as Barrande has remarked, look to some special law to explain these remarkable changes in life forms.[12] We shall see this more clearly when we consider the present distribution of organic beings and find how slight is the relation between the physical conditions of various countries and the nature of their inhabitants.

This great fact of the parallel succession of the forms of life throughout the world is explicable on the theory of natural selection. New species are gradually formed by having some advantage over older forms and the forms that are already dominant, or have some advantage over the other forms in their own country, give birth to the greatest number of new varieties, which I think of as incipient species. We have distinct evidence for this, in that the plants that are the most common and the most widely diffused have produced the greatest number of new varieties. It is also natural that the dominant, varying, and far-spreading species that have already invaded, to a certain extent, the territories of other species should have the best chance of spreading still further and of giving rise in new countries to other new varieties and new species. The process of diffusion would often be very slow, depending on climate changes and geographical changes, on strange accidents, and on the gradual acclimatization of new species to the various climates through which they might have to pass; but, over time, the dominant forms would generally succeed in spreading and would ultimately prevail. The diffusion would, it is probable, be slower for the terrestrial inhabitants of distinct continents than for the marine inhabitants of the continuous sea. We might therefore expect to find—as indeed we do find—a less strict degree of parallelism in the succession of terrestrial productions than with those of the sea.

It seems to me, then, that the parallel—and taken in a large sense—simultaneous succession of the same forms of life throughout the world accords well with the principle of new species having been formed by dominant species spreading widely and varying, with the new species thus produced being themselves dominant, owing to their having had some advantage over their already dominant parents as well as over other species, and again spreading, varying, and producing new forms. The old forms that were beaten and consequently forced to yield their places to the new, victorious forms will generally be allied in groups, from inheriting some inferiority in common. Thus, as new and improved groups spread throughout the world, old groups disappear

---

[12] See footnote 32.

from the world, and the succession of forms everywhere tends to correspond both in their first appearance and final disappearance.

There is one other remark connected with this subject that is worth making. I have previously explained, in Chapter 10, my reasons for believing that most of our great geological formations that are so rich in fossils were deposited during periods of subsidence and that blank time intervals of vast duration, as far as fossils are concerned, occurred whenever the sea bed was either stationary or rising, and likewise when sediment was not being thrown down quickly enough to embed and preserve organic remains, protecting them from decomposition and from being eaten. During these long and missing geological intervals, I suppose that the inhabitants of each region underwent a considerable amount of unrecorded modification and extinction and that there was much migration from other parts of the world. As we have reason to believe that large areas are affected by the same movement, it seems likely that strictly contemporaneous formations have often been accumulated over very wide areas in the same quarter of the world; but we are very far from having any right to conclude that this has invariably been the case and that large areas have invariably been affected by the same movements. When two formations have been deposited in two regions during nearly, but not exactly, the same period, we should find in both—from the causes explained in the foregoing paragraphs—the same general succession in the forms of life. But the species would not exactly correspond, for there will have been a little more time in the one region than in the other for modification, extinction, and immigration.

I suspect that cases of this nature occur in Europe. Mr. Joseph Prestwich,[13] in his admirable Memoirs on the Eocene deposits of England and France, is able to draw a close general parallelism between the successive stages in those two countries; but when he compares certain stages in England with those in France, although he finds in both countries a curious agreement in the numbers of the species belonging to the same genera, yet the species themselves differ in a manner very difficult to account for considering the proximity of the two areas—unless, indeed, we assume that an isthmus once separated two seas inhabited by distinct but contemporaneous faunas. Lyell has made similar observations on some of the later Tertiary formations. Barrande has also shown that there is a striking general parallelism in the successive Silurian deposits of Bohemia and Scandinavia; nevertheless, he finds a surprising amount of difference in the species that were living in the two areas. But this is to be expected if the several formations in these regions were not deposited during the same exact periods—a formation in one region often corresponding with a blank interval in the other—and if in both regions the species have gone on slowly changing during the accumulation of the several geological formations and during the long intervals of time between them. In this case, the several geological formations in the two regions could be arranged in the same order, in accordance with the general succession of the

---

[13] Prestwich eventually became Professor of Geology at Oxford University, in 1874.

forms of life, and the order would falsely appear to be strictly parallel; nevertheless the species would not all be the same in the apparently corresponding stages in the two regions.

On the Affinities of Extinct Species to Each Other, and to Living Forms

Let us now consider the mutual affinities of extinct and living species. All fall into a few grand classes, a fact that is easily explained on the principle of descent. The more ancient any form is, the more, as a general rule, it differs from living forms. But, as Buckland long ago remarked, extinct species can all be classed either within still existing groups or between them. That the extinct forms of life help to fill in the gaps between existing genera, families, and orders is certainly true, but as this statement has often been ignored or even denied, it may be worth making some remarks on this subject and to give some examples.

If we confine our attention either to the living or to the extinct species of any one taxonomic class of organisms, the series is far less perfect than if we combine both into one general system. In the writings of Professor Owen, we continually meet with the expression of generalized forms, as applied to extinct animals, and in the writings of Agassiz we read of prophetic or synthetic types; these terms imply that such forms are, in fact, intermediate or connecting links. Another distinguished paleontologist, M. Gaudry,[14] has shown in the most striking manner that many of the fossil mammals that he discovered in Attica, near Athens, Greece, serve to close up the gaps between the members of existing genera. Similarly, although Cuvier ranked the ruminants and pachyderms as two of the most distinct orders of mammals, so many fossil links have now been disentombed that Owen has had to alter the whole classification and has now placed certain pachyderms in the same suborder with cattle, deer, and other ruminants; he dissolves, for example, by gradations the apparently wide gap between the morphology of pigs and camels. The hoofed quadrupeds (ungulates) are now divided into even-toed or odd-toed groups,[15] but the fossilized *Macrauchenia* of South America[16] connects to a certain extent these two grand divisions. No one will deny that the members of the genus *Hipparion*[17] are intermediate between the existing horse and certain other existing ungulate forms. And what a wonderful connecting link in the chain of mammals is the rodent-like *Typotherium*

---

[14] Gaudry was a Frenchman who worked on European and South American mammals, including fossil ungulates.

[15] Whales and other cetaceans evolved from even-toed ungulates; for even-toed ungulate species, including pigs and hippopotamus, body weight is evenly borne by the third and fourth toes. With horses, zebras, donkeys, and other odd-toed ungulates, body weight is borne mostly by the third toe.

[16] This is a genus of long-necked and long-toed fossil mammals; Darwin collected the type specimens of this group during his voyage on the HMS Beagle.

[17] These are extinct horses that lived for more than 22 million years, from about 23 million years ago until about 781,000 years ago.

from South America,[18] as the name given to it by Professor Gervais expresses, and which cannot be placed in any existing order. Members of the aquatic Sirenia ("sea cows")—a taxonomic order that includes the manatees and dugongs—form a very distinct group of mammals, and one of the most remarkable peculiarities in existing dugong and lamentin species is the entire absence of hind limbs, without even a rudiment being left; but the extinct *Halitherium*, a primitive sea cow, had, according to Professor Flower, an ossified thigh-bone "articulated to a well-defined acetabulum in the pelvis," and it thus makes some approach to the characteristics of ordinary hoofed quadrupeds, to which members of the mammalian order Sirenia are in other respects allied. The cetaceans (i.e., whales, dophins, and porpoises) are widely different from all other mammals, but members of the genera *Zeuglodon* and *Squalodon*[19] from the Tertiary, which have been placed by some naturalists in an order by themselves, are considered by Professor Huxley to undoubtedly be cetaceans "and to constitute connecting links with the aquatic Carnivora."[20]

Even the wide interval between birds and reptiles has been shown by Professor Huxley to be partially bridged over, in the most unexpected manner, by the ostrich and the extinct *Archeopteryx* on the one hand, and, on the other hand, by *Compsognathus*, a genus of small, bipedal dinosaurs, a group that includes the most gigantic of all terrestrial reptiles.

Turning to the invertebrates, the French geologist and paleontologist Joachim Barrande asserts—and a higher authority could not be named—that he is every day taught that, although fossilized Paleozoic animals can certainly be classed under existing groups, those groups were not so distinctly separated from each other at that ancient period as they now are.

Some writers have objected to the idea that any extinct species—or group of species—should be considered as being intermediate between any two living species or groups of species. If by this term they mean that an extinct form is directly intermediate in all its characters between two living forms or groups, then that objection is probably valid. But in a natural classification many fossil species certainly have characteristics that stand between those of living species and some extinct genera between living genera, even between genera belonging to distinct families. The most common case, especially with respect to such very distinct groups as fish and reptiles, seems to be that, supposing them to be distinguished at the present day by a score of characters, the ancient members are separated by somewhat fewer characters so that the two groups formerly made a somewhat nearer approach to each other than they now do.

It is commonly believed that the more ancient a form is, the more at least some of its characters tend to connect groups now widely separated from each other. This remark no doubt will only apply to those groups that have undergone much change

---

[18] The members of this genus lived from about 55 million years ago to about 2 million years ago.

[19] These are primitive whales. The genus *Zeuglodon* is now known as *Basilosaurus* ("king lizard"). These animals lived for about 6 million years, from about 40 million years ago to 34 million years ago. Fossils of *Squaldon* are between about 33 million and 14 million years old.

[20] The Carnivora is an order that today includes about 280 species of placental mammals.

in the course of geological ages. It would be difficult to prove the truth of the proposition, however, for every now and then even a living animal, such as the South American lungfish *Lepidosiren*, is discovered having affinities directed toward very distinct groups. Yet, if we compare the older Reptiles and Batrachians, the older Fish, the older Cephalopods, and the Eocene Mammals, with the recent members of the same classes, we must admit that there is general truth in the remark.

Now let us see how far these several facts and inferences accord with the theory of descent with modification. As the subject is somewhat complex, I must request the reader to tur,n to the diagram in Chapter 4 (Figure 4.11). We may suppose that the numbered letters in italics represent genera, while the dotted lines diverging from them represent the species in each genus. The diagram is much too simple, showing such a small number of genera and species, but this is not important for us here. The horizontal lines may represent successive geological formations, and all the forms beneath the uppermost line may be considered to have gone extinct. The three existing genera, a14, q14, p14, will form a small family; b14 and f14, a closely allied family or subfamily; and o14, i14, m14, a third family. These three families, together with the many extinct genera on the several lines of descent diverging from the parent form (A) will form an order; all will have inherited something in common from their ancient ancestor. On the principle of the continued tendency to divergence of character, which was formerly illustrated by this diagram, the more recent any form is, the more it will generally differ from its ancient ancestor. Hence, we can understand the rule that the most ancient fossils differ most from currently existing forms. We must not, however, assume that divergence of character is a necessary contingency; it depends solely on the descendants from a species being thus enabled to seize on many and different places in the economy of nature. Therefore it is quite possible, as we have seen in the case of some Silurian forms,[21] that a species might go on being slightly modified in relation to its slightly altered conditions of life and yet still retain the same general characteristics throughout a vast period. This is represented in the diagram by F14.

As I noted previously, all the many forms—both extinct and recent—that have descended from (A) combine to form a single order, and this order, from the continued effects of extinction and the gradual divergence of character, has gradually become divided into several subfamilies and families, some of which are supposed to have perished at different geological periods and some to have endured to the present day.

By looking at the diagram, we can see that if many of the extinct forms supposed to be embedded in the successive geological formations were discovered at several points low down in the series, the three existing families on the uppermost line would be rendered less distinct from each other. If, for instance, the genera a1, a5, a10, f8, m3, m6, m9, were dug up, these three families would be so closely linked together that they probably would have to be united into a single great family, in nearly the same

---

[21] The Silurian is part of the Paleozoic era and lasted from about 444 million years ago until about 419 million years ago (see Table 10.1).

manner as has occurred with ruminants and certain pachyderms. Yet anyone would be partly justified in refusing to consider the extinct genera as being intermediate between the others, which thus link together the living genera of three families, for they are not directly intermediate but are rather intermediate only by a long and circuitous course through many widely different forms. If many extinct forms were to be discovered above one of the middle horizontal lines or geological formations—for instance, above No. VI—but none from beneath this line, then only two of the families (those on the left hand a14, etc., and b14, etc.) would have to be united into one family, and two families would remain that would be less distinct from each other than they were before the discovery of the fossils. So again, if we suppose that the three families formed of the eight genera (a14 to m14) on the uppermost line differ from each other by half a dozen important characteristics, then the families that existed at the period marked VI would certainly have differed from each other by a smaller number of characters for they would at this early stage of descent have diverged to a lesser degree from their common ancestor. This would explain why ancient and extinct genera are often to a greater or lesser degree intermediate in character between their modified descendants or between their collateral relations.

Under nature, the process will be far more complicated than is represented in the diagram: the groups will have been more numerous, they will have endured for extremely unequal lengths of time, and they will each have been modified to various degrees. As we possess only the last volume of the geological record, and that in a very broken condition, we have no right to expect, except in rare cases, to fill up the wide intervals in the natural system and thus to unite distinct families or orders. All that we have a right to expect is that those groups that have, within known geological periods, undergone much modification should in the older formations make some slight approach to each other, so that the older members should differ less from each other in some of their characters than do the existing members of the same groups—and this, by the concurrent evidence of our best paleontologists, is frequently the case.

Thus, the theory of descent with modification explains the main facts with respect to the mutual affinities of the extinct forms of life to each other and to still living forms in a satisfactory manner. Those facts are wholly inexplicable on any other view.

On this same theory, it is evident that the fauna during any one great geological period in the Earth's history will be intermediate in general character between that which preceded and that which succeeded it. Thus the species that lived at the sixth great stage of descent in the diagram are the modified offspring of those that lived at the fifth stage and are the parents of those that became still more modified at the seventh stage; hence they could hardly fail to be nearly intermediate in character between the forms of life above and below on the diagram. We must, however, allow for the entire extinction of some preceding forms and, in any one region, must also allow for the immigration of new forms from other regions and for a large amount of modification during the long and blank intervals between the successive geological formations. Subject to these allowances, the fauna of each geological period is undoubtedly intermediate in character between the preceding and succeeding faunas. I need give only one example: namely, the manner in which the fossils of the Devonian system, when this system was first discovered, were at once recognized by paleontologists as intermediate in character between those of the overlying Carboniferous and underlying Silurian geological systems (see Table 10.1). But each fauna is not necessarily exactly intermediate, as unequal intervals of time have elapsed between consecutive formations.

The characteristics of the fauna of each geological period as a whole are nearly intermediate between those of the faunas from preceding and succeeding periods. It is no real objection to the truth of that general statement that certain genera offer exceptions. For instance, consider that when Dr. Falconer arranged the species of mastodons and elephants in two different series—the first according to their mutual similarities and the second according to their periods of existence—the two arrangements did not agree. But remember that the species that are extreme in character are not necessarily the oldest or the most recent, nor are those that are intermediate in character also intermediate in age. Supposing for an instant, in this and other such cases, that the record of the first appearance and disappearance of the species was complete, which is far from the case, we have no reason to believe that forms successively produced will necessarily endure for corresponding lengths of time. A very ancient form may occasionally have lasted much longer than a form subsequently produced elsewhere, especially in the case of terrestrial productions inhabiting separated regions. To compare small things with great things, consider this: if the principal living and extinct races of the domestic pigeon were arranged in serial affinity, that arrangement would not accord closely with the order in time of their production and even less with the order of their disappearance—for one thing, the parent rock pigeon still lives. Moreover, many varieties between the rock pigeon and the carrier pigeon have become extinct, and carrier pigeons that are extreme in the important character of beak length originated earlier than did the short-beaked tumblers, which are at the opposite end of the series in this respect.

Closely connected with the statement that the organic remains from an intermediate geological formation are in some degree intermediate in character is the widely accepted fact that fossils from two consecutive geological formations are far more closely related to each other than are the fossils from two remote formations. Professor Pictet, for example, notes the general resemblance of the organic remains from the several stages of the Chalk formation, even though the species in each layer are distinct. This fact alone, from its generality, seems to have shaken Pictet's original belief in the immutability of species. Anyone who is acquainted with how existing species are presently distributed over the globe will not attempt to account for the close resemblance of distinct species in closely consecutive geological formations by the physical conditions of the ancient areas having remained nearly the same. Remember that the forms of life, at least those inhabiting the sea, have changed almost simultaneously throughout the world, and therefore under the most different climates and conditions. Consider the prodigious vicissitudes of climate during the

Pleistocene period, which includes the whole glacial epoch, and note how little the specific forms of the inhabitants of the sea were affected.

On the theory of descent with gradual modification, the full meaning of the fossil remains found in closely consecutive geological formations being closely related to each other is obvious, although though they are typically ranked as distinct species. As the accumulation of the deposits in each geological formation has often been interrupted, and as long blank intervals of time have intervened between successive geological formations, we should not expect to find (see Chapter 10) in any one or in any two geological formations all of the intermediate varieties between the species that appeared at the start and at the close of these periods. But we ought to find after intervals—very long as measured by years but only moderately long as measured geologically—closely allied forms, or, as they have been called by some authors, "representative species"—and these assuredly we do find. We find, in short, such evidence of the slow and scarcely detectable modifications of specifies as we have any right to expect.

On the State of Development of Ancient Compared with Living Forms

We saw in Chapter 4 that the degree of differentiation and specialization of the various parts in living beings, once those beings have reached maturity, is presently the best standard that we have of judging their degree of perfection. We have also seen that, because the specialization of parts is generally advantageous to each being, so natural selection will tend to render the organization of each being more specialized and more perfect over time, and in this sense "higher"; of course it can also leave many creatures with simple and unimproved structures that are well-fitted for simple conditions of life, and in some cases will even degrade or simplify the organization in ways that leave such degraded beings better fitted for their new walks of life. In another and more general manner, new species will generally become superior to their predecessors; indeed, to succeed in the struggle for life they must beat all the older forms that they come into close competition with. We may therefore conclude that if under a nearly similar climate the Eocene inhabitants of the world could be put into competition with the inhabitants existing today, the Eocene organisms would be beaten and exterminated by the latter, as would the Mesozoic inhabitants by those in the Eocene. The Paleozoic inhabitants would similarly be beaten and exterminated by the Mesozoic forms.[22] Thus, by this fundamental test of victory in the battle for life, as well as by the standard of the specialization of organs, modern forms ought, on the theory of natural selection, to stand higher—and be more generally complex—than

---

[22] Indeed, it now seems that the evolution of teleost fishes drove all trilobite and many brachiopod species into extinction and may also have driven selection for the loss of external shells among cephalopods.

ancient forms. Is this the case? Most paleontologists would answer "Yes," and it seems that this answer must be admitted to be true, though difficult of proof.

That certain brachiopod species[23] have been only slightly modified from an extremely remote geological period is no valid objection to this conclusion; nor is the finding that the shells of certain terrestrial and freshwater snail and bivalve species have remained nearly the same from the time when—as far as is known—they first appeared. And it is similarly not an insuperable difficulty that members of the Foraminifera[24] have not, as Dr. Carpenter has insisted, become any more complex in organization since even the Laurentian (i.e., Precambrian) epoch; some organisms would surely have had to remain fitted for simple conditions of life, and what could be better fitted for this end than these lowly organized protozoans? These objections would be fatal to my view if a continuous advance in organization was a necessary contingent. They would likewise be fatal if the foraminiferans that I just mentioned could be shown to have first come into existence during the Laurentian epoch, or if the brachiopods mentioned above first appeared during the Cambrian formation; for, in this case, there would not have been enough time for the development of these organisms up to the standard which they had then reached. Once any organisms have advanced up to any given point, there is no necessity, on the theory of natural selection, for their further continued process, although they will, during each successive age, have to be slightly modified if they are to hold their places in relation to slight changes in their conditions. The foregoing objections hinge on the question of whether we really know how old the world is and at what period the various forms of life first appeared, and this may well be disputed.[25]

The problem of whether or not organization on the whole has become more complex over time is in many ways excessively intricate. The geological record, which is clearly and routinely imperfect, does not extend far enough back in time to show with unmistakable clearness thatthe general organization of organisms has largely advanced within the known history of the world. Even at the present day, looking to members of the same taxonomic class, naturalists are not unanimous about which forms ought to be ranked as the most highly developed: thus, some consider the selaceans or sharks to be the highest, most perfectly developed fish, based on some important structural relationships to reptiles, while others look at teleosts as the most highly developed fishes. The ganoids (a group of fish with hard bony scales, as with

---

[23] Brachiopods are a group of marine organisms with shells that resemble those of bivalves, although the shell symmetry and internal anatomy of these two groups differ considerably. This group includes the genus *Lingula*, which has been mentioned several times earlier in this chapter.

[24] Foraminiferans are single-celled, mostly marine, amoeba-like organisms that secrete single or multi-lobed shells, which are usually composed of calcium carbonate. Most foraminiferan species today are benthic, living in or on the sea bottom—even at the deepest parts of the ocean—but some are planktonic. There are about 10,000 foraminiferan species living today, but about 40,000 species have been found so far in the fossil record, starting about 540 million years ago at the start of the "Cambrian explosion."

[25] Indeed, we now know more about these issues with the help of radiometric age dating techniques. The planet Earth is now known to be more than 4 billion years old, far older than anyone believed during Darwin's time. And, as mentioned again in footnote 31, fossils of bacteria have now been found in geological layers that are more than 3 billion years old.

sturgeon) stand intermediate in their characteristics between the selaceans and teleosts; at the present day, there are far more teleost species on our planet, but at one time selaceans and ganoids alone existed. And, in this case, depending on the standard of what constitutes "highness," so will it be said that fishes have either advanced or retrogressed in their degree of organization over time.

To attempt to compare members of distinct types in the scale of highness seems hopeless; who will decide whether a cuttlefish[26] is higher in organization than a bee— that insect which the great Von Baer believed to be "in fact more highly organized than a fish, although upon another type"? In the complex struggle for life, it is quite credible that crustaceans, although not very high in their own group (the phylum Arthropoda) might beat the cephalopods, which are the highest, most complex molluscs; and that such crustaceans, though not so highly developed, would stand especially high in the scale of invertebrate animals if judged by the most decisive of all trials: the law of battle. In addition to these inherent difficulties in deciding which life forms are the most advanced in organization, we should not only compare the highest members of a class at any two geological periods—though undoubtedly this is one and perhaps the most important element in striking a balance—but we also ought to compare all the members, high and low, at two time periods. At an ancient epoch the highest and lowest molluscoidal animals—namely, cephalopods and brachiopods— swarmed in large numbers; at the present time, however, both groups are greatly reduced while others, intermediate in their organization, have largely increased; in consequence, some naturalists maintain that molluscs were formerly more highly developed than they are at present. But a stronger case can be made for the opposite conclusion by considering the vast reduction in the number of brachiopod species over evolutionary time and the fact that our existing cephalopods, including the octopus, though few in the number of species, are more highly organized than their ancient representatives.[27]

We ought also to compare the relative proportional numbers, at any two time periods, of the high and low classes of organisms throughout the world. If, for instance, at least 50,000 kinds of vertebrate animals exist on Earth today, and if we knew that at some former period only 10,000 kinds existed, we ought to look at this increase in number in the highest class as something that implies a great displacement of lower forms, as a decided advance in the organization of the world. We thus see how hopelessly difficult it is to compare with perfect fairness, under such extremely complex relations, the standard of organization of the imperfectly known faunas of successive geological periods.

---

[26] Cuttlefish are cephalopods, a group that also contains the octopus and squid, along with the chambered nautilus. Cuttlefish have an interesting internal shell, commonly called *cuttlebone*. Cuttlebones are commonly used as calcium-rich dietary supplements for a wide range of pets, including birds, chinchillas, reptiles, and hermit crabs. Remarkably, cephalopods are in the same phylum as the snails and bivalves, the Mollusca.

[27] Indeed, the octopus has an amazingly large and complex brain and a remarkably complex visual system and range of behaviors.

We shall appreciate this difficulty more clearly by looking at certain existing faunas and floras. From the extraordinary manner in which European species have recently spread over New Zealand and have seized on places that must have previously been occupied there by indigenous species, we must believe that if all the animals and plants of Great Britain were to be set free in New Zealand, a multitude of British forms would in the course of time become thoroughly naturalized there as well, and would exterminate many of the native species. On the other hand, from the fact that hardly a single inhabitant of the southern hemisphere has become wild in any part of Europe, we may well doubt whether, if all the species unique to New Zealand were set free in Great Britain, any considerable number would be able to seize on places now occupied by our native plants and animals. Viewed in this way, the productions of Great Britain stand much higher in the scale of complexity than those of New Zealand. Yet the most skillful naturalist, by simply examining the species in these two countries, could not have foreseen this result.

Agassiz and several other highly competent judges insist that ancient animals resemble, to a certain extent, the embryos of recent animals belonging to the same classes, and that the geological succession of extinct forms is nearly parallel with the embryological development of existing forms. This view accords admirably well with our theory. In Chapter 14, I shall attempt to show that the adult form of an organism differs from that of its embryo because of variations having supervened at a not early age and having been inherited by the descendants at a corresponding age. This process, while it leaves the embryo almost unaltered, continually adds, in the course of successive generations, more and more difference to the adult. Thus the embryo comes to be left as a sort of picture, preserved by nature, of the former and less modified condition of the ancestral species.[28] This view may be true and yet may never be capable of being proven. Seeing, for instance, that the oldest known mammals, reptiles, and fishes strictly belong to their proper classes, it would be in vain to look for animals having the common embryological character of the Vertebrata until beds rich in fossils were discovered far beneath the lowest Cambrian strata—a discovery of which the chance is small.

On the Succession of the Same Types Within the Same Areas, During the Later Tertiary Periods

Mr. William Clift showed many years ago that the fossil mammals that were found in Australian caves were closely related to the living marsupials of that continent. In South America, a similar relationship is readily apparent, even to an uneducated eye,

---

[28] Sorry, Charles. We now understand that while the similarities in embryological development seen in members of a given taxonomic group do indeed reflect a common ancestry, embryological characteristics do not illustrate what adult forms looked like earlier in their evolution.

in the gigantic pieces of armor (e.g., those of the armadillo) found in several parts of La Plata, Argentina. Similarly, Professor Owen has shown in the most striking manner that most of the fossil mammals that are buried there in such numbers are clearly related to present-day South American species. This relationship is even more clearly seen in the wonderful assortment of fossil bones collected from the caves of Brazil by the Danish naturalists Peter Wilhelm Lund and Peter Clausen.

I was so much impressed with these facts that I strongly insisted, both in 1839 and 1845, on this "law of the succession of types" and on "this wonderful relationship in the same continent between the dead and the living." Professor Owen has subsequently extended the same generalization about relationships to the mammals of the Old World. We see the same law in his restorations of the extinct and gigantic birds of New Zealand, and we also see it in the birds of Brazilian caves. Mr. Woodward has shown that the same law holds good with seashells, although, from the wide distribution of most molluscs, it is not as well displayed by them. Other cases could be added, such as the relationships between the extinct and living land snails of Madeira, and between the extinct and living brackish water snails of the Aralo-Caspian Sea.

Now, what does this remarkable law of the succession of the same types within the same areas mean? He would be a bold person who, after comparing the present climate of Australia with parts of South America—continents that share the same latitude—would attempt to account, on the one hand, through dissimilar physical conditions, for the dissimilarity of the inhabitants of these two continents, and, on the other hand, through similarity of conditions for the uniformity of the same types in each continent during the later tertiary periods. Nor can anyone pretend that it is an immutable law that marsupials should have been chiefly or solely produced in Australia or that members of the order Edentata[29] and other American types should have been solely produced in South America. For we know that in ancient times Europe was occupied by numerous marsupials and that, in the publications alluded to earlier, it was shown that in America the law of distribution of terrestrial mammals was formerly different from what it is now. The terrestrial mammals of North America formerly resembled—and quite strongly— those presently found in the southern half of the continent, and those in the southern half were formerly more closely allied to the northern half than they are at present. In a similar manner we know, from Falconer and Cautley's discoveries, that the mammals of Northern India were formerly more closely related to those of Africa than they are at the present time. Analogous facts could be given in relation to the distribution of marine animals.

On our theory of descent with modification, we can immediately explain the great law of the long enduring—but not immutable—succession of the same types within the same areas over long periods of time; for the inhabitants of each part of the world will obviously tend to leave behind in that part, during the next succeeding period of time, closely allied, although to some degree modified, descendants. If the

---

[29] A former group of placental mammals that included the aardvarks, anteaters, armadillos, and tree sloths.

inhabitants of one continent formerly differed greatly from those of another continent, so will their modified descendants still differ to nearly the same manner and degree. But after very long intervals of time and after great geographical changes that will have permitted much intermigration between areas that were formerly separated from each other, the feebler forms will yield to the more dominant forms, and there will be nothing immutable in the distribution of organic beings.

It may be asked in ridicule whether I suppose that the Megatherium[30] and other huge allied monsters that formerly lived in South America have left behind them the sloth, armadillo, and anteater as their degenerate descendants. This cannot for an instant be admitted. These huge animals have become wholly extinct and have left no descendants. But in the caves of Brazil there are many extinct species that are very similar in size and in all other characteristics to the species still living in South America; some of these fossils may have been the actual ancestors of today's living species. It must not be forgotten that, on our theory, all the species of any one genus have descended from some one ancestral species; thus if six genera, each having eight species, were to be found in one geological formation, and in a succeeding, more recent geological formation there be six other allied or representative genera, each with the same number of species, then we may conclude that generally only one species of each of the older genera has left modified descendants, which constitute the new genera containing the several species; the other seven species of each old genus would have gone extinct and left no progeny. Or, and this will be a far commoner case, two or three species in only two or three of the six older genera will have given rise to members of the new genera, with the other species and the other old genera having become utterly extinct. In failing orders, with the genera and species decreasing in numbers over time, as is the case with the order Edentata of South America, still fewer genera and species will leave behind modified blood descendants.

Summary of the Preceding and Present Chapters

In this and the previous several chapters I have attempted to show the following: (1) that the geological record is extremely imperfect; (2) that only a small portion of the globe has so far been geologically explored with care; (3) that only certain classes of organisms have been largely preserved as fossils; (4) that the number of specimens and of species that have been preserved in our museums is absolutely as nothing compared with the number of generations that must have passed away long ago, even within a single geological period; (5) that, owing to subsidence being almost

---

[30] *Megatherium* was a genus of ground sloths that were the size of today's elephants. These animals, unique to South America, lived from about 5.3 million years ago until the end of the most recent Ice Age, nearly 12,000 years ago. Their extinction may well have been caused by the growing numbers of human hunters (see Figure 11.2).

necessary for the accumulation of deposits rich in fossil species of many kinds and being thick enough to outlast future degradation, great intervals of time must have elapsed between most of our successive geological formations; (6) that there has probably been more extinction during the periods of subsidence and more variation during the periods of elevation, and, during the latter, the geological record will have been the least perfectly kept; (7) that each single geological formation has not been continuously deposited; (8) that the duration of each geological formation is probably short compared with the average duration of specific forms; (9) that migration has played an important part in the first appearance of new forms in any one area and geological formation; (10) that widely ranging species have varied most frequently and have most often given rise to new species; (11) that varieties have at first been only local; and last, (12) that although each species must have passed through numerous transitional stages, the periods during which each underwent modification, though many and long as measured by years, have probably been short in comparison with the periods during which each remained unchanged. **Taken together, these causes will to a large extent explain why—though we do find many links—we do not find interminable varieties that connect together all extinct and existing forms by the finest graduated steps.** It should also be constantly borne in mind that any fossilized variety that might be found linking two forms would be ranked as a new and distinct species, unless the whole chain could be perfectly restored; we cannot pretend that we have any sure criterion enabling us to discriminate between species and varieties.

Anyone who rejects this view of the imperfection of the geological record will rightly reject the whole theory. For some readers may ask in vain "Where are the numberless transitional links that must formerly have connected the closely allied or representative species that are found in the successive stages of the same great geological formation?" Some may refuse to believe in the immense intervals of time that must have elapsed between our consecutive geological formations; some may overlook the importance of migration, when the formations of any one great region, as those of Europe, are considered; some may urge the apparent—but often falsely apparent—sudden appearance of whole groups of species in the fossil record. Some may ask "Where are the remains of those infinitely numerous organisms that must have existed long before the Cambrian system was deposited?" We now know that at least one animal did then exist.[31] But I can answer this last question only by supposing that where our oceans now extend they have extended for an enormous period, and where our oscillating continents now stand they have stood since the commencement of the Cambrian system,[32] but that, long before that epoch, the world presented a

[31] As noted earlier, in footnote 25, well-defined fossils of bacteria have now been found in deposits that are more than 3 billion years old, in Australia. And fossils of multicellular, soft-bodied animals have now been found from a variety of sites in various parts of the world from as far back as 635 million years ago, well before the start of the Cambrian. So, yes, indeed, life—including multicellular life—did in fact exist well before the Cambrian "explosion" that took place about 540 millions years ago.

[32] We now know that entire continents slowly move in different directions, changing ocean widths and the distances between major land masses. The movement is caused by the uprising of molten magma from deep within the Earth, through weak areas ("deep-sea rifts") along the ocean floor. These ideas of plate

widely different aspect and that the older continents, formed of formations older than any known to us, exist now only as remnants in a metamorphosed condition, or lie still buried under the ocean.

Passing from these difficulties, the other great leading facts in paleontology agree admirably with the theory of descent with modification through variation and natural selection. We can thus understand how it is that new species come in slowly and successively; how species of different classes do not necessarily change at the same time, or at the same rate, or to the same degree; and yet in the long run that all undergo modification to some extent. **The extinction of old forms is the almost inevitable consequence of the development of new forms.** We can now also understand why, when a species has once disappeared, it never reappears. Groups of species increase in numbers slowly, and they endure for unequal periods of time, for the process of modification is necessarily slow and depends on many complex contingencies. The dominant species belonging to large and dominant groups tend to leave many modified descendants that go on to eventually form new subgroups and larger groups. As these are formed, the species of the less vigorous groups—from their inferiority inherited from a common ancestor—tend to become extinct together and to leave no modified offspring on the face of the Earth. But the utter extinction of a whole group of species has sometimes been a slow process, resulting from the survival of a few descendants having lingered in protected and isolated situations. **Once a group has wholly disappeared, however, it never reappears: the link of generation has been broken.**

We can also understand how it is that dominant forms that spread widely and yield the greatest number of varieties tend to populate the world with allied, but modified, descendants, and how it is that those descendants will generally succeed in displacing those groups that are their inferiors in the struggle for existence. Hence, after long intervals of time, the productions of the world will appear to have changed simultaneously.

And we can understand how it is that all the forms of life, whether ancient or recent, make together only a few grand classes. We can understand, from the continued tendency to diverge in character, why the more ancient a form is, the more it generally differs from those now living, and why ancient and extinct forms often tend to fill up gaps between existing forms, sometimes blending two groups that were previously classed as distinct into one, but more commonly bringing them only a little closer together. The more ancient a form is, the more often it stands in some degree intermediate between groups now distinctly different; for the more ancient a form is, the more nearly it will be related to, and consequently resemble, the common ancestor of groups that subsequently became widely divergent. Extinct forms are seldom

tectonics and continental drift were not fully accepted until the 1960s, although they were proposed a number of times beginning in the 1500s. Such "continental drift" may well explain Darwin's previous observation that various forms of life in the geological record seem to have changed simultaneously in different parts of the world: those different parts were more closely connected at those times and became distantly separated later, through continental drift.

directly intermediate between existing forms, but rather are intermediate only by a long and circuitous course through other extinct and different forms. We can clearly see why the organic remains of closely consecutive geological formations are closely allied: they are closely linked together by generation. And we can clearly see why the remains of an intermediate geological formation are intermediate in character.

All of these things make good sense once we accept the idea of gradual change through natural selection. The inhabitants of the world at each successive period in its history have beaten their predecessors in the race for life and are, in that sense, higher in the scale, and their structure has generally become more specialized; this may account for the common belief held by so many paleontologists that organization on the whole has progressed over time. Extinct and ancient animals resemble to a certain extent the embryos of the more recent animals belonging to the same classes, and this wonderful fact receives a simple explanation according to our views. The succession of the same types of structure within the same geographical areas during the later geological periods ceases to be mysterious and is intelligible on the principle of inheritance.

If, then, the geological record is really as imperfect as many believe it to be, the main objections to the theory of natural selection are greatly diminished or disappear altogether. On the other hand, it seems clear to me that all the chief laws of paleontology plainly proclaim that species have been produced by ordinary generation, with old forms having been gradually supplanted by new and improved forms of life, the unavoidable products of variation and the subsequent survival of the fittest.

Key Issues to Talk and Write About

1. If you had written this chapter, what are two key issues that you would want to have readers think and write about?

12
Geographical Distribution

The more one looks at how species are distributed around the world, the more questions one comes up with. For example, why are the same species so often found living in many different places that are widely separated from each other (e.g., by mountains, valleys, or an ocean) instead of being limited to one particular area? The degree to which the same or similar species are found in widely separated areas is not explained by the climatic conditions in those areas; indeed, widely separated regions with very similar climates often support completely different groups of organisms. So what is the explanation? And how do we explain why areas that are separated by physical barriers that prevent migration between them sometimes bear similar species but sometimes have especially great differences in species composition? And why do we find the same or similar terrestrial species living on the mainland and also on nearby islands? In this chapter, Darwin explains all of these intriguing observations at many levels, using facts and logical thinking. For one thing, it all makes sense if each species originated in a single location. Their later distributions will then be related to the degree to which the members of a given species have been able to disperse over long distances or have been separated from other members for long periods of time. Isolated groups of a single species are subject to different selective pressures and may gradually become more and more different from each other through natural selection and, eventually, even become different species. Geological forces seem to have played a major role in creating or removing barriers to dispersal. In particular, over long periods of time, there have been many changes in the height of the land relative to the height of the adjacent sea, creating many opportunities, by a variety of means, for dispersing offspring into many distant places or dispersing seeds over great distances. Darwin also summarizes evidence indicating that various regions of our planet have long ago experienced marked, long-lasting changes in climate. These changes caused, for example, migration to new areas during periods of cooling, fueled by the ability of cold-adapted organisms to spread to new areas as temperatures fall and later, as conditions begin to warm, fueled by the need of those organisms to live under colder conditions. If we assume that each species has arisen at a single location, then where they are currently found must be related to the degree to which they have been able to spread to other areas or to which they have become isolated from other members of the species.

Thus the present geographical distribution of species is related to geological events and gradual changes in climate, along with patterns of dispersal and the impact of natural selection.

The Readable Darwin. Second Edition. Jan A. Pechenik, Oxford University Press. © Oxford University Press 2023.
DOI: 10.1093/oso/9780197575260.003.0013

In considering the distribution of living organisms over the face of the globe, **the first great fact that strikes us is that neither the similarities nor the dissimilarities of the inhabitants of various regions can be wholly accounted for by differences in climate or other physical conditions**; almost every author who has studied the subject has recently come to this very same conclusion. The case of America alone would almost suffice to prove its truth. If we exclude the Arctic and northern temperate regions, all authors agree that one of the most fundamental divisions in the geographical distribution of species is that between the New and Old Worlds.[1] And yet, if we travel over the vast American continent from the central parts of the United States to its extreme southern point, we meet with the most diversified environmental conditions: humid districts, arid deserts, lofty mountains, grassy plains, forests, marshes, lakes and great rivers, and covering an enormous range of temperatures. Indeed, there is hardly a climate or condition in the Old World that cannot be paralleled in the New—at least so closely as the same species generally require. **Notwithstanding this general parallelism in the physical conditions of Old and New Worlds, yet how widely different are their species compositions!** How can we explain that? And in the southern hemisphere, if we compare large tracts of land in Australia, South Africa, and western South America between latitudes 25 and 35 degrees, we shall find regions that are extremely similar in all their physical conditions, and yet it would not be possible to point out three faunas and floras that are more utterly dissimilar than these. Or, again, we may compare the species of South America that are found south of latitude 35 degrees with those found north of latitude 25 degrees—areas that are exposed to considerably different physical conditions; remarkably, the species found in those two regions are incomparably more closely related to each other than they are to the species of Australia or Africa that are living under nearly the same climate conditions. Analogous facts could be given with respect to the inhabitants of the sea. So, clearly, the similarities and differences among species living in different lands have little to do with climate conditions.

   A second great fact which strikes us in our general review is that barriers of any kind, or any other obstacles to free migration, are related in a close and important manner to the differences seen among the organisms living in different regions. We see this in the great difference in nearly all the terrestrial organisms living in the New and Old Worlds, except in the northern parts where the land masses almost join, and where, under a slightly different climate long ago, there might have been free migration for the northern temperate forms, as there now is for the strictly arctic organisms. We see this same fact in the great difference between the inhabitants of Australia, Africa, and South America living at the same latitudes and thus under similar climates; these countries are almost as much isolated from each other as is possible, so that migration between them should be essentially impossible. Moreover, we see the same fact within each continent: we find different species living on the opposite

---

[1] Here Darwin is referring to North America and Europe, for example.

sides of lofty and continuous mountain ranges, and of great deserts, and even of large rivers. However, as mountain chains, deserts, and rivers are not as impassable, or as likely to have endured so long as the oceans that separate the continents from each other, the differences seen in species composition are not nearly as great as those characterizing distinct continents.

We find the same law at work in the sea. The marine inhabitants of the eastern and western shores of South America are very distinct, with extremely few shelled species, crustacean species, or echinoderm species in common.[2] In contrast, Dr. Gunther has recently shown that about 30% of the fish species are the same on the opposite sides of the isthmus of Panama; but this fits in, too, as naturalists now believe that the isthmus was formerly open, allowing free exchange of individuals from one side to the other.

Heading further west from the west coast of America, there is a wide stretch of open ocean, the Pacific Ocean, with not a single island available as a halting place for emigrants; here we have a barrier of another kind—a barrier of open ocean—and as soon as this is passed we meet a totally different fauna in the eastern islands of the Pacific. Thus we have three distinct marine faunas ranging northward and southward in parallel lines, not far from each other and living under corresponding climate conditions; but because they are separated from each other by impassable barriers, either of land or open sea, they are almost wholly distinct.

On the other hand, proceeding still further westward from the eastern islands of the tropical parts of the Pacific Ocean, we encounter no impassable barriers and have innumerable islands as potential halting places, or continuous coasts, until, after traveling over a hemisphere, we come to the shores of Africa; over this vast space we meet with no well-defined and distinct marine faunas. **Again, it is clear that barriers to dispersal generally result in very different species compositions, whereas a lack of such barriers results in the same species being found over very long distances.**

Although so few marine animal species are common to the above-named three approximate faunas of Eastern and Western America and the eastern Pacific islands, yet many fish species range all the way from the Pacific Ocean into the Indian Ocean, and many shelled mollusc species[3] are common to both the eastern islands of the Pacific and the eastern shores of Africa on almost exactly opposite meridians of longitude.

A third great fact is that species found on the same continent or in the same sea are clearly related to each other, although the species themselves are distinct at different points and stations. This is a law of the widest generality, and every continent offers innumerable instances. Nevertheless, the naturalist, in traveling, for instance, from north to south, never fails to be struck by the manner in which successive groups of organisms, specifically distinct though clearly closely related, replace each other. He hears from closely allied—yet distinct—kinds of birds singing very similar songs and sees their nests similarly constructed, but not quite alike, and with

---

[2] By "shelled species," Darwin is referring to species of gastropods, bivalves, and brachiopods; crustaceans include crabs and shrimp, while echinoderms include sea urchins, brittlestars, and seastars.

[3] These species presumably all have microscopic, long-lived, widely dispersing larval stages.

Figure 12.1 *Rhea* is a genus of flightless birds native to South America. The birds in this genus are distantly related to the emu and the ostrich.

their eggs colored in nearly the same manner. The plains near the Straits of Magellan are inhabited by one species of flightless bird in the genus *Rhea* (the "American ostrich") (Figure 12.1) and northward on the plains of La Plata by another species of the same genus, and not by a true ostrich[4] or emu, like those inhabiting Africa and Australia at the same latitude. On these same plains of La Plata we see the agouti and bizcacha.[5] These two rodents (Order Rodentia) have nearly the same habits as our British hares and rabbits, but they plainly display an American type of structure. We ascend the lofty peaks of the Cordillera mountain ranges of South America, and we find an alpine species of bizcacha; we look to the waters, and we do not find the beaver or muskrat, but instead find the coypu (the "river rat," *Myocastor coypus*) and the capybara (*Hydrochoerus hydrochaeris*), large rodents of the South American type. Innumerable other instances could be given. If we look to the islands off the

---

[4] True ostriches belong to the genus *Struthio*.
[5] These are two families of rabbit-like rodents in the family Chinchillidae.

American shore, however much they may differ in geological structure, the inhabitants are essentially American, though they may all be peculiar species. If we look back to past ages, as shown in Chapter 11, we find American types then prevailing on the American continent and in the American seas. We see in these facts some deep organic bond, throughout space and time, over the same areas of land and water, independent of physical conditions. The naturalist must be dull indeed, who is not led to inquire what this bond is.

The bond is simple inheritance, that cause which alone, as far as we positively know, produces organisms quite like each other, or, as we see in the case of varieties, nearly alike. The dissimilarity among organisms that inhabit different regions may be attributed largely to gradual modification through variation and natural selection. The degrees of dissimilarity will depend on (1) the extent to which migrations of the more dominant forms of life from one region into another have been effectively prevented, at periods more or less remote; (2) on the nature and number of the former immigrants; and (3) on the impact that the inhabitants have had on each other in leading to the preservation of different modifications, as the relations between organisms in the struggle for life are the most important of all relations. Thus the high importance of physical barriers comes into play simply by checking migration. The issue of time comes in as well, for the process of modification through natural selection is very slow. Widely ranging species, abounding in individuals that have already triumphed over many competitors in their own widely extended homes, will have the best chance of seizing on new places when they spread out into new regions. In their new homes they will be exposed to new conditions, both physical and ecological, and will frequently undergo further modification and improvement. They will thus become still further victorious and will go on to produce groups of modified descendants. On this principle of inheritance with modification we can understand how it is that sections of genera, entire genera, and even entire families can be found only within a particular area, as is so commonly and notoriously the case.

As noted in the previous chapter, there is no evidence of the existence of any law of necessary development. As the degree of variability varies independently within each species, it will be taken advantage of by natural selection only so far as it benefits each individual in its complex struggle for life; thus the amount of modification in different species will not be uniform. If a number of species, after having long competed with each other in their old home, were to migrate together into a new and afterward isolated place, they would be little liable to modification—neither migration nor isolation in themselves can alter anything. The principles of natural selection come into play only by bringing organisms into new relationships with each other and, to a lesser degree, bringing them into areas with different physical conditions. We saw in the previous chapter that some forms have retained nearly the same characteristics from an enormously remote geological period; thus, certain species have not become greatly or at all modified over time, even though they have migrated over vast spaces.

According to these views, it is obvious that members of the same genus, even if they are now inhabiting the most distant quarters of the world, must have originated at the same location as they are descended from the same ancestor. In the case of those species that have undergone, during whole geological periods, little modification, there is not much difficulty in believing that they have all originally migrated from the same region, for during the vast geographical and climatical changes that have supervened since ancient times, almost any amount of migration is possible. But in many other cases, in which we have reason to believe that the various species of a genus have been produced within comparatively recent times, there is great difficulty on this point. It is also obvious that the individuals of the same species, though now inhabiting distant and isolated regions, must have originated from some one place where their parents were first produced; as I have explained earlier, it is difficult to believe that individuals that are essentially identical should have been produced from parents of distinctly different species.

Single Centers of Supposed Creation

We are thus brought to a question that has been much discussed by naturalists, namely, whether species have been created at one or at more than one point of the Earth's surface. Undoubtedly there are many cases of extreme difficulty in understanding how any one species could possibly have migrated from some one point to the several distant and isolated points where they are now found. Nevertheless, the simplicity of the view that each species was first produced within a single region captivates the mind. Whoever rejects it rejects the *vera causa* of ordinary generation with subsequent migration and calls in the agency of a miracle.

It is universally admitted that in most cases each species inhabits one continuous area and that when any plant or animal species inhabits two points so distant from each other, or with an interval of such a nature that the space could not have been easily passed over by migration, the fact is given as something remarkable and exceptional. The impossibility of migrating across a wide sea is clearer in the case of terrestrial mammals than perhaps with any other organic beings; accordingly, we do indeed find no inexplicable instances of the same mammal species inhabiting distant points of the world. No geologist feels any difficulty in finding the same quadruped species in Great Britain and the rest of Europe, for these land masses were no doubt once united. But if the same species can be produced independently at two separate locations, why do we not find a single mammal species common to both Europe and Australia or to both Europe and South America? The physical conditions of life are nearly the same at these different locations, allowing a multitude of European animals and plants to have now become naturalized in America and in Australia; and some of the aboriginal plants are identically the same at these distant points of the northern and southern hemispheres. So why don't we find any particular mammal species living in both Europe and Australia?

The answer, I believe, is that mammals have not been able to migrate across the wide and broken interspaces, whereas some plants, from their varied means of dispersal, have been able to do so. **The great and striking influence of barriers of all kinds on species distributions is intelligible only on the view that most species have been produced on one side of a barrier and have not been able to migrate to the opposite side.** Some few families, many subfamilies, very many genera, and a still greater number of species and varieties within genera are confined to a single region, and it has been observed by several naturalists that the most natural genera—or those genera in which the species are most closely related to each other—are generally confined to the same country or, if they have a wide range, that their range is continuous. What a strange anomaly it would be, then, if a directly opposite rule were to prevail when we go down one step lower in the series—namely, to the individuals of the same species—and these had not been, at least at first, confined to some single region!

Hence, it seems to me, as it has to many other naturalists, that **the view of each species having been produced in one area alone, and its members having subsequently migrated from that area as far as their powers of migration and subsistence under past and present environmental conditions permitted, is the best explanation of current and former species distributions.** Yet, many cases undoubtedly occur in which we cannot explain how the same species could have passed from one location to the other. But the geographical and climatical changes that have certainly occurred within recent geological times must have often caused the formerly continuous range of many species to become discontinuous. Thus we are reduced to consider whether the exceptions to continuity of range are so numerous and of so grave a nature that we ought to give up the compelling belief that each species had to have been produced within one area and has then migrated thereafter as far as it could.

It would be hopelessly tedious to discuss all the exceptional cases in which the same species are now found living at distant and separated points. Nor do I for a moment pretend that any explanation could even be offered in many instances. But, after some preliminary remarks, I will discuss a few of the most striking classes of facts: namely, (1) the existence of the same species on the summits of distant mountain ranges and at distant points in the Arctic and Antarctic regions, (2) the wide distribution of freshwater organisms (see Chapter 13), and (3) the occurrence of the same terrestrial species on islands and on the nearest mainland, even when those islands are separated from the mainland by hundreds of miles of open sea.

If the existence of the same species at distant and isolated points of the Earth's surface can, in many instances, be explained by assuming that each species has migrated from a single birthplace, then, considering our ignorance with respect to former climate and geographical changes and to the various occasional means of transport from one place to another, it seems safest to believe that a single birthplace for each species is the law.

Here is another important issue that we should consider on this subject: How likely is it that the several species found within a genus—that must, on our theory, all have descended from a common ancestor—can have migrated from some one particular area, undergoing gradual modification during their migration? When most of the related species inhabiting one region are different from those inhabiting another region, though closely related to them, if it can be shown that migration from the one region to the other has probably occurred at some time in the past, our general view will be much strengthened: the explanation is obvious on the principle of descent with modification. A volcanic island, for instance, upheaved and formed some few hundreds of miles from a continent, would probably receive from it in the course of time a few colonists. And the descendants of those early colonists, though modified, would still be related by inheritance to the inhabitants of that continent. Cases of this nature are common, and are, as we shall see later, inexplicable on the theory of independent creation. This view of the relation of the species of one region to those of another does not differ much from that advanced by Mr. Alfred Wallace, who concludes that "every species has come into existence coincident both in space and time with a pre-existing closely allied species." And it is now well known that he attributes this coincidence to descent with modification.

The question of whether or not there are single or multiple centers of species creation differs from another—though related—question: Are all the individuals of the same species descended from a single pair of individuals, or from a single hermaphrodite,[6] or whether, as some authors suppose, from many individuals simultaneously created? For organisms that are able to replicate themselves without ever mating, if such organisms actually exist, each species would then be descended from a succession of modified varieties that have supplanted each other but that have never blended with other individuals or varieties of the same species, so that, at each successive stage of modification, all the individuals of the same form will have descended from a single parent. But with all organisms that routinely unite for each birth, which is of course the general rule, or even with those that mate only occasionally, the individuals of the same species inhabiting the same area will be kept nearly uniform by interbreeding; in such a case, many individuals will go on simultaneously changing, and the whole amount of modification at each stage will *not* be due to descent from a single parent. To illustrate what I mean, consider our English racehorses. Even though they differ from the horses of every other breed, they do not owe their difference and superiority to descent from any single pair of ancestors but rather to continued care in the selecting and training of many individuals during each generation.

Before discussing the three classes of facts listed above, which I have selected as presenting the greatest amount of difficulty on the theory of "single centers of creation," I must say a few words on the means of dispersal.

Means of Dispersal

As Sir Charles Lyell and other authors have ably treated this subject of dispersal. I will give here only the briefest abstract of the most important facts.

---

[6] Here Darwin is referring to a single individual that can function simultaneously as both a male and a female.

Change of climate must have had a powerful influence on migration. A region now impassable to certain organisms because of the nature of its climate might have been a high road for migration in the past, when the climate was different, something that I will soon discuss in some detail. Changes of level in the land must also have been highly influential. For example, suppose that a narrow isthmus now separates two marine faunas; submerge it, or let it formerly have been submerged, and the two faunas will now blend together, or may formerly have so blended. Where the sea now extends, land may at a former period have connected islands or possibly even continents together and thus have allowed terrestrial productions to pass from one place to the other. No geologist disputes that there have been great changes of level within the period of existing organisms. Indeed, Edward Forbes has insisted that all the islands in the Atlantic must have been recently been connected with Europe or Africa, and Europe must have been connected likewise with America. Other authors have thus hypothetically bridged over every ocean and united almost every island with some mainland.[7] If, indeed, the arguments used by Forbes are to be trusted, we must then admit that scarcely a single island exists today that has not recently been joined to some continent. This view cuts the Gordian knot[8] of how the same terrestrial species have become dispersed to the most distant locations and removes many a difficulty; but, to the best of my judgment, we are not justified in thinking that such enormous geographical changes have occurred within the period of existing species. It seems to me that although we do indeed have abundant evidence of great oscillations in the level of the land or sea, we lack similarly compelling evidence of such vast changes in the position and extension of our continents as to have recently united them to each other and to the several intervening oceanic islands. I freely admit the former existence of many islands that are now buried beneath the sea, islands that may have served as stopping places for plants and for many animals during their migration. In the coral-producing oceans, such sunken islands are now marked by rings of coral or atolls standing over them.[9] Whenever it is fully admitted—as it will someday be—that each species has proceeded from a single birthplace, and when, in the course of time, we know something definite about the means of distribution, we shall be enabled to speculate with security on the former extension of the land. But I do not believe that it will ever be proven that within the recent period most of our continents, which now stand quite separate from each other, have been continuously or almost continuously united with each other and with the many existing oceanic islands in the past.[10]

---

[7] As noted earlier, Darwin did not know about continental drift, an idea that did not become generally accepted until the 1950s and 1960s.

[8] This metaphor refers to solving a seemingly impossible problem by thinking "outside the box."

[9] Many years before publishing *The Origin of Species*, Darwin came up with a novel idea about the formation of coral atolls, an idea that turned out to be correct. He postulated that coral reefs would initially form around the edges of volcanic islands; if those islands later gradually sank at an appropriate rate while the coral population continued to grow, you would eventually end up with a circular ring of coral, with a lagoon—and no land—in the middle.

[10] Not only didn't Darwin know anything about genetics, he also didn't know that there was another mechanism for formerly linking continents together. The idea that all continents were once connected and have subsequently drifted apart was first officially suggested by Alfred Wegener in 1912. His theory of

Several facts in distribution—such as the great difference in the species composition of marine faunas on the opposite sides of almost every continent, the close relation of the fossilized tertiary inhabitants of several lands and even seas to their present inhabitants, and the degree of relationship between the mammals inhabiting islands and those of the nearest continent, being in part determined (as we shall hereafter see) by the depth of the intervening ocean—these and other such facts are opposed to the admission of such prodigious geographical revolutions within the recent period, as are necessary on the view advanced by Forbes and admitted by his followers. The nature and relative proportions of the inhabitants of oceanic islands are likewise opposed to the belief of their former continuity of continents. Nor does the almost universally volcanic composition of such islands favor the admission that they are the wrecks of sunken continents; if they had originally existed as continental mountain ranges, at least some of the islands would have been formed, like other mountain summits, of granite, metamorphic schists, old fossiliferous and other rocks, instead of consisting of mere piles of volcanic matter.

I must now say a few words about what are called "accidental means" of distribution, but which more properly should be called "occasional means of distribution." I shall here confine myself to plants. In botanical works, this or that plant is often stated to be ill-adapted for wide dissemination, but the greater or lesser facilities for transport across the sea may be said to be almost wholly unknown.[11] Until I tried, with Mr. Berkeley's aid, a few experiments of my own, it was not even known how far seeds could resist the injurious action of seawater. To my surprise I found that out of 87 kinds of plants that I tested, the seeds from 64 plants germinated after being immersed in seawater for 28 days, and a few even survived an immersion of 137 days. It deserves notice that the seeds from certain orders of plants were far more injured than the seeds from other orders: nine members of the family Leguminosae were tested, and only one was able to resist saltwater immersion. Even worse, seeds from seven species of the related orders Hydrophyllaceae and Polemoniaceae were all killed by a month's immersion in the seawater. For convenience sake I chiefly tested small seeds without the capsules or fruit, and, as all of these sank in a few days, they could not have been floated across wide spaces of the sea, even if they were not injured by saltwater. Afterward I tried some larger fruits, capsules, etc., and some of those floated for a long time.

It is well known what a difference there is in the buoyancy of green and seasoned timber, and it occurred to me that floods would often wash into the sea dried plants or branches with seed capsules or fruit attached to them. Hence I was led to dry the stems and branches of 94 plants with ripe fruit and to place them on seawater. Most sank quickly, but some which floated for only a very short time while they were still

continental drift, based largely on the observation that continents could be fitted together remarkably well, as in a jigsaw puzzle, was initially received with considerable skepticism; it became widely accepted in the 1960s, leading to our present theory of plate tectonics.

[11] Here Darwin begins a long and fascinating discussion about the potential for seeds to be dispersed for great distances by ocean currents without losing their ability to germinate if they eventually reach land.

fresh and green floated for a much longer time after being dried. For instance, ripe hazelnuts sank immediately, but when dried they floated for 90 days, and they later germinated successfully after being planted. Similarly, an asparagus plant with ripe berries floated for only 23 days, but after being dried it floated for 85 days and its seeds afterward germinated. Similarly, although the ripe seeds of *Helosciadium* sank within 2 days, when dried they floated for more than 90 days and also germinated afterward. Altogether, out of the 94 dried plants that I worked with, 18 of them floated for at least 28 days, and some of those 18 floated for a very much longer period. So that as 64 of the 87 kinds of seeds (73.6%) germinated after being immersed in seawater for 28 days, and as 18 of the 94 distinct species (19.1%) with ripe fruit (but not all the same species as in the foregoing experiment) floated for more than 28 days after being dried, we may conclude, as far as anything can be inferred from these scanty facts, that the seeds of 14% of the plant species of any country might be floated by sea currents for nearly one month without losing their ability to germinate. According to Johnston's *Physical Atlas*, the average rate of the several Atlantic Ocean water currents is 33 miles per day, with some currents running almost twice as fast, at a rate of 60 miles per day. On this average, the seeds of 14% of the plants belonging to one country might be floated across 924 miles of ocean to another country, and, if blown by an inland gale to a favorable spot of land, those seeds would still be able to germinate.

Subsequent to my work, M. Charles Martens, in Belgium, tried similar experiments, but in a much better manner, placing the seeds in a box in the actual ocean, so that they were alternately wet and exposed to the air like naturally floating plants. He tried seeds from 98 different plants, most of which were different species from the ones that I tested, but he chose many large fruits, and likewise seeds, from plants that live near the sea; this would have favored both the average length of their flotation and their resistance to the injurious action of seawater. Of the 98 different plant seeds that he tested, 18 of them (18.4%) floated for 42 days without losing their ability to germinate. Note that he did not dry the plants or branches with the fruit before conducting this experiment; if he had, as we have seen, some of the seeds would have floated much longer.

To be fair, I do not doubt that plants exposed to waves would float for less time than those protected from violent movement, as in our experiments. Therefore, it would perhaps be safer to assume that the seeds of about 10% of the plants of a flora, after having been dried, could float successfully across a space of sea 900 miles in width without losing the ability to germinate. The fact that larger fruits often float longer than the smaller fruits is interesting, as plants with large seeds or fruit which, as Alph. de Candolle has shown, generally have restricted ranges, could hardly be transported by any other means.

Seeds may occasionally also be transported in another manner. Drift timber is thrown up on most islands, even on those in the midst of the widest oceans. And the native peoples of the coral islands in the Pacific Ocean procure stones for their tools solely from the roots of such drifted trees, these stones being a valuable royal tax. I have found that when irregularly shaped stones are embedded in the roots of trees, small parcels of earth are very frequently enclosed in their interstices and behind them, enclosed so perfectly that not a single particle could be washed away during the longest transport. Indeed, out of one small portion of earth thus completely enclosed by the roots of an oak about 50 years old, three dicotyledonous plants germinated; I am absolutely certain of the accuracy of this observation. And I can also show that the carcasses of birds, when floating on the sea, sometimes escape being immediately devoured. Many kinds of seeds long retain their vitality in the crops of such floating birds: although peas and vetches are killed by even a few days' immersion in seawater, nearly all germinated when taken out of the crop of a pigeon that had floated on artificial seawater for 30 days, much to my surprise.

In addition, living birds can hardly fail to be highly effective agents in transporting seeds. I could give many facts showing how frequently birds of many kinds are blown by gales for vast distances across the ocean. We may safely assume that under such circumstances these birds would often be flying at 35 miles an hour; indeed, some authors have given a far higher estimate of this flying speed. Although I have never seen an instance of nutritious seeds passing through the intestines of a bird, hard seeds of fruit are known to pass uninjured through even the digestive organs of a turkey. In the course of two months, I picked up 12 kinds of seeds in my garden from the excrement of small birds, and these seemed perfect; indeed, some of them germinated for me. But the following fact is even more important: the crops of birds do not secrete gastric juice, and do not, as I know by trial, injure in the least the germination of seeds. Now, after a bird has found and devoured a large supply of food, it is positively asserted that all the grains do not pass into the gizzard for 12 or even 18 hours, during which time a bird might easily be blown as far away as 500 miles. Hawks are known to look out for tired birds, and the contents of their torn crops might thus readily get scattered. Some hawks and owls bolt their prey whole, and after an interval of from 12 to 20 hours, disgorge pellets, which, as I know from experiments made in the Zoological Gardens, include seeds capable of germination. Indeed, some seeds of the oat, wheat, millet, canary, hemp, clover, and beet germinated after having been from 12 to 21 hours in the stomachs of different birds of prey; two seeds of beet, in fact, grew after having been thus retained for 2 days and 14 hours.

Fish may also be involved in transporting seeds from place to place, and in a very interesting but indirect way. Freshwater fish, I find, eat seeds of many land and water plants; as fish are frequently devoured by birds, those seeds might thus be transported from place to place. I therefore forced many kinds of seeds into the stomachs of dead fish, and then gave their bodies to fishing-eagles, storks, and pelicans. These birds, after an interval of many hours, either rejected the seeds in pellets or released them in their excrement; although the seeds of some plant species were always killed by this process, the seeds of other species retained the power of germination.

Locusts are sometimes also blown to great distances from the land. I myself caught one 370 miles from the coast of Africa, and I have heard of others being caught at even greater distances. Even more impressively, the Rev. R. T. Lowe informed Sir Charles Lyell that in November, 1844, swarms of locusts visited the island of Madeira

Figure 12.2 A swarm of locusts. A single swarm can contain tens of millions of individuals.

in countless numbers, as thick as the flakes of snow in the heaviest snowstorm and extending upward as far as could be seen with a telescope (Figure 12.2). During two or three days they slowly careered round and round in an immense ellipse, at least five or six miles in diameter, and at night alighted on the taller trees, which were completely coated with them. They then disappeared over the sea as suddenly as they had appeared and have not visited the island since. Now, it is believed by some farmers in parts of Natal, though probably on insufficient evidence, that injurious seeds are introduced into the farmers' grassland in the dung left by the great flights of locusts that often visit that country. In consequence of this belief, Mr. Weale mailed me a small packet of the dried pellets, out of which I extracted several seeds, with the help of a microscope, and raised from them seven plants belonging to two species of grasses, each belonging to a different genus. Hence a swarm of locusts, such as that which visited Madeira, might readily introduce several kinds of plants onto an island lying far from the mainland.

Birds can also be direct and effective agents of seed transport: although the beaks and feet of birds are generally clean, earth sometimes adheres to them. In one case I removed 61 grains of dry argillaceous earth, and in another case 22 grains, from the foot of a partridge; in that material I found a pebble as large as the seed of a vetch. Here is an even better case: the leg of a woodcock was sent to me by a friend with a little cake of dry earth attached to the shank, weighing only nine grains[12]; this contained a seed of the toad-rush (*Juncus bufonius*), which germinated and flowered after I planted it. Similarly, Mr. Swaysland, of Brighton, who has paid close attention to

[12] A troy grain refers to a weight of 64.8 milligrams (mg), based on the mass of a single seed of a cereal. So here Darwin is talking about something that weighs 583 mg., or about 0.58 grams (g).

our migratory birds over the last 40 years, informs me that he has often shot wagtails (family Motacillidae), wheatears, and whinchats (*Saxicola rubetra*) on their first arrival on our shores, before they had alighted, and has several times noticed little cakes of earth attached to their feet.

Indeed, many similar facts could be given showing how generally it is that soil adhering to the legs of birds contains embedded seeds. For instance, Professor Newton sent me the leg of a red-legged partridge (*Caccabis rufa*) that had been wounded and could not fly, with a ball of hard earth adhering to it that weighed 6.5 ounces. The earth had been kept for three years, but when broken, watered, and placed under a bell glass, no less than 82 plants sprung from it: these consisted of 12 monocotyledons, including the common oat, and at least one kind of grass, and of 70 dicotyledons, which consisted (judging from the young leaves) of at least three distinct species. With such facts before us, can we doubt that the many birds that are annually blown by gales across great spaces of ocean and which annually migrate—for instance, the millions of quails that migrate across the Mediterranean Sea each year—must occasionally transport from one place to another a few seeds embedded in dirt adhering to their feet or beaks? I shall return to this subject later.

There are also other likely means of seed transport, means that do not directly involve living animals. As icebergs are known to be sometimes loaded with earth and stones, and have even carried brushwood, bones, and the nest of a land-bird, it can hardly be doubted that they must occasionally also have transported seeds from one part of the Arctic and Antarctic regions to another, as suggested by Charles Lyell, and, during the glacial period that ended some 15,000 years ago, from one part of the now temperate regions to another. In the Azores, which harbor a large number of plants common to Europe, in comparison with the species on the other islands of the Atlantic that are found nearer to the mainland, and (as remarked by Mr. H.C. Watson) from their somewhat northern character in comparison with the latitude, I suspected that the Azores had been partly stocked by ice-borne seeds during the glacial epoch.[13] At my request, Sir Charles Lyell wrote to M. Hartung to inquire whether he had observed erratic boulders on these islands, and he answered that he had indeed found large fragments of granite and other rocks that do not otherwise occur in the archipelago. Hence we may safely infer that icebergs formerly landed their rocky burdens on the shores of these mid-ocean islands; it is at least possible that they brought thither the seeds of northern plants.

Considering that these several known means of seed transport—along with other means that surely remain to be discovered—have been active year after year for tens of thousands of years or more, it would, I think, be a marvelous fact if many plant species had not thus become widely transported over substantial distances.

These means of transport are sometimes called accidental, but this is not strictly correct: the currents of the sea are not accidental, nor is the direction of prevalent

---

[13] This most recent glacial period ("Ice Age") began about 110,000 years ago and ended about 15,000 years ago.

gales of wind. Even so, it should be observed that scarcely any means of transport would carry seeds for very great distances without them dying in the process; for seeds do not retain their vitality when exposed for a great length of time to the detrimental effects of seawater. Nor could they be long carried in the crops or intestines of birds without losing their viability. These means, however, would suffice for occasional transport across tracts of sea some hundred miles in breadth, or from island to island, or from a continent to a neighboring island, although not from one distant continent to another.

The floras of distant continents would not by such means become mingled, but rather would remain as distinct as they now are. The water currents, from their course, could never bring seeds from North America to Britain, though they might (and do in fact) bring seeds from the West Indies to our western shores, where, even if they were not killed by their very long immersion in saltwater, they could not endure our climate. Almost every year, one or two land birds are blown across the whole Atlantic Ocean, from North America to the western shores of Ireland and England. But the seeds could be transported by these rare wanderers by only one means: namely, by dirt adhering to their feet or beaks, which is in itself a rare event. Even in this case, how small would be the chance of a seed falling on favorable soil and coming to maturity!

Although it would be very difficult to prove this, it would be a great error to argue that just because a well-stocked island like Great Britain has not, as far as is known, received any immigrants from Europe or any other continent within the last few centuries, a poorly stocked island, though standing more remote from the mainland, could not receive successful colonists by similar means. Out of a hundred kinds of seeds or animals transported to an island, even if far less well-stocked than Britain, perhaps not more than one would be so well fitted to its new home as to become a successful invader. But this is no valid argument against what could be achieved by occasional means of transport during the long periods of geological time, while the island was being upheaved and before it had become fully stocked with inhabitants. On almost bare land, with few or no destructive insects or birds living there, nearly every viable seed that chanced to arrive, if fitted for the climate, would germinate and survive. Thus these mechanisms, over long periods of time, could well account for the presently documented wide distributions of many plants at widely separated locations.

Dispersal During the Glacial Period

The fact that the exact same species of many plants and animals are found on the summits of mountains that are separated from each other by hundreds of miles of lowlands—on which such alpine species could not possibly exist—is one of the most striking cases known of the same species living at distant points without any apparent possibility of their having migrated from one point to the other. It is indeed remarkable to see so many plants of the same species living on the various snowy regions of the Alps or Pyrenees and in the extreme northern parts of Europe. But it is far more remarkable that the plants on the White Mountains of New Hampshire, in the United States of America, are all identical with those of Labrador in northern Canada, more than 3,500 miles to the north, and nearly exactly the same, as we hear from Professor Asa Gray at Harvard, with those on the loftiest mountains of Europe. Even as long ago as 1747, such facts led Gmelin to conclude that the same species must have been independently created at many distinct points; we might still believe this today had not Agassiz and others called vivid attention to the glacial period, which, as we shall immediately see, affords a simple explanation of these facts.[14] Indeed, we have evidence of almost every conceivable kind, organic and inorganic, that, within a very recent geological period, central Europe and North America both suffered under a cold, arctic climate. The ruins of a house burnt by fire do not tell their tale more plainly than do the mountains of Scotland and Wales, with their scored flanks, polished surfaces, and perched boulders, of the icy streams with which their valleys were filled not long ago. So greatly has the climate of Europe changed since then that gigantic moraines in Northern Italy, left many years ago by old glaciers, are now clothed by vines and maize. Similarly, erratic boulders and scored rocks plainly reveal a former cold period throughout a large part of the United States.

The former influence of the glacial climate on the distribution of the inhabitants of Europe, as explained by Edward Forbes, is substantially as I shall explain below. But we shall follow the changes more readily by imagining a new glacial period that comes on slowly, lingers for some long period of time, and then passes away, just as formerly occurred. As the cold came on, and as each more southern zone became better and better fitted for the inhabitants of the north, these northern species would take the places of the former inhabitants of the more southern, temperate regions. At the same time, those inhabitants of the more southern regions would travel further and further southward, unless they were stopped by barriers, in which case they would perish as it continued getting colder. The mountains would gradually become covered with snow and ice, and their former alpine inhabitants would descend to the plains. By the time that the cold had reached its maximum, we should have an arctic fauna and flora covering the central parts of Europe as far south as the Alps and Pyrenees and even stretching into Spain. The now temperate regions of the United States would likewise be covered by arctic plants and animals during such an ice age, and these would be nearly the same species as those living in Europe, since the present circumpolar inhabitants, which we suppose to have everywhere traveled southward, are remarkably uniform all around the world.

Much later, as the warmth eventually returned, the arctic forms would retreat northward, closely followed up in their northward movement by species from the more temperate regions. And as the snow melted from the bases of the mountains, the arctic forms would seize on the cleared and thawed ground, always ascending higher and higher up the mountain as the warmth increased and the snow still further

[14] The realization that there had been long glacial periods in the past was a fairly new idea, first appearing around 1840.

disappeared, while their brethren were pursuing their northward journey. Hence, when the warmth had fully returned, the same species that had for some considerable time lived together on the European and North American lowlands would again be found in the arctic regions of the Old and New Worlds, on many isolated mountain-summits far distant from each other.

Thus we can understand why we see so many of the same plant species at points so immensely remote as the mountains of the United States and those of Europe. We can thus also understand why the alpine plants of each mountain range are most closely related to the arctic forms living due north or nearly due north of them: for the first migration when the cold came on and the remigration that followed as the warmth returned would generally have been due south and then due north. The alpine plants of Scotland, for example, as remarked by Mr. H. C. Watson, and those of the Pyrenees, as remarked by Ramond, are more especially related to the plants of northern Scandinavia; those of the United States to Labrador; and those of the mountains of Siberia to the arctic regions of that country. These views, grounded as they are on the perfectly well-ascertained occurrence of a former glacial period, seem to me to explain in so satisfactory a manner the present distribution of the alpine and arctic species of Europe and America that, when in other regions we find the same species on distant mountain-summits, we may almost conclude, even without additional evidence, that a colder climate must have formerly permitted their migration across the intervening lowlands, regions that have now become too warm for their existence.

As the arctic forms moved first southward as the region cooled and then later moved back to the north as they warmed back up, they will not have been exposed during their long migrations to any great diversity of temperature. And, as they will have all migrated in a body together, their mutual interactions will not have been much disturbed. Hence, in accordance with the principles inculcated in this volume, these forms will not have been liable to much modification. But with the alpine species, left isolated from the moment of the returning warmth, first at the bases and ultimately on the summits of the mountains, the case will have been somewhat different; for it is not likely that all the same arctic species will have been left on mountain ranges far distant from each other and have survived there ever since; they will also, in all probability, have become mingled with ancient alpine species, which must have existed on the mountains before the start of the glacial epoch and which, during the coldest period, will have been temporarily driven down to the plains; they will also have been subsequently exposed to somewhat different climatical influences. Their interactions will thus have been in some degree disturbed, so that they will have consequently been liable to modification[15]—and they have indeed been modified: if we compare the present alpine plants and animals of the several great European mountain ranges, one with another, though many of the species remain identically the

---

[15] And of course we now know that the isolation of small populations can itself result in genetically based modification over time.

same, some exist as varieties, some as doubtful forms or subspecies, and some as distinct yet closely related species representing each other on the several ranges.

Here I have assumed that, at the start of our imaginary new glacial period, the arctic species were as uniformly distributed around the polar regions as they are at the present day. But we must also assume that many subarctic and some few temperate forms were the same around the world, for some of the same species now exist both on the lower mountain slopes and on the plains of North America and Europe. How can I account for this degree of uniformity of the subarctic and temperate forms around the world at the start of the real glacial period? At the present day, the subarctic and northern temperate species found in both the Old and New Worlds are separated from each other by the whole Atlantic Ocean and by the northern part of the Pacific. During the glacial period, when the inhabitants of the Old and New Worlds lived further southward than they do at present, they must have been still more completely separated from each other by even wider spaces of ocean. Thus it may well be asked how the same species could then or previously have entered the two continents. The explanation, I believe, lies in the nature of the climate before the start of the glacial period.[16] At that time, during the newer Pliocene period, most of the world's inhabitants were specifically the same as they are now, and we have good reason to believe that the climate was warmer than it is at present. Hence, we may suppose that the organisms now living under latitude 60 degrees lived further north during the Pliocene period, under the Polar Circle at latitude 66–67 degrees, and that the present arctic productions then lived on the broken land still nearer to the pole. Now, if we look at a terrestrial globe, we see under the Polar Circle that there is almost continuous land from western Europe through Siberia, and from there to eastern America. This continuity of the circumpolar land, with the consequent freedom for intermigration under a more favorable climate, will account very nicely for the supposed uniformity of the subarctic and temperate productions of the Old and New Worlds during a period before the start of the glacial epoch.

Believing, from reasons that I alluded to previously, that our continents have long remained in nearly the same relative position[17] though subjected to great oscillations of level, I am strongly inclined to extend the above view and to infer that during some earlier and still warmer period, such as the older Pliocene period, a large number of the same plants and animals inhabited the almost continuous circumpolar land and that these plants and animals, both in the Old and New Worlds, began to slowly migrate southward as the climate became colder, long before the commencement of the glacial period. We now see, I believe, their descendants, mostly in a modified condition, in the central parts of Europe and the United States. On this view we can understand why there is a clear relationship between the species now found living in

---

[16] As noted earlier, this period began about 110,000 years ago. The Pliocene mentioned in the next paragraph of the text extends from about 5.3 to 2.6 million years ago (see Table 10.1).

[17] Again, recall that Darwin was not aware of continental drift, which provides yet another mechanism through which populations of a given species will have become widely separated over time and then likely evolve in very different directions.

North America and Europe, a relationship that is highly remarkable considering the distance between the two areas and their separation by the whole Atlantic Ocean. We can further understand the singular fact remarked on by several observers that the species of Europe and America during the later tertiary stages[18] were more closely related to each other than they are at the present time, for, during these warmer periods, the northern parts of the Old and New Worlds will have been almost continuously united by land, which served as a bridge for the intermigration of their inhabitants, a bridge that later became impassable by cold.

During the slowly decreasing warmth of the Pliocene period, as soon as the species that had inhabited both the New and Old Worlds migrated south of the Polar Circle, they will have been completely cut off from each other. This separation, as far as the more temperate productions are concerned, must have taken place long ago. As the various plant and animal species migrated southward, they will have become mingled in the one great region with the native American productions and would have had to compete with them; in the other great region, they will have had to compete with those of the Old World. Consequently we have here everything favorable for much modification, selection, and diversification to take place—and for far more modification than with the alpine productions, left isolated within a much more recent period, on the several mountain ranges and on the arctic lands of Europe and North America. Thus it is that when we compare the now living productions of the temperate regions of the New and Old Worlds, we find very few identical species, but we find within every great class many forms, which some naturalists rank as geographical races and others as distinct species, along with a host of closely allied or representative forms that are ranked by all naturalists as distinct species.

As on the land, so it has been in the waters of the sea, with a slow southern migration of a marine fauna, which, during the Pliocene or even a somewhat earlier period, was nearly uniform along the continuous shores of the Polar Circle. This will account, on the theory of descent with modification, for many closely related forms now living in marine areas that are today completely separated from each other. Thus, I think, we can understand the presence of some closely related, still existing and extinct tertiary forms on the eastern and western shores of temperate North America, as well as the still more striking fact of many closely related crustaceans (as described in Dana's admirable work) as well as some fish and other marine animals, inhabiting both the Mediterranean Sea and the Sea of Japan—these two areas being now completely separated by the breadth of a whole continent and by wide spaces of ocean.

These cases of close relationship in species either now or formerly inhabiting the seas on the eastern and western shores of North America, the Mediterranean, Japan, and the temperate lands of North America and Europe, are inexplicable on the theory of divine creation. Moreover, we cannot argue that such species have been created

---

[18] We now know that the time that Darwin refers to here extends from about 23 million years ago to almost 2 million years ago. During this long period, North America and Eurasia formed a continuous land-mass, resulting from a substantial drop in sea level.

alike in correspondence with the nearly similar physical conditions characterizing the different areas; for if we compare, for instance, certain parts of South America with parts of South Africa or Australia, we see countries that are closely similar in all their physical conditions and yet find their inhabitants utterly dissimilar.

Alternate Glacial Periods in the North and South

But let us now return to our more immediate subject. I am convinced that Forbes's view about the role of climate change on the distribution of species in Europe may be largely extended. In Europe we meet with the plainest evidence of the glacial period, from the western shores of Britain to the Ural Mountain range of eastern Russia and southward to the Pyrenees, on the border between Spain and France. We may infer, from the frozen mammals and the nature of the mountain vegetation, that Siberia was similarly affected. In the Lebanon, according to Dr. Hooker, perpetual snow formerly covered the central axis and fed glaciers which rolled 4,000 feet down the valleys. The same observer has recently found a great moraine[19] at a low level on the Atlas range in North Africa. Along the Himalaya, at points 900 miles apart, glaciers have left the marks of their former low descent, and in Sikkim, in northwest India, Dr. Hooker saw maize growing on ancient and gigantic moraines. Southward of the Asiatic continent, on the opposite side of the Equator, we know—from the excellent researches of Drs. J. Haast and Hector—that in New Zealand immense glaciers formerly descended to a low level; the same plants found by Dr. Hooker on widely separated mountains in this island tell the same story of a former cold period. From facts communicated to me by the Rev. W. B. Clarke, it appears that there are also traces of former glacial action on the mountains of the southeastern corner of Australia.

And now let us consider America: in North America, ice-borne fragments of rock have been observed on the eastern side of the continent as far south as latitudes 36 and 37 degrees and have also been found on the shores of the Pacific, where the climate is now so different, as far south as latitude 46 degrees. Erratic boulders have also been noticed on the Rocky Mountains. In the Cordillera mountain range of South America, nearly under the Equator, glaciers once extended far below their present level. In central Chile, I examined a vast mound of detritus with great boulders crossing the Portillo valley, which—and there can hardly be any doubt about this—once formed a huge glacial moraine. Similarly, Mr. D. Forbes informs me that he found in various parts of the Cordillera, from latitude 13 to 30 degrees south, at the height of about 12,000 feet, deeply furrowed rocks resembling those with which he was familiar in Norway and likewise great masses of detritus, including grooved pebbles, another clear sign of former glacial movement. Along this whole space of the Cordillera, true

---

[19] Moraines are glacially formed accumulations of rocks, sediment, and other debris.

glaciers do not now exist even at much more considerable heights. Further south, on both sides of the continent, from latitude 41 degrees to the southern-most extremity, we have the clearest evidence of former glacial action in numerous immense boulders transported far from their parent source.

So we have before us these several facts: (1) from the glacial action having extended all around the northern and southern hemispheres; (2) from the period having been in a geological sense recent in both hemispheres; (3) from its having lasted in both hemispheres for a very long time, as may be inferred from the amount of work accomplished; and, (4) from glaciers having recently descended to a low level along the whole line of the Cordillera mountain range of South America. It therefore at one time appeared to me that we could not avoid the conclusion that the temperature of the whole world had been simultaneously lowered during the glacial period. But Mr. Croll, in a series of admirable memoirs, has now attempted to show that a glacial condition of climate is the result of various physical causes brought into operation by an increase in the eccentricity of the Earth's orbit. Although all of these causes tend toward the same end, the most powerful appears to be the indirect influence of the eccentricity of the orbit upon oceanic currents. According to Mr. Croll, cold periods on Earth regularly recur every 10,000 or 15,000 years and these, at long intervals, are extremely severe, owing to certain contingencies, of which the most important, as Sir C. Lyell has shown, is the relative position of the land and water. Mr. Croll believes that the last great glacial period occurred about 240,000 years ago and endured, with slight alterations of climate, for about 160,000 years.[20] With respect to the more ancient glacial periods, several geologists are convinced, from direct evidence, that such occurred during the Miocene and Eocene formations, not to mention still more ancient formations. But the most important result for us, arrived at by Mr. Croll, is that whenever the northern hemisphere passes through a cold period, the temperature of the southern hemisphere is actually raised, with the winters rendered much milder, chiefly through changes in the direction of the ocean currents. Conversely, it will be the opposite with the northern hemisphere, whenever the southern hemisphere passes through a glacial period. This conclusion throws so much light on the geographical distribution of organisms that I am strongly inclined to trust in it. But I will first give some facts that demand an explanation.

In South America, Dr. Hooker has shown that besides there being a great many closely related species, between 40 and 50 of the flowering plants of Tierra del Fuego—a considerable part of its scanty flora—are common to both North America and Europe, enormously remote as these areas in opposite hemispheres are from each other. Similarly, on the lofty mountains of equatorial America a great many peculiar species belonging to European genera are found. Similarly, on the Organ Mountains of Brazil, Gardner has found some few temperate European, Antarctic, and Andean genera that do not exist in the low intervening hot countries. On the Silla of Caraccas,

---

[20] As noted previously, we now put the start of the most recent ice age at 110,000 years ago, and it actually lasted for about 95,000 years.

Venezuela, the illustrious Humboldt long ago found species belonging to genera characteristic of the Cordillera.

On the mountains of Abyssinia, in Africa, we find several forms characteristic of those found in Europe, as well as some few representatives of the flora of the Cape of Good Hope. At the Cape of Good Hope we find a few European species, believed not to have been introduced there by man, and on the mountains we find several representative European forms that have not been discovered in the intertropical parts of Africa. Dr. Hooker has also lately shown that several of the plants living on the upper parts of the lofty island of Fernando Po and on the neighboring Cameroon Mountains in the Gulf of Guinea are closely related to those living on the mountains of Abyssinia and likewise to those of temperate Europe. It now also appears, as I hear from Dr. Hooker, that the Rev. R. T. Lowe has now discovered some of these same temperate plants on the mountains of the Cape Verde Islands. This extension of the same temperate forms, almost under the Equator, across the whole continent of Africa, and to the mountains of the Cape Verde archipelago, is one of the most astonishing facts ever recorded in the distribution of plants. How can we explain this?

On the Himalayan mountains and on the isolated mountain ranges of the peninsula of India, on the heights of Ceylon, and on the volcanic cones of Java, many plants occur either identically the same or representing each other, and, at the same time, representing plants of Europe not found in the intervening hot lowlands. A list of the plant genera collected on the loftier peaks of Java, in Indonesia, raises a strikingly similar picture of a collection made on a hillock in Europe. Still more striking is the fact that a number of forms peculiar to Australia are also represented by certain plants growing on the summits of the mountains of Borneo, in the Malay Archipelago, some 2,100 miles north of Australia. Some of these Australian forms, as I hear from Dr. Hooker, extend along the heights of the peninsula of Malacca and are thinly scattered, on the one hand, over India and, on the other hand, as far north as Japan.

Dr. F. Müller has discovered several European species on the southern mountains of Australia; other European species, not introduced by man, occur on the lowlands. Similarly, a long list can be given, as I am informed by Dr. Hooker, of European genera that are also found in Australia although not in the intermediate torrid regions. In the admirable "Introduction to the Flora of New Zealand," Dr. Hooker gives analogous and striking facts in regard to the plants of that large island.

Hence, we see some of the same plant species, or varieties of the same species, growing on the more lofty mountains of the tropics in all parts of the world and on the temperate plains of the northern and southern hemispheres. It should, however, be observed that these plants are not strictly arctic forms; for, as Mr. H. C. Watson has remarked, "in receding from polar toward equatorial latitudes, the alpine or mountain flora really become less and less Arctic." Besides these identical and closely related forms, many species inhabiting the same widely sundered areas belong to genera that are no longer found in the intermediate tropical lowlands.

Although these brief remarks apply to plants alone, some few analogous facts could also be given in regard to terrestrial animals and for marine species. For example, let me quote a statement by the highest authority, Professor Dana, that "it is certainly a wonderful fact that New Zealand should have a closer resemblance in its crustacea to Great Britain, its antipode, than to any other part of the world." Sir J. Richardson, also, speaks of the reappearance on the shores of New Zealand, Tasmania, etc., of northern forms of fish, and Dr. Hooker informs me that 25 species of marine algae are found in both New Zealand and Europe but have not been found in the intermediate tropical seas.

So we have some remarkable facts before us: namely, the presence of temperate forms on the highlands across the whole of equatorial Africa and along the Peninsula of India to Ceylon and the Malay Archipelago and, in a less well-marked manner, across the wide expanse of tropical South America. Thus it appears almost certain that at some former period, no doubt during the most severe part of a glacial period, the lowlands of these great continents were everywhere inhabited under the Equator by a considerable number of temperate forms. At this period the equatorial climate at the level of the sea was probably about the same as that now experienced at the height of from 5,000 to 6,000 feet at the same latitude, or perhaps even rather cooler. During this, the coldest period, the lowlands under the Equator must have been clothed with a mingled tropical and temperate vegetation, like that described by Hooker as growing luxuriantly at the height of from 4,000 to 5,000 feet on the lower slopes of the Himalayas, but with perhaps a still greater preponderance of temperate forms. So again, in the mountainous island of Fernando Po, in the Gulf of Guinea, Mr. Mann found temperate European forms beginning to appear at the height of about 5,000 feet. On the mountains of Panama, at the height of only 2,000 feet, Dr. Seemann found the vegetation like that of Mexico, "with forms of the torrid zone harmoniously blended with those of the temperate."

Now let us see whether Mr. Croll's conclusion (i.e., that when the northern hemisphere suffered from the extreme cold of the great glacial period, the southern hemisphere was actually warmer) throws any clear light on the present apparently inexplicable distribution of various organisms in the temperate parts of both hemispheres and on the mountains of the tropics. The glacial period, as measured by years, must have been very long; and, when we remember over what vast spaces some naturalized plants and animals have spread within just a few centuries, this period will have been ample for any amount of migration. As the cold became more and more intense, we know that arctic forms gradually invaded the temperate regions. And, from the facts just given, there can hardly be a doubt that some of the more vigorous, dominant, and widest-spreading temperate forms invaded the equatorial lowlands. The former inhabitants of these hot lowlands would at the same time have migrated southward to the tropical and subtropical regions, for the southern hemisphere was at this period warmer. Much later, near the end of the glacial period, as both hemispheres gradually recovered their former temperatures, the northern temperate forms living on the lowlands under the Equator would have been gradually driven to their former homes or have been destroyed, being replaced by the equatorial forms returning from the south. However, some of the northern temperate forms would almost certainly have gradually ascended any adjoining high land as the climate continued to warm, where, if sufficiently lofty, they would have long survived like the arctic forms on the mountains of Europe. They might have survived even if the climate was not perfectly fitted for them, for the change of temperature must have been very slow, and plants undoubtedly possess a certain capacity for acclimatization, as shown by their transmitting to their offspring different constitutional powers of resisting heat and cold.

In the regular course of events, the southern hemisphere would in its turn also be subjected to a severe glacial period, with the northern hemisphere rendered warmer; the southern temperate forms would then invade the equatorial lowlands. The northern forms that had before been left on the mountains during periods of extreme cold would now descend and mingle with the southern forms. These latter, when the warmth returned, would return to their former homes, leaving some few species on the mountains and carrying southward with them some of the northern temperate forms that had descended from their mountain fastnesses.[21] Thus, we should have some few species being identical in the northern and southern temperate zones and on the mountains of the intermediate tropical regions. But the species left during a long time on these mountains, or in opposite hemispheres, would have to compete with many new forms and would be exposed to somewhat different physical conditions; hence, they would be eminently liable to modification and would generally now exist as varieties of the original species or as representative new species—and this is indeed the case. We must also bear in mind the occurrence in both hemispheres of former glacial periods; for these will account, in accordance with the same principles, for the many quite distinct species inhabiting the same widely separated areas and belonging to genera not now found in the intermediate torrid zones.

It is a remarkable fact, strongly insisted on by Hooker in regard to America and by Alph. de Candolle in regard to Australia, that many more identical or slightly modified species have migrated from the north to the south than from the south to the north. We do see, however, a few southern, Australian forms on the mountains of Borneo and Abyssinia. I suspect that this preponderant migration from the north to the south is due to the greater extent of land in the north and to the northern forms having existed in their own homes in greater numbers and having consequently been advanced through natural selection and competition to a higher stage of perfection, or dominating power, than the southern forms. And thus, when the two sets became commingled in the equatorial regions during the alternations of the glacial periods, the northern forms were the more powerful and were able to hold their places on the mountains, later migrating southward with the southern forms; but not so the southern in regard to the northern forms.

In the same manner, at the present day, we see that very many European productions cover the ground in La Plata, New Zealand, and to a lesser degree in Australia,

---

[21] Now there's a word I've never heard before! A "fastness" is a remote and inaccessible mountainous area, basically a mountain "refuge." The term seems to have been first used by the poet William Wordsworth, in the early nineteenth century.

and have outcompeted the native species. In contrast, extremely few southern forms have become naturalized in any part of Europe, even though hides, wool, and other objects likely to carry seeds have been largely imported into Europe during the last two or three centuries from La Plata and during the last 40 or 50 years from Australia. The Neilgherrie Mountains in India, however, offer a partial exception; for here, as I hear from Dr. Hooker, Australian forms are rapidly sowing themselves and becoming naturalized. Before the last great glacial period, no doubt the intertropical mountains were stocked with endemic alpine forms; but these have almost everywhere yielded to the more dominant forms generated in the larger areas and more efficient workshops of the north. In many islands, the native productions are nearly equaled or even outnumbered by those which have become naturalized, and this is the first stage toward the extinction of those native species. Mountains are essentially islands on the land, and their inhabitants have yielded to those produced within the larger areas of the north, just in the same way as the inhabitants of real islands have everywhere yielded and are still yielding to continental forms brought there through human agency.

The same principles apply to the distribution of terrestrial animals and of marine species in the northern and southern temperate zones and on the intertropical mountains. During the height of the glacial period, when the ocean currents were widely different from what they are now, some of the inhabitants of the temperate seas might have reached the Equator; of these, a few would perhaps at once have been able to migrate southward by keeping to the cooler currents, while others might have remained and survived in the colder depths until the southern hemisphere was in its turn subjected to a glacial climate, permitting their further progress. In nearly the same manner, according to Forbes, isolated spaces inhabited by arctic productions exist to the present day in the deeper parts of the northern temperate seas.

I am far from supposing that the above arguments remove all the difficulties regarding the distribution and affinities of the identical and allied species that now live so widely separated in the north and south and sometimes on the intermediate mountain ranges. The exact lines of migration cannot be indicated. We cannot explain why only certain species and not others have migrated, or why certain species have been modified and have given rise to new forms while others have remained unaltered. We cannot hope to explain such facts until we can explain why one species and not another becomes naturalized by our actions in a foreign land, or why one species ranges twice or thrice as far and is twice or thrice as common as another species within their own homes.

Various special difficulties also remain to be solved; for instance, the occurrence, as shown by Dr. Hooker, of the same plants at points so enormously remote as Kerguelen Land, New Zealand, and Fuegia, although icebergs, as suggested by Lyell, may have played an important role in their dispersal. The existence at these and other distant points of the southern hemisphere of species which, although distinct, belong to genera exclusively confined to the south, is a more remarkable case. Some of these species are so distinct that we cannot suppose that there has been time since the commencement of the last glacial period for their migration and subsequent modification to the necessary degree. The facts seem to indicate that distinct species belonging to the same genera have migrated in radiating lines from a common center; and I am inclined to look in the southern, as in the northern hemisphere, to a former and warmer period, before the commencement of the last glacial period, when the Antarctic lands, now covered with ice, supported a highly peculiar and isolated flora. It may be suspected that before this flora was exterminated during the last glacial epoch, a few forms had been already widely dispersed to various points of the southern hemisphere by occasional means of transport and by the aid, as halting-places, of now sunken islands. Thus the southern shores of America, Australia, and New Zealand may have become slightly tinted by the same peculiar forms of life.[22]

Sir Charles Lyell has speculated, in a striking passage—and in language almost identical with mine—on the effects of great alternations of climate throughout the world on the geographical distribution of animals and plants. And we have now seen that Mr. Croll's conclusion that successive glacial periods in the one hemisphere coincide with warmer periods in the opposite hemisphere, together with the admission of the slow modification of species, explains a multitude of facts in the distribution of the same and related forms of life in all parts of the globe. The living waters have flowed during one period from the north and during another from the south, and in both cases have reached the Equator; but the stream of life has flowed with greater force from the north than in the opposite direction and has consequently more freely inundated the south. Just as the tides leave their drift in horizontal lines, rising higher on the shores where the tide rises highest, so have the living waters left their living drift on our mountain summits, in a line gently rising from the Arctic lowlands to a great latitude under the Equator. The various beings thus left stranded on mountain tops may be compared with the savage races of man, driven up and surviving in the mountain fastnesses[23] of almost every land, which serves as a record, full of interest to us, of the former inhabitants of the surrounding lowlands.

Key Issues to Talk and Write About

1. How does Darwin explain why we see so many identical species on the tops of distant mountains?
2. Describe how one of his examples about species distributions could now be explained by continental drift. What was Darwin's explanation?
3. Make a list of the experiments that Darwin talks about in this chapter, that he himself conducted.

---

[22] Some animal and plant species that are now found in South America and Australia were once also living in Antarctica, a result of all of those land masses (together with what was to become North America, Europe, India, and Africa) once forming a single, huge supercontinent called Pangaea. Pangaea began breaking apart about 175 million years ago.

[23] As noted earlier, a "fastness" is a refuge.

4. Try to summarize the essential content of this chapter in a single paragraph.
5. Find out two interesting things about one of the people that Darwin mentions in this chapter. Choose from the following:
   Reverend William Branwhite Clarke
   James Croll
   James Dwight Dana
   Alphonse de Candolle
   Edward Forbes
   Joseph Hooker
   James Hector
   Charles Lyell
   Louis François Ramond
   Sir John Richardson
   Alfred Russell Wallace
   Hewett Cottrell Watson

13

Geographical Distribution, Continued

In this chapter, Darwin continues his examination of species distributions. He argues that they are difficult to explain through their origins by divine creation but that they can be explained in a convincing, logical way through dispersal mechanisms—including the dispersal of eggs and hatchlings—and their limitations; gradual rises of land areas over time; and the processes of competition, natural selection, and speciation. Of particular interest, Darwin explains why freshwater species are more widely distributed than might be expected from the physical separation of lakes and rivers; why so many island plant and animal species are unique to particular islands while others are also found on the mainland; why most islands have no terrestrial mammal, frog, or toad species; why different groups of islands are homes to very different species even when soil and environmental conditions are very similar; why species that are endemic to particular islands are so similar to those on nearby islands but are also clearly related to those on the nearest continent or on the nearest large island; why islands that are separated from each other by deep water are more likely to have different mammal species than those that are separated by shallow water; and why, in general, different areas having the same physical conditions are so often inhabited by very different species. Darwin's theory of gradual evolution by natural selection explains all of these issues, with evidence from a large variety of sources. And, as always, Darwin freely admits his uncertainties.

Freshwater Plant and Animal Productions

As lakes and river systems are separated from each other by barriers of land, it might have been thought that freshwater productions would not have ranged widely within the same country and, as the sea is apparently a still more formidable barrier, that they would never have extended across the oceans to distant countries. But the case is exactly the reverse. Not only do many freshwater species belonging to a variety of different taxonomic classes have an enormous range, but allied species also prevail in a remarkable manner throughout the world. When first collecting samples in the fresh waters of Brazil, I well remember feeling much surprise at the similarity of the freshwater insects and snails compared with those of Britain (and at the dissimilarity of the surrounding terrestrial beings).

The Readable Darwin. Second Edition. Jan A. Pechenik, Oxford University Press. © Oxford University Press 2023.
DOI: 10.1093/oso/9780197575260.003.0014

But the wide-ranging distributions of freshwater productions can, I think, in most cases be explained by their having become fitted, in a manner highly useful to them, for short and frequent migrations from pond to pond—or from stream to stream—within their own countries; opportunities for wide dispersal would follow from this capacity as an almost necessary consequence. We can here consider only a few cases; of these, some of the most difficult to explain are presented by fish. It was formerly believed that the same freshwater species never existed on two continents that were distant from each other. But Dr. Gunther has lately shown that the fish species *Galaxias attenuatus* inhabits a very wide range of locations: Tasmania, New Zealand, the Falkland Islands, and the mainland of South America. This is a wonderful case, and it probably indicates previous dispersal from an Antarctic center during a former warm period. This case, however, is rendered in some degree less surprising by the species of this genus having the power of crossing by some unknown means considerable spaces of open ocean: thus there is one species common to New Zealand and to the Auckland Islands, even though those locations are separated by a distance of about 230 miles. Freshwater fish often range widely on the same continent, and as if capriciously—for in two adjoining river systems some of the species may be the same and some wholly different.

At least occasionally, freshwater fish are probably transported by what may be called accidental means. Thus fishes are commonly picked up and dropped while still alive at distant points by whirlwinds, and it is known that fish eggs (the "ova") retain their vitality for a considerable time after their removal from the water. The widespread dispersal of these fish may, however, be mainly attributed to changes in the level of the land within the recent period, causing rivers to flow into each other. Instances could also be given of this having occurred during floods, without any change of level. The large differences of the fish on the opposite sides of most mountain ranges, which are continuous and consequently must, from an early period, have completely prevented the merging of the river systems on the two sides, leads to the same conclusion. Some freshwater fish belong to very ancient forms, and in such cases there will have been ample time for great geographical changes and, consequently, time and means for much migration. Moreover, Dr. Gunther has recently been led by several considerations to infer that with fishes the same forms have a long endurance. Moreover, marine fish can with care be slowly accustomed to live in fresh water; indeed, according to Achille Valenciennes, there is hardly a single group of fish of which all the members are confined to fresh water. Thus, a marine species belonging to a freshwater group might travel far along the shores of the sea and could then probably become adapted without much difficulty to the fresh waters of a distant land.

Some freshwater snail species have very wide ranges, and allied species which, on our theory, are descended from a common parent and must have dispersed from a single source, in fact prevail throughout the world. Their distribution at first perplexed me much as their ova are not likely to be transported by birds, and the ova, as well as the adults, are immediately killed by seawater. I could not even understand

Figure 13.1 A member of the diving beetle genus *Dyticus*.

how some naturalized species[1] have spread rapidly throughout the same country. But I have observed two facts—and many others no doubt will be discovered—that throw some light on this subject. When ducks suddenly emerge from a pond covered with duckweed,[2] I have twice seen these little plants adhering to their backs; in removing a little duckweed from one aquarium to another in my home, I have sometimes unintentionally stocked the one aquarium with freshwater snails from the other.

But another agency is perhaps even more effective in transporting snail eggs: I suspended the feet of a duck in an aquarium where many ova of freshwater snails were hatching, and I found that many of the extremely minute and just-hatched snails crawled onto the duck's feet and clung to them so firmly that when the foot was taken out of the water they could not be jarred off, although at a somewhat more advanced age they would voluntarily drop off. Moreover, these just-hatched snails, though aquatic in their nature, could survive perfectly well on the duck's feet for 12 to 20 hours in damp air; in this length of time a duck or heron might fly at least 600 or 700 miles and, if blown across the sea to an oceanic island or to any other distant point, would be sure to alight on a pool or rivulet, thus transporting the young snails in the process. As another example, Sir Charles Lyell informs me that the freshwater, predatory diving beetle *Dyticus* (Figure 13.1) has been caught with a small, freshwater air-breathing snail (genus *Ancylus*) firmly adhering to it. In addition, a freshwater diving beetle of the same family, a member of the genus *Colymbetes*, once flew on board the *Beagle* when we were 45 miles away from the nearest land; how much farther it might have been blown by an even more favorable gale no one can tell.

---

[1] Naturalized species are species that invade or are moved into a new area and are then able to survive and reproduce successfully there.

[2] Duckweed is a member of the freshwater plant genus *Lemna*. They are the smallest flowering plants known, each consisting of a single leaf no longer than 63 mm long (about ¼ of an inch). They are commonly seen floating in dense concentrations at the surfaces of ponds and lakes.

With respect to plants, it has long been known what enormous ranges many fresh-water and even marsh species have, both over continents and to the most remote oce-anic islands. This is strikingly illustrated, according to Alph. de Candolle, in those large groups of terrestrial plants that have very few aquatic members; for the latter seem immediately to acquire, as if in consequence, a wide range. I think this must be explained by a favorable means of dispersal. I have previously mentioned that various amounts of soil and mud occasionally adhere to the feet and beaks of birds. Wading birds, which frequent the muddy edges of ponds, if suddenly flushed, would be the most likely to have muddy feet. Birds of this order wander farther than those of any other and are occasionally found on the most remote and barren islands of the open ocean. They would not be likely to alight on the surface of the sea, so any dirt on their feet would not be washed off as they travel; when gaining the land, they would be sure to fly to their natural freshwater haunts.

I do not believe that botanists are aware how charged the mud of ponds is with seeds: I have tried several little experiments, but will here give only the most striking case. One February, I took three tablespoons of mud from three different points be-neath the water, on the edge of a little pond; this mud when dry weighed only 6.75 ounces. I kept this sample covered up in my study for six months, pulling up and counting each plant as it grew: there were 537 plants, and of many kinds. And yet the viscid mud that all of these plants sprouted in was all contained in a single breakfast cup! Considering these facts, I think it would be an inexplicable circumstance if water birds did not in fact transport the seeds of freshwater plants to unstocked ponds and streams situated at very distant points. The same mechanism of transport may have come into play with the eggs of some of the smaller freshwater animals.

Other and unknown agencies have probably also played a part. I have previ-ously stated that freshwater fish eat some kinds of seeds, though they reject many other kinds after having swallowed them; even small fish swallow seeds of mod-erate size, such as those of the yellow waterlily and members of the pondweed genus *Potamogeton*. Herons and other birds, century after century, have gone on daily devouring fish; they then take flight and go to other waters or are blown across the sea; and we have seen that when ejected many hours afterward in pellets or in the bird's excrement, the seeds still retain their power of germination. When I saw the great size of the seeds produced by that fine waterlily, a member of the genus *Nelumbium*, and remembered Alph. de Candolle's remarks on the distribution of this plant, I thought that the means of its dispersal must remain inexplicable, but Audubon states that he has in fact found the seeds of the great southern waterlily (probably, according to Dr. Hooker, *Nelumbium luteum*) (Figure 13.2) in a heron's stomach. Now this bird must often have flown with its stomach thus well-stocked to distant ponds and then, getting a hearty meal of fish, analogy makes me believe that it would have rejected the seeds in the pellet in a fit state for germination.

In considering these several means of distribution, we should remember that when a pond or stream is first formed, for instance on a rising islet, it will be unoccupied, so that a single seed or egg will have a good chance of succeeding if transported there.

Figure 13.2 *Nelumbium luteum*, now known as the American lotus, *Nelumbo lutea*.

Although there will always be a struggle for life between the inhabitants of the same pond, however few in kind, yet as the number of inhabitants even in a well-stocked pond is small in comparison with the number of species inhabiting an equal area of land, there will probably be less competition between them than between terrestrial species; an intruder from the waters of a foreign country would consequently have a better chance of seizing on a new place in these waters than would terrestrial colonists. We should also remember that many freshwater productions are low in the scale of nature. We have reason to believe that such beings become modified more slowly than those of the higher forms; this will give time for the migration of aquatic species. We should also not forget the probability that many freshwater forms that formerly ranged continuously over immense areas will later have become extinct at intermediate points. But the wide distribution of freshwater plants and of the lower animals, whether retaining the same identical form or in some degree modified, apparently depends largely on the wide dispersal of their seeds and eggs by animals—especially by freshwater birds, which have great powers of flight and naturally travel from one piece of water to another.

On the Inhabitants of Oceanic Islands

We now come to the last of the three classes of facts that I have selected as presenting the greatest amount of difficulty in explaining present animal and plant distributions, on the view that not only all the individuals of the same species have migrated from some one area, but that allied species, although now inhabiting the most distant points, have proceeded from a single area, the birthplace of their early ancestors.

I have already given my reasons for disbelieving in continental extensions within the period of existing species on so enormous a scale that all the many islands of the several oceans were thus stocked with their present terrestrial inhabitants. This view removes many difficulties, but it does not agree with all the facts in regard to the productions of islands. In the following remarks I shall not confine myself to the mere question of dispersal, but shall consider some other cases that bear on the truth of our two theories: independent creation versus descent with modification.

Few animal and plant species inhabit oceanic islands compared with those on equal continental areas: Alphonse de Candolle admits this for plants, and Wollaston for insects. New Zealand, for instance, with its lofty mountains and diversified stations, extending over 780 miles of latitude, together with the outlying islands of Auckland, Campbell, and Chatham, contain altogether only 960 kinds of flowering plants; if we compare this moderate number with the species that swarm over equal areas of land in Southwestern Australia or at the Cape of Good Hope, we must admit that some cause, independently of different physical conditions, must have given rise to so great a difference in number. Even the uniform county of Cambridge has 847 plant species, and the little island of Anglesea has 764 species; it must be admitted, though, that a few ferns and a few introduced plants are included in those numbers, and the comparison in some other respects is also not quite fair. We have evidence that the barren island of Ascension aboriginally possessed less than six flowering plant species; and yet many species have now become naturalized on it, as they have in New Zealand and on every other oceanic island that can be named. In St. Helena there is reason to believe that the naturalized plants and animals have nearly or even completely exterminated many native productions. He who admits the doctrine of the divine creation of each separate species will have to admit that a sufficient number of the best adapted plants and animals were not created for oceanic islands: indeed, humans have unintentionally stocked them far more fully and perfectly than did nature.

Although relatively few species are found living on oceanic islands, the proportion of endemic species (i.e., those found nowhere else in the world) is often extremely large. If we compare, for instance, the number of endemic land snails found in Madeira, or of birds endemic to the Galapagos Archipelago, with the number of species found on any continent and then compare the area of the island with that of the continent, we shall see that this is true. This fact might have been theoretically expected, for, as already explained, species occasionally arriving after long intervals of time in the new and isolated district, and having to compete with new associates, would be eminently liable to modification over time and would often produce groups of modified descendants. But it by no means follows that just because nearly all the species of one class are peculiar on any particular island, those of another class, or of another section of the same class, will be peculiar as well. This difference seems to depend partly on the species that have not become modified having immigrated together in a body, so that their mutual relations have not been much disturbed, and partly on the frequent arrival of unmodified immigrants from the mother country with which the island forms have intercrossed. It should be borne in mind that the offspring of such crosses would certainly gain in vigor, so that even an occasional cross would produce more effect than might have been anticipated.

I will give a few illustrations of my previous remarks. For one thing, in the Galapagos Islands, there are 26 terrestrial bird species, of which 21 (or perhaps 23) are unique to those islands. In contrast, of the 11 marine bird species found in the Galapagos, only two are unique, and it is obvious that marine birds could arrive at these islands much more easily and frequently than land birds. Bermuda, on the other hand, which lies at about the same distance from North America as the Galapagos Islands do from South America, and which has a very peculiar soil, does not possess a single endemic land bird. How can we explain this? Well, we know from Mr. J. M. Jones's admirable account of Bermuda that very many North American birds occasionally or even frequently visit this island. Similarly, almost every year, as I am informed by Mr. Edward V. Harcourt, many European and African birds are blown to Madeira; this island is inhabited by 99 different species, of which only one is peculiar to that island, although it is very closely related to a European form. Three or four other species are found only on this one island and in the Canaries. Thus, the islands of Bermuda and Madeira have been stocked from the neighboring continents with birds, which for long ages have struggled there together and have become mutually co-adapted. Hence, when settled in their new homes, each species will have been kept by the others to its proper place and habits, and will consequently have been but little liable to modification. Any tendency to modification will also have been checked by interbreeding with unmodified immigrants that would be arriving frequently from the mother country. Madeira is also inhabited by a wonderful number of peculiar land snails, whereas not one single species of marine snail is peculiar to its shores: now, though we do not know how the marine species are dispersed, yet we can see that their eggs or larvae, perhaps attached to seaweed or floating timber or to the feet of wading birds, might be transported across 300 or 400 miles of open sea far more easily than land snails.[3] The different orders of insects inhabiting Madeira present nearly parallel cases.

Oceanic islands are sometimes deficient in entire taxonomic classes of certain animals, and their places are occupied instead by the members of other classes. In the Galapagos Islands, for example, reptiles, and in New Zealand gigantic wingless birds, take—or recently took—the place of mammals. Although New Zealand is here spoken of as an oceanic island, it is in some degree doubtful whether it should be ranked as such; it is of large size and is not separated from Australia by a profoundly deep sea; from its geological character and the direction of its mountain ranges, the Rev. W. B. Clarke has lately maintained that this island, as well as New Caledonia, should be considered as appurtenances of Australia. Turning to plants, Dr. Hooker has shown that, in the Galapagos Islands, the proportional numbers of the different orders are very different from what they are elsewhere. All such differences in number, and the

---

[3] As it turns out, many marine snails produce microscopic larvae that can be dispersed great distances in ocean currents for weeks or months before metamorphosing to the juvenile stage.

absence of certain whole groups of animals and plants, are generally accounted for by supposed differences in the physical conditions of the islands, but this explanation is not a little doubtful. Facility of immigration seems to have been fully as important as the nature of the environmental conditions.

Many remarkable little facts could be given with respect to the inhabitants of oceanic islands. For instance, in certain islands not occupied by a single species of mammal, some of the endemic plants nevertheless have beautifully hooked seeds; yet it is well known that these hooks serve to transport the seeds in the wool or fur of quadrupeds, none of which are found on these islands! But a hooked seed might occasionally be carried to an island by other means; the resulting plant could then become gradually modified, over many generations, to form an endemic species that still retained its hooked seeds even though the hooks would now form a useless appendage, like the shriveled wings found under the soldered wing-covers of many island beetles.

As another example, islands often possess trees or bushes belonging to taxonomic orders that elsewhere include only herbaceous species. Now trees, as Alphonse de Candolle has shown, generally have—whatever the cause may be—confined ranges. Hence trees would be unlikely to reach distant oceanic islands; thus, a herbaceous plant that had no chance of successfully competing with the many fully developed trees growing on a continent might, if it became established on an island, gain an advantage over other herbaceous plants on that island by growing taller and taller and overtopping them. In this case, natural selection would tend to add to the stature of the plant, to whatever order it belonged, and thus first convert it into a bush and then into a tree.

Absence of Batrachians and Terrestrial Mammals on Oceanic Islands

With respect to the absence of whole orders of animals on oceanic islands, Bory St. Vincent long ago remarked that batrachians (a group of amphibians that includes the frogs and toads, members of the order Anura) are never found on any of the many islands with which the great oceans are studded. I have taken pains to verify this assertion and have found it to be true, with the exception of New Zealand, New Caledonia, the Andaman Islands, and perhaps the Solomon Islands and the Seychelles. But I have already remarked that it is doubtful whether New Zealand and New Caledonia ought to be classed as oceanic islands, and this is still more doubtful with respect to the Andaman and Solomon groups and the Seychelles. This general absence of frogs, toads, and newts on so many true oceanic islands cannot be accounted for by their physical conditions; indeed, it seems that islands are peculiarly fitted for these animals. Indeed, frogs have been introduced into Madeira, the Azores, and Mauritius and have multiplied so much as to have become a nuisance. But as these animals and their spawn are immediately killed by seawater (with the exception, as far as is known, of one Indian species), there would be great difficulty in their being successfully transported across the sea; we can see, therefore, why they do not exist on strictly oceanic islands. But why, on the theory of creation, they should not have been created there, it would be very difficult to explain.

Mammals offer another and similar case. I have carefully searched reports from the oldest exploratory voyages and have not found a single instance, free from doubt, of a terrestrial mammal (excluding domesticated animals kept by the natives) inhabiting an island situated more than 300 miles from a continent or great continental island; and many islands situated much closer are equally barren. The Falkland Islands, which are inhabited by a wolf-like fox, come nearest to an exception. However, this group cannot be considered as oceanic as it lies on a bank in connection with the mainland at a distance of about 280 miles; moreover, icebergs formerly brought boulders to its western shores, and these icebergs may have formerly transported foxes, as now frequently happens in the arctic regions.

Yet it cannot be said that small islands will not support at least small mammals, for small mammals occur on such very small islands in many parts of the world, when they are near a continent; and hardly an island can be named on which our smaller quadrupeds[4] have not become naturalized and greatly multiplied. It cannot be said, on the ordinary view of creation, that there has not been time for the divine creation of mammals, for many volcanic islands are sufficiently ancient, as shown by the stupendous degradation which they have suffered and by their tertiary strata. There has also been time for the production of endemic species belonging to other classes, and on continents new species of mammals are known to appear and disappear at a quicker rate than other and lower animals.

Although terrestrial mammals do not occur on oceanic islands, aerial mammals do occur on almost every island. Indeed, New Zealand possesses two bats found nowhere else in the world, and Norfolk Island, the Viti Archipelago, the Bonin Islands, the Caroline and Marianne Archipelagoes, and Mauritius all possess their peculiar bats. Why, it may be asked, has the supposed creative force produced bats and no other mammals on remote islands? On my view this question can easily be answered, for no terrestrial mammal can be transported across a wide space of sea, but bats can fly across. In fact, bats have been seen wandering by day far over the Atlantic Ocean, and two North American species, either regularly or occasionally, are known to visit Bermuda, which lies about 600 miles from the mainland. I hear from Mr. Tomes, who has specially studied this family, that many bat species have enormous ranges and are found on different continents and on far distant islands. Hence, we have only to suppose that such wandering species have been modified in their new homes in relation to their new position, and we can understand the presence of endemic bats on oceanic islands, despite the absence of all other terrestrial mammals.

---

[4] Quadrepeds are animals with four feet.

Another interesting relation exists between the depth of the sea separating islands from each other—or from the nearest continent—and the degree of affinity of their mammalian inhabitants. Mr. Windsor Earl has made some striking observations on this point—observations that have since been greatly extended by Mr. Alfred Wallace's admirable researches—in regard to the great Malay Archipelago, which is traversed near Celebes by a space of deep ocean, and this separates two widely distinct mammalian faunas. On either side, the islands stand on a moderately shallow submarine bank and are inhabited by the same or by closely allied quadrupeds. I have not as yet had time to follow up on this subject in all quarters of the world, but, as far as I have gone, the relation holds good. For instance, Britain is separated by a shallow channel from Europe, and the mammal species are the same on both sides; and so it is with all the islands near the shores of Australia. The West Indian Islands, on the other hand, stand on a deeply submerged bank, nearly 1,000 fathoms in depth, and here, although we do find American forms, the species and even the genera are quite distinct. As the amount of modification that animals of all kinds undergo depends partly on the lapse of time, and as the islands that are separated from each other—or from the mainland—by shallow channels are more likely to have been continuously united within a recent period than the islands separated by deeper channels, we can understand how it is that a relation exists between the depth of the sea separating two mammalian faunas and the degree of their affinity, a relation that is quite inexplicable on the theory of independent acts of creation.

The foregoing statements in regard to the inhabitants of oceanic islands—namely, (1) the small number of species, with a large proportion consisting of endemic forms (the members of certain groups, but not those of other groups in the same class, having been modified); (2) the absence of certain whole orders, as of batrachians and of terrestrial mammals, notwithstanding the presence of aerial bats; and (3) the singular proportions of certain orders of plants, herbaceous forms having been developed into trees, etc.—seem to me to accord better with the belief in the efficiency of occasional means of transport, carried on during a long course of time, than with the belief in the former connection of all oceanic islands with the nearest continent. On this latter view, it is probable that the members of the various classes would have immigrated more uniformly, and, from the species having entered in a body, their mutual relations would not have been much disturbed; consequently, they would either have not been modified, or all of the species would have been modified in a more equable manner.

I do not deny that there are many and serious difficulties in understanding how many of the inhabitants of the more remote islands, whether still retaining the same specific form or a form subsequently modified after its arrival, have reached their present homes. But the probability of other islands having once existed as halting places for ships, of which not a wreck now remains, must not be overlooked. I will specify one difficult case. Almost all oceanic islands, even the most isolated and smallest, are inhabited by land snails—generally by endemic species, but sometimes by species that are also found elsewhere, striking instances of which have been given by

Dr. A. A. Gould in relation to the Pacific. Now it is notorious that land snails are easily killed by seawater; their eggs, at least such as I have tried, sink in it and are killed. Yet there must be some unknown but occasionally efficient means for their transport. Would the just-hatched young sometimes adhere to the feet of birds roosting on the ground and thus get transported? It occurred to me that land snails, when hibernating and having a membranous diaphragm over the mouth of the shell, might be floated in chinks of drifted timber across moderately wide arms of the sea. And I find that several species in this state withstand a seven-day emersion in seawater without any sign of injury. One land snail species, *Helix pomatia*, after having been thus treated, and again hibernating, was put into seawater for 20 days and recovered completely. During this length of time the snail might have been carried by a marine current of average swiftness to a distance of 660 geographical miles. As this *Helix* species has a thick calcareous operculum on its foot, I removed it, and when it had formed a new membranous one, I again immersed it in seawater for 14 days; once again, it recovered and crawled away. Baron Aucapitaine has since tried similar experiments. He placed 100 land snails (belonging to 10 different species) in a box pierced with holes and immersed it for a fortnight in the sea. Out of the 100 snails so treated, 27 recovered. The presence of an operculum seems to have been of importance, as 11 of the 12 specimens of the operculum-bearing species *Cyclostoma elegans* revived after this treatment. It is remarkable, seeing how well the *Helix pomatia* snails resisted submersion in seawater in my studies, that not one of the 54 specimens belonging to four other *Helix* species recovered after similar treatments by Aucapitaine. It is, however, not at all probable that land snails have often been thus transported; the feet of birds offer a more probable method of transport.

On the Relations of the Inhabitants of Islands to Those of the Nearest Mainland

The most striking and important fact for us is the affinity between the species that inhabit islands to those of the nearest mainland, without being actually the same. Numerous instances could be given. The Galapagos Archipelago, situated under the Equator, lies about 500 to 600 miles from the shores of South America. In this Galapagos Archipelago, almost every product of the land and of the water bears the unmistakable stamp of the American continent. There are 26 different land birds. Of these 21—or perhaps even 23—are ranked as distinct species and would commonly be assumed to have been created here; yet the close affinity of most of these birds to American species is manifest in every character in their habits, gestures, and tones of voice. So it is with the other animals and with a large proportion of the plants as well, as shown by Dr. Hooker in his admirable Flora of this archipelago. The naturalist, looking at the inhabitants of these volcanic islands in the Pacific—which lie distant several hundred miles from the continent—feels that he is standing on American land. Why should this be so? Why should the species that are supposed to have been created in the Galapagos Archipelago—and nowhere else—bear so plainly the stamp of affinity to those created in America? There is nothing in the conditions of life, in the geological nature of the islands, in their height or climate, or in the proportions in which the several classes are associated together that closely resembles the conditions of the South American coast. In fact, there is a considerable dissimilarity in all these respects. On the other hand, there is a considerable degree of resemblance in the volcanic nature of the soil, in the climate, height, and size of the islands, between the Galapagos and the Cape Verde Archipelagos—but what an entire and absolute difference there is in their inhabitants! The inhabitants of the Cape Verde Islands are related to those of Africa, as those of the Galapagos are to America. Facts such as these admit of no sort of explanation on the ordinary view of independent creation. On my view, however, it is obvious that the Galapagos Islands would be likely to receive colonists from America, whether by occasional means of transport or (though I do not believe in this doctrine) by formerly continuous land, and the Cape Verde Islands from Africa; such colonists would be liable to modification, the principle of inheritance still betraying their original birthplace.

Many analogous facts could be given: indeed, it is an almost universal rule that the endemic productions of islands are related to those of the nearest continent or of the nearest large island. The exceptions are few, and most of them can be explained. Thus, although Kerguelen Land stands nearer to Africa than to America, the plants are very closely related to those of America, as we know from Dr. Hooker's account; but on the view that this island has been mainly stocked by seeds brought with earth and stones on icebergs, drifted by the prevailing currents, this anomaly disappears. The endemic plants of New Zealand are much more closely related to plants found in Australia, the nearest mainland, than to any other region, and this is what might have been expected; but New Zealand's endemic plants are also plainly related to those of South America, which, although it is the next nearest continent, is so enormously remote that the fact becomes an anomaly. But this difficulty partially disappears on the view that New Zealand, South America, and the other southern lands have been stocked in part from a nearly intermediate though distant point, namely, from the Antarctic islands when they were clothed with vegetation during a warmer tertiary period, before the commencement of the last glacial period. The affinity, which, though feeble, I am assured by Dr. Hooker is real, between the flora of the southwestern corner of Australia and of the Cape of Good Hope, is a far more remarkable case; but this affinity is confined to the plants and will, no doubt, someday be explained.[5]

The same law that has determined the relationship between the inhabitants of islands and the nearest mainland is sometimes displayed—on a small scale, but in a most interesting manner—within the limits of the same archipelago. Thus, although each separate island of the Galapagos Archipelago is occupied—and the fact is a marvelous one—by many distinct species, these species are clearly related to each

---

[5] It has now been explained! These botanical similarities owe to continental drift, a phenomenon that was first proposed by Alfred Wegener in 1912, but that was not widely accepted until the 1960s.

other in a very much closer manner than to the inhabitants of the American continent or of any other quarter of the world. This is what might have been expected, for
islands situated so near to each other would almost necessarily receive immigrants
from the same original source and from each other. But how is it that many of the
immigrants have been differently modified, though only in a small degree, in islands
situated within sight of each other, despite their having the same geological nature,
the same height, the same climate, and so forth? This long appeared to me a great
difficulty, but it arises in chief part from the deeply seated error of considering the
physical conditions of a country to be the most important selective force acting on
animals and plants. But in fact **it cannot be disputed that the nature of the other species with which each has to compete is at least as important—and in fact generally
a far more important element of success.** Now if we look to the species that inhabit
the Galapagos Archipelago and are likewise found in other parts of the world, we find
that they differ considerably in the several islands. This difference might indeed have
been expected if the islands have been stocked by occasional means of transport—a
seed, for instance, of one plant having been brought to one island, and that of another
plant to another island, though all proceeding from the same general source. Hence,
when in former times an immigrant first settled on one of the islands, or when it subsequently spread from one to another, it would undoubtedly be exposed to different
environmental selective forces in the different islands, since it would have to compete
with a different set of organisms; a plant, for instance, would find that the ground
best-fitted for it was already occupied by somewhat different species in the different
islands and would be exposed to the attacks of somewhat different enemies. If, then,
it varied, natural selection would probably favor different varieties in the different islands. Some other species, however, might spread and yet retain the same character
throughout the group, just as we see some species spreading widely throughout a continent and remaining the same.

The really surprising fact in this case of the Galapagos Archipelago, and to a lesser
degree in some analogous cases, is that each new species, after being formed in any
one island, did not spread quickly to the other islands. But these islands, although
they are all within sight of each other, are separated by deep arms of the sea that in
most cases are wider than the British Channel, and there is no reason to suppose that
these islands have at any former period been continuously united. The currents of
the sea are rapid and deep between the islands, and gales of wind are extraordinarily
rare; thus the islands are far more effectually separated from each other than they
appear on a map. Nevertheless, some of the species—both of those found in other
parts of the world and of those confined to the archipelago—are common to the several islands, and we may infer from the present manner of distribution that they have
spread from one island to the others. But I think that we often take an erroneous
view of the probability of closely allied species invading each other's territory when
put into free intercommunication. Undoubtedly, if one species has any advantage
over another, it will in a very brief time wholly or in part supplant it; but, if both are
equally well-fitted for their own places, both will probably hold their separate places

for almost any length of time. Being familiar with the fact that many species that have been transported to new locations by humans have subsequently spread with astonishing rapidity over wide areas, we are apt to infer that most species would thus spread—but we should remember that the species that become naturalized in new countries are not generally closely related to the aboriginal inhabitants, but are very distinct forms belonging in a large proportion of cases to distinct genera, as shown by Alphonse de Candolle. In the Galapagos Archipelago, even many of the birds differ on the different islands, even though they are so well adapted for flying from island to island. For example, three closely allied species of mocking-thrush are each confined to its own island. Now let us suppose that the mocking-thrush of Chatham Island gets to Charles Island, which has its own mocking-thrush: Why should the Chathma Island mocking-thrush succeed in establishing itself there? We may safely infer that Charles Island is well stocked with its own species, for annually more eggs are laid and young birds hatched than can possibly be reared, and we may infer that the mocking-thrush peculiar to Charles Island is at least as well fitted for its home as is the species peculiar to Chatham Island. Sir C. Lyell and Mr. Wollaston have communicated to me a remarkable fact bearing on this subject: namely, that Madeira and the adjoining islet of Porto Santo possess many distinct but representative species of land snails, some of which live in crevices of stone. Although large quantities of stone are annually transported from Porto Santo to Madeira, yet this latter island has not become colonized by the Porto Santo snail species. Nevertheless, both islands have been colonized by some European land snails, which no doubt must have had some advantage over the indigenous species.

From these considerations, I think we need not greatly marvel at the fact that endemic species inhabiting the several islands of the Galapagos Archipelago have not all spread from island to island. In addition, on the same continent, prior occupation has probably also played an important part in checking the commingling of the species that inhabit different districts with nearly the same physical conditions. Thus, the southeast and southwest corners of Australia have nearly the same physical conditions and are united by continuous land and yet they are inhabited by a vast number of distinct mammal, bird, and plant species; and so it is, according to Mr. Bates, with the butterflies and other animals inhabiting the great, open, and continuous valley of the Amazons.

The same principle that governs the general character of the inhabitants of oceanic islands—namely, their relation to the source from which colonists could have been most easily derived, together with their subsequent modification—is of the widest application throughout nature. We see this on every mountain summit, and in every lake and marsh. Alpine species, for example, are related to those of the surrounding lowlands, excepting in as far as the same species have become widely spread during the glacial epoch: thus in South America we find alpine hummingbirds, alpine rodents, alpine plants, etc., all strictly belonging to American forms, and it is obvious that a mountain, as it became slowly upheaved, would be colonized from the surrounding lowlands. So it is with the inhabitants of lakes and marshes, excepting in so far as a greater facility of transport has allowed the same forms to prevail throughout large portions of the world. We see the same principle in the character of most of the blind animals inhabiting the caves of America and of Europe. Other analogous facts could be given. **It will, I believe, be found universally true that wherever in two regions—regardless of how far apart they are—many closely allied or representative species occur, there will likewise be found some identical species; and wherever many closely allied species occur, we will find many forms that some naturalists rank as distinct species and others as mere varieties, with these doubtful forms showing us the steps in the process of modification.**

The relationship between the power and extent of migration in certain species—either at the present or at some former period of time—and the existence of closely allied species at remote points of the world is shown in another and more general way. Mr. Gould remarked to me long ago that, in those genera of birds that range all over the world, many of the species have very wide ranges. I can hardly doubt that this rule is generally true, though difficult to prove. Among mammals, we see it strikingly displayed in bats and to a lesser degree in the Felidae and Canidae.[6] We see the same rule in the distribution of butterflies and beetles. So it is with most of the inhabitants of fresh water, for many of the genera in the most distinct classes range over the world, and many of the species have enormous ranges. It is not meant that all, but that some of the species have very wide ranges in the genera that range very widely. Nor is it meant that the species in such genera have, on an average, a very wide range, for that will largely depend on how far the process of modification has gone. For instance, two varieties of the same species inhabit America and Europe, and thus the species has an immense range; but, if variation were to be carried a little further, the two varieties would be ranked as distinct species, and their ranges would be greatly reduced. Still less is it meant that species that have *the capacity* of crossing barriers and ranging widely, as in the case of certain powerfully winged birds, will *necessarily* range widely; we should never forget that to range widely implies not only the ability to cross barriers, but also the more important power of being victorious in distant lands in the struggle for life with foreign associates. But, according to the view that all the species belonging to any particular genus, though distributed to the most remote points of the world, are descended from a single ancestor, we ought to find—and I believe as a general rule we do indeed find—that at least some of the species range very widely.[7]

We should bear in mind that many genera in all classes of organisms are of ancient origin and that the species in this case will have had ample time for dispersal and subsequent modification. There is also reason to believe, from geological evidence, that within each great class the lower organisms change at a slower rate than the higher;

---

[6] Cats are members of the mammalian family Felidae, while the family Canidae includes dogs, wolves, foxes, and other dog-like carnivores.

[7] As noted earlier, plate tectonics, which Darwin knew nothing about since it was only widely validated in the 1960s, probably also played an important role in distributing ancient organisms over much of our planet.

they will consequently have had a better chance of ranging widely and of still retaining the same specific character. This fact, together with that of the seeds and eggs of most lowly organized forms being very minute and better fitted for long-distance transport, probably accounts for a law that has long been observed, and that has lately been discussed by Alphonse de Candolle in regard to plants: namely, that the lower any group of organisms stands, the more widely it ranges.

I have now discussed a number of important relationships: (1) that lower organisms range more widely than the higher organisms; (2) that some of the species of widely ranging genera also range widely themselves; (3) the finding that alpine, lacustrine, and marsh productions are generally related to those that live on the surrounding lowlands and dry lands; (4) the striking relationship between the inhabitants of islands and those of the nearest mainland; and (5) the even closer relationship of the distinct inhabitants found living on the islands of the same archipelago. These relationships are inexplicable on the ordinary view of the independent creation of each species but are easy to explain if we admit that there has been colonization from the nearest or readiest source together with the subsequent adaptation of the colonists to their new homes.

Summary of Chapters 12 and 13

In Chapter 12 and in this chapter, I have endeavored to show that if we make due allowance for our ignorance of the full effects of changes of climate and of the level of the land, which have certainly occurred within the recent period, and of other changes that have probably occurred; if we remember how ignorant we are with respect to the many curious means of occasional transport; and if we bear in mind—and this is a very important consideration—how often a species may have ranged continuously over a wide area and then have become extinct in the intermediate tracts, then it is not difficult to believe that all the individuals of the same species, wherever found, are descended from common parents. Indeed, we are led to this conclusion, which has been arrived at by many naturalists under the designation of "single centers of creation," by various general considerations, more especially (1) from the importance of barriers of all kinds and (2) from the analogical distribution of subgenera, genera, and families.

With respect to distinct species belonging to the same genus, which on our theory have spread from one parent source, if we make the same allowances as before for our ignorance and remember that some forms of life have changed very slowly, enormous periods of time having been thus granted for their migration, the difficulties are far from insuperable—although, as in the case of individuals of the same species, they are often great.

To illustrate the effects that climate change has had on the distribution of various taxonomic groups, I have attempted to show how important a part the last glacial period (which affected even the equatorial regions) has played and which, during the alternations of the cold in the north and the south, has allowed the productions of opposite hemispheres to mingle and left some of them stranded on the summits of mountains in all parts of the world. And, as a way of showing how diversified are the means of occasional transport from one place to another, I have discussed at some little length the means of dispersal of freshwater productions.

If we can admit that, in the long course of time, all the individuals of the same species—and likewise of the several species belonging to the same genus—have all proceeded from some one source, then all the grand leading facts of geographical distribution are explicable on the theory of migration, together with subsequent modification and the multiplication of new forms. We can thus understand the high importance of physical barriers, whether of land or water, in not only separating but in apparently forming the several zoological and botanical provinces. We can thus understand why related species are often concentrated within the same areas and how it is that under different latitudes—in South America, for instance—the inhabitants of the plains and mountains, and of the forests, marshes, and deserts, are linked together in so mysterious a manner and are likewise linked to the extinct beings that formerly inhabited the same continent. Bearing in mind that the mutual relation of organism to organism is of the highest importance, we can see why two areas that have nearly the same physical conditions should often be inhabited by very different forms of life: for, according to the length of time that has elapsed since the colonists entered one of the regions, or both; according to the nature of the communication that allowed certain forms and not others to enter, either in greater or lesser numbers; according or not as those that entered happened to come into more or less direct competition with each other and with the original inhabitants; and according as the immigrants were capable of varying more or less rapidly, there would ensue in the different regions, independently of the physical conditions in those regions, infinitely diversified conditions of life. Indeed, there would be an almost endless amount of organic action and reaction, and we should find some groups of beings greatly modified and some only slightly modified, some developed in great force, and some existing in scanty numbers—and this we do find in the several great geographical provinces of the world.

On these same principles we can understand, as I have endeavored to show, why oceanic islands should have few inhabitants but that, of these, a large proportion should be endemic or peculiar; and why, in relation to the means of migration, one group of beings should have all its species peculiar and another group, even within the same class, should have all its species the same with those in an adjoining quarter of the world. We can see why whole groups of organisms, such as frogs, toads, and terrestrial mammals, should be absent from oceanic islands, while the most isolated islands should possess their own peculiar species of aerial mammals or bats. We can see why, on islands, there should be some relation between the presence of mammals, in a more or less modified condition, and the depth of the sea between such islands and the mainland. We can clearly see why all the inhabitants of an archipelago, though they may be specifically distinct on the several islets, should all be closely related to each other and should likewise be related, but less closely, to those of the nearest continent or some other source whence the immigrants might have been derived. We can see why, if there exist very closely allied or representative species in two areas, however distant from each other, some identical species will almost always be found in both of them.

As the late Edward Forbes often insisted, there is a striking parallelism in the laws of life throughout time and space, with the laws governing the succession of forms in past times being nearly the same as those presently governing the differences in different areas. We see this in many facts. For example, the endurance of each species and group of species is continuous in time; indeed, the apparent exceptions to the rule are so few that they may fairly be attributed to our not having as yet discovered in an intermediate deposit certain forms that are absent in it, but which occur above and below. And, similarly, in space, it certainly is the general rule that the area inhabited by a single species, or by a group of species, is continuous; the exceptions, which are not rare, may—as I have attempted to show—be accounted for by former migrations under different circumstances, or through means of occasional transport, or by the species having become extinct in the intermediate areas. Both in time and space, species and groups of species have their points of maximum development. Groups of species living during the same period of time, or living within the same area, are often characterized by trifling features in common, as of sculpture or color. In looking to the long succession of past ages, as in looking to distant provinces throughout the world, we find that species in certain classes differ little from each other, while those in another class, or only in a different section of the same order, differ greatly from each other. In both time and space, the lowly organized members of each class generally change less than the highly organized, but there are in both cases marked exceptions to that rule. According to our theory, these several relations throughout time and space are clearly intelligible: for whether we look to the allied forms of life which have changed during successive ages, or to those that have changed after having migrated into distant quarters, in both cases they are connected by the same bond of ordinary generation; in both cases the laws of variation have been the same, and modifications have been accumulated by the same means: by natural selection.

Key Issues to Talk and Write About

1. Try rewriting these sentences, to make them clearer or less wordy (or both).
    a. "The Galapagos Archipelago, situated under the equator, lies at a distance of between 500 and 600 miles from the shores of South America."
    b. "New Zealand in its endemic plants is much more closely related to Australia, the nearest mainland, than to any other region."

c. "For alpine species, excepting in as far as the same species have become widely spread during the glacial epoch, are related to those of the surrounding lowlands; thus we have in South America, alpine hummingbirds, alpine rodents, alpine plants, etc., all strictly belonging to American forms; and it is obvious that a mountain, as it became slowly upheaved, would be colonized from the surrounding lowlands."

d. "The species of all kinds which inhabit oceanic islands are few in number compared with those on equal continental areas."

2. What studies that are described in this chapter did Darwin do himself?

3. Briefly describe the mechanisms that Darwin suggests as being able to explain why particular species of fish, land snails, and plants often have surprisingly wide geographical ranges.

4. How does Darwin explain why some species are unique to single islands but nevertheless seem similar to those of neighboring islands?

14

Evidence for Mutual Affinities Among Organic Beings

Evidence from Morphology, Embryology, and Rudimentary Organs

In this chapter, Darwin presents strong evidence and logical thinking in support of gradual evolutionary change linking the members of all taxonomic group into a single genealogical series. Why are all of the smaller taxonomic groups of animals and plants, such as species and families, naturally placed within larger groups, such as orders, classes, and phyla? Why are the members of different classes of organisms so different from each other? Why can all living and extinct organisms be grouped together within the same few major taxonomic categories? Why are embryological characteristics so often used in grouping organisms? Why are identical larval stages so often found in such a wide variety of adult organisms? Why are atrophied or only partially developed, currently functionless organs so useful in placing organisms into their proper groups, and why are characteristics of seemingly minor physiological importance to an organism often so useful in classification? Why are some body parts that different animals use for greatly different purposes constructed on the same basic pattern? The most rational explanation for these and related questions is simple, says Darwin: the explanation is that all organisms living today have descended from common ancestors, with a great deal of change and extinction occurring over long periods of time. Thus our natural classification reflects a "community of descent."

From the most remote period in the history of the world, people have noticed that living organisms resemble each other in descending degrees, so that groups of organisms lie within other groups of organisms. The group Vertebrata, for example, contains seven other major groups, including the bony fish (Osteichthyes), the frogs and salamanders (Amphibia), the birds (Aves), and mammals (Mammalia, including humans). And each of those groups contains other groups. The Mammalia, for example, contains at least two major subgroups: marsupials and placental mammals. This classification of organisms, with groups contained within other groups, is not arbitrary, unlike the way that the stars in the sky have been grouped to form constellations. The existence of these animal and plant groupings would have been of little

The Readable Darwin. Second Edition. Jan A. Pechenik, Oxford University Press. © Oxford University Press 2023.
DOI: 10.1093/oso/9780197575260.003.0015

significance if one group had been adapted to live exclusively on land and another for life in water, or one to feed on flesh and another to feed on vegetable matter, and so forth. But the actual groupings are quite different: indeed, it is notorious how commonly the members of even any one subgroup differ in their lifestyles.

In Chapter 2 (on variation) and Chapter 4 (on natural selection), I attempted to show that within each country it is the most wide-ranging, much diffused, and common species—that is, the dominant species belonging to the larger genera in each class of organisms—that show the most variation in traits among individuals. It is these varieties that will ultimately become converted into new and distinct species; thus I have called these varieties "incipient species." And these, on the principles of inheritance, tend to eventually produce other new and dominant species. Consequently, the groups that are now large and which generally include many dominant species tend to go on increasing in size over time. I further attempted to show that from the varying offspring of each species always trying to occupy as many and as different ecological niches[1] as possible in nature, they tend constantly to diverge in character. This latter conclusion is supported by observing the great diversity of forms which, in any small area, come to compete mostly closely with each other and by certain facts in naturalization.

I also attempted to show that there is a steady tendency for the forms that are increasing in number and diverging in character to eventually supplant and exterminate the preceding, less divergent and less improved forms. If you turn to Figure 4.11, which illustrates the action of these several principles that I have already discussed in detail, you will see that the modified descendants of any one ancestor inevitably become broken up over time into groups within groups. In the diagram, each letter (e.g., a, q, or b) on the uppermost line may represent one genus that includes a number of species, and all of the first eight genera along this upper line form together one taxonomic class, for all are descended from a single ancient parent (a member of original genus A) and, consequently, have inherited something in common. But the three genera on the left hand side of the figure (a-14, q-14, p-14) also have, on this same principle, much in common and form a subfamily distinct from that containing the next two genera to the right (b-14 and f-14), which diverged from a common parent at the fifth stage of descent in the diagram.

The members of these first five genera shown in the left side of the diagram also have much in common with each other, though less than when grouped in subfamilies, and they form a family distinct from that containing the three genera (o-14, e-14, and m-14) still farther to the right in the diagram, which diverged from the other descendants of parent A at an even earlier period.

Moreover, all of the genera that descended from ancestor (A) form a taxonomic order distinct from that which includes the six genera descended from ancestor (I). So here we have many species descended from a single ancestor grouped into genera,

---

[1] In the original version of this chapter, Darwin referred to what we now call "niches" as "places in the economy of nature."

and the genera are distributed among various subfamilies, families, and orders, all under one great class. The grand fact of the natural subordination of living organisms into groups within groups which, from its familiarity, does not always strike us sufficiently, is in my judgment thus explained.

No doubt we could classify organic beings in many ways, as with all other objects, either artificially by single characters or more naturally by a number of characters. We know, for instance, that minerals can be thus arranged, as can the chemical elements. In the case of minerals and elements, of course, the grouping has nothing to do with genealogical succession, and no cause can presently be assigned for their falling into groups. But with organic beings the case is different, and the view that I have just given explains their natural arrangement in groups within groups very well. Indeed, no other explanation of this grouping has ever been attempted.

Naturalists, as we have seen, try to arrange the species, genera, and families in each taxonomic class based on what is called the "Natural System." But what is meant by such a "natural system"? Some authors look at it merely as a scheme for arranging together those living objects that are most like each other and for separating those that are most unlike each other, or as an artificial method of enunciating, as briefly as possible, various general propositions—giving in a single sentence the characteristics that all mammals have in common, for instance, and in another sentence stating the characteristics that all carnivores have in common, and then that all members of the dog genus have in common. And then, just by adding another sentence, they can give a full description for each kind of dog. The ingenuity and utility of this system are indisputable.

But many naturalists think that something more is meant by this Natural System; they believe that it reveals the plan of the Creator. But unless it is be specified what is meant by "the plan of the Creator," it seems to me that nothing is added to our knowledge. Expressions such as that famous one by the Swedish taxonomist Carl Linnaeus, namely that "the characteristics do not make the genus, but that the genus gives the characteristics," seem to imply that some deeper bond is included in our classification system than mere resemblances. I believe that this is indeed the case and that community of descent—the only known potential cause of close similarity between groups of organic beings—is the bond that is partially revealed to us by our classifications.

Now let us consider the rules followed in classification and the difficulties that are encountered if one believes that classification either gives some unknown "plan of creation" or is simply a scheme for making general statements and placing together the forms which most resemble each other.

In ancient times, it was thought that those parts of the structure that determined the habits of life and the general place of each being in the economy of nature would be very important in classification, but nothing can be more false. No one regards the external similarity of a mouse and a shrew to be of any importance in classification, or that between a dugong and a whale, or between a whale and a fish. Those resemblances, though so intimately connected with the whole life of the being, are ranked as merely "adaptive or analogical characters." We shall return to this issue later. It may even be given as a general rule that the *less* any part of an animal or plant's organization is concerned with specialized habits, the *more important* it becomes for classification. For example, Sir Richard Owen, in speaking of the dugong, says, "The generative organs, being those which are most remotely related to the habits and food of an animal, I have always regarded as affording very clear indications of its true affinities. We are least likely in the modifications of these organs to mistake a merely adaptive for an essential character." With plants, how remarkable it is that the organs of vegetation, on which their nutrition and life depend, are of such little significance in suggesting relationships for classification, whereas the reproductive organs, along with their products, in the form of seeds and embryos, are of paramount importance! So again, in formerly discussing certain morphological characters that are not functionally important to an organism, we have seen that they are often of the highest service in classification. This depends on their constancy throughout many allied groups; and their constancy chiefly depends on any slight deviations *not* having been preserved and accumulated by natural selection, which acts only on characters that affect survival and reproductive success.

That the mere physiological importance of an organ does not determine its value for classification is almost proven by the fact that, in allied groups, in which the same organ seems to have nearly the same physiological value in all members, its value in classification is widely different. No naturalist can have worked with any group without being struck with this fact, which has been fully acknowledged in the writings of almost every author. It will suffice to quote the highest authority, the Scottish botanist Robert Brown, who, in speaking of certain organs in the plant family Proteaceae[2] says that their importance in determining what genera they belong to "like that of all their parts, not only in this, but, as I apprehend in every natural family, is very unequal, and in some cases seems to be entirely lost." Again, in another work he says that the genera belonging to the plant family Connaraceae "differ in having one or more ovaria, in the existence or absence of albumen, in the imbricate or valvular aestivation. Any one of these characters singly is frequently of more than generic importance, though here even, when all taken together, they appear insufficient to separate members of the genus *Cnestis* from those of the genus *Connarus*."[3] Let me give an example among insects: in one great division of the large insect order Hymenoptera, which includes the ants, wasps, and bees, the antennae, as Westwood has remarked, are extremely constant in structure, while in another division they differ much, and the differences are of quite subordinate value in classification. Yet no one will say that the antennae found in these two divisions of organisms belonging to the same order are of unequal physiological importance. **Any number of instances could be given of an important organ differing in how useful it is for purposes of classification within the same group of organisms.**

[2] These flowering plants—about 1,000 species—are found mostly in tropical and subtropical parts of the southern hemisphere.

[3] These are two genera of tropical flowering plants found in the family Connaraceae.

Similarly, no one will say that rudimentary or atrophied organs are of high physiological or vital importance to the organism that has them; yet, undoubtedly, organs in this condition are often of much value in classification. No one will dispute that the rudimentary teeth in the upper jaws of young ruminants[4], and certain rudimentary bones of the leg, are extremely useful in demonstrating the close affinity between ruminants (e.g., cows, sheep, and deer) and pachyderms (e.g., elephants and the rhinoceros). Similarly, Robert Brown has strongly insisted that the position of the rudimentary florets of grasses is of the highest importance in classifying them.

Numerous instances could be given of characters derived from parts that must be considered of very trifling physiological importance but that are universally admitted as being most useful in defining major groups. Here are some examples: whether or not there is an open passage from the nostrils to the mouth, the only character that, according to Sir Richard Owen, absolutely distinguishes fishes from reptiles; the inflection of the angle of the lower jaw in kangaroos, wallabies, and other marsupials; the manner in which the wings of insects are folded; mere color in certain algae; mere pubescence (soft down or fine short hairs on the leaves and stems of plants) on parts of the flower in grasses; and the nature of the skin-covering among vertebrates—whether it is hair or feathers, for example. If the duck-billed platypus had been covered with feathers instead of hair, this external and trifling character would have been considered by naturalists to be an important aid in determining how closely related this strange creature is to birds.

How important such trifling characters are for purposes of classification depends mainly on their being correlated with many other characters of more or less importance. The value indeed of an aggregate of characters is very evident in natural history. Thus it has often been remarked that a species may depart from its allies in several characters, both of high physiological importance and of almost universal prevalence, and yet leave us in no doubt where that species should be ranked. **Indeed, a classification founded on any single character, however important it may be, has always failed;** for no part of the organization is invariably constant. The importance of an aggregate of characters, even when none are especially important by themselves, explains the aphorism enunciated by Linnaeus, namely, that the characters do not give the genus, but the genus gives the character, for this seems founded on the appreciation of many trifling points of resemblance too slight to be defined. Certain tropical and subtropical flowering plants belonging to the family Malpighiaceae bear both perfect and degraded flowers; in the latter, as the French botanist de Jussieu has remarked, "The greater number of the characters proper to the species, to the genus, to the family, to the class, disappear, and thus laugh at our classification." When the asphead tree *Aspicarpa* produced in France, during several years, only these degraded flowers, which departed so wonderfully in a number of the most important points of

---

[4] Ruminants are mammals such as cows, goats, and sheep that have a specialized stomach capable of fermenting ingested food before it is digested. There are about 150 ruminant species alive today around the world.

structure from the accepted representatives of the order, yet M. Richard sagaciously saw, as A.-H. de Jussieu observes, that this genus should nevertheless be retained among the Malpighiaceae. This case well illustrates the spirit of our classifications.

When naturalists are at work, they do not trouble themselves about the physiological value of the characters they are using in defining a group or in allocating any particular species to a particular group. If they find a character that is nearly uniform and that is common to a great number of forms, but not common to others, they use it as one of high value in classifying; if common to some lesser number, they use it as of subordinate value. This principle has been broadly confessed by some naturalists to be the true one for classifying organisms, and by none more clearly than by that excellent botanist, Augustin Saint-Hilaire. If several trifling characters are always found in combination, though no apparent bond of connection can be discovered between them, special value is set on them. In most groups of animals, important organs, such as those for propelling the blood or for aerating it, or those for propagating the race, are found to be nearly uniform among species and are considered to be especially useful in classification; but in some groups, all of these most important vital organs are found to offer characters of quite subordinate value. Thus, as Fritz Müller[5] has lately remarked, all members of the ostracod genus *Cypridina*[6] are furnished with a heart, while in the closely allied genera *Cypris* and *Cytherea*, there is no such organ; similarly, one species of *Cypridina* has well-developed gills, while another species in the same genus lacks them.

We can see why characters derived from the embryo should be of equal importance with those derived from the adult, for a natural classification of course includes organisms at all ages and stages of development. But it is by no means obvious, on the ordinary view, why the structure of the embryo should ever be more important for this purpose than that of the adult, which alone plays its full part in the economy of nature. Yet it has been strongly urged by those great naturalists, Henri Milne Edwards and Louis Agassiz, that embryological characters are the most important of all in deducing relationships, and this doctrine has very generally been admitted as true. There are, of course some exceptions owing to the adaptive characters of larvae not having been excluded; in order to show this, Fritz Müller, for example, tried using such characters alone in arranging the various members of the great class of crustaceans, and the arrangement did not prove a natural one.

But there can be no doubt that early embryonic characters are of the highest value for classification, and not only with animals but with plants as well. Thus the main divisions of flowering plants are in fact founded on differences in embryonic characteristics—on the number and position of the cotyledons, for example, and on the mode of development of the rudimentary shoot of an embryonic plant (the

[5] Müller was a famous German biologist who spent much of his life studying natural history in the forests of Brazil. The phenomenon of "Müllerian mimicry" is named after him.

[6] Ostracods are small (typically about 1 mm long), aquatic crustaceans, a major arthropod group that also includes such larger organisms as krill, crabs, lobsters, and shrimp.

(A)   (B)

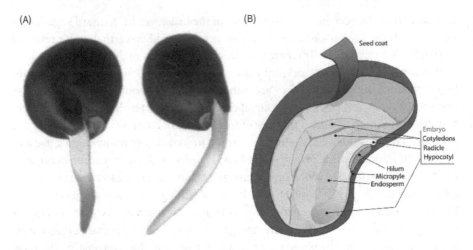

Figure 14.1 A plant radicle, emerging from the seed.

plumule) and the first part of the plant that emerges from the seed (the radicle: Figure 14.1).[7] We can immediately see why these characters possess so high a value in classification for the natural system is genealogical in its arrangement. Indeed, our classifications are often plainly influenced by chains of affinities. Nothing can be easier than to define a number of characters common to all birds; but with crustaceans, any such definition has hitherto been found impossible. There are crustaceans at the opposite ends of the series that have hardly a character in common; yet the species at both ends, from being plainly allied to others within the grouping, and these to others, and so onwards, can be recognized as unequivocally belonging to this and to no other class of arthropods.

Geographical distribution has often been used—though perhaps not quite logically—in classification, particularly in very large groups of closely related forms. The Dutch zoologist Coenraad Temminck insists on the utility or even necessity of this practice for certain groups of birds, and it has been followed by several entomologists and botanists as well.

Finally, with respect to the comparative value of the various groupings of species, such as orders, suborders, families, subfamilies, and genera, they seem to be, at least at present, almost arbitrary. Indeed, several of the best botanists have strongly insisted on their arbitrary value. I could easily give examples among plants and insects of groups that were first ranked by experienced naturalists as only genera and which were later raised to the rank of a subfamily or family; this has not been done because further research has detected important structural differences that were initially

---

[7] The cotyledon is an embryonic leaf, while the radicle grows downward into the soil, serving to anchor the seed.

overlooked, but rather because numerous allied species, with slightly different grades of difference, have been subsequently discovered.

All of the foregoing rules and aids and difficulties in classification may be explained, if I do not greatly deceive myself, on the view that our natural system is founded on descent with modification—that the characters which naturalists consider as showing true affinity between any two or more species are those which have been inherited from a common parent. That is, **all true classification is genealogical, and community of descent is the hidden bond that naturalists have been unconsciously seeking, and not some unknown plan of creation, or the mere putting together and separating of objects that are more or less alike.**

But I must explain my meaning more fully. Although I believe that the *arrangement* of the groups within each class, in due subordination and relation to each other, must be strictly genealogical in order to be natural, I also believe that the *amount* of difference in the several branches or groups, though allied to the same degree in blood with their common ancient ancestor, may differ greatly depending on how much modification they have undergone over time. This is expressed by the various species being ranked under different genera, families, sections or orders. To better understand my meaning, please refer to Figure 4.11. We will suppose that the letters A to L along the bottom of the figure represent allied genera that existed during the Silurian epoch,[8] and that all descended from some still earlier form. In three of these genera (A, F, and I), a species has transmitted modified descendants to the present day, represented by the 15 genera (a-14 to z1-4) on the uppermost horizontal line. Now, all of these modified descendants from a single species are related in blood or descent to the same degree; they may metaphorically be called cousins to the same millionth degree. And yet they differ widely and to different degrees from each other. The forms descended from ancestor A, now broken up into two or three families of species, constitute a distinct order from those descended from ancestor I, whose descendants are now broken up into two families. Nor can the existing species descended from ancestor A be placed in the same genus with the parent A, or those from ancestral parent I with parent I. But the members of existing genus F14 may be supposed to have been only slightly modified, and it will then rank with the parent genus F, just as some few still-living organisms belong to genera found in Silurian deposits. Thus it is that the comparative value of the differences between these organic beings, which are all related to each other to the same degree in blood, has come to be widely different. Nevertheless, their genealogical arrangement remains strictly true, not only at the present time, but also at each successive period of descent. All the modified descendants from ancestor A will have inherited something in common from their common parent, as will all the descendants from ancestor I; so will it be with each subordinate branch of descendants at each successive stage. If, however, we suppose any descendant of ancestor A or of ancestor I to have become so much modified as to have lost all traces

---

[8] The Silurian geological period ran from about 444 million years ago to the beginning of the Devonian period, about 419 million years ago.

of its parentage, its true place in the natural system will be lost to us, as seems to have occurred with some few existing organisms. All the descendants of genus F, along its whole line of descent, are supposed to have been but little modified, and they still form a single genus. But this genus, though much isolated, will still occupy its proper intermediate position.

The representation of the groups as given in the two-dimensional diagram in Figure 4.11, is really much too simple. The branches ought to have diverged in all directions. If the names of the groups had been simply written down in a linear series, the representation would have been even less natural, and it is notoriously impossible to represent in a series, on a flat surface, the affinities that we discover in nature among the members of the same group. Thus, the natural system is genealogical in its arrangement, like a pedigree. But the amount of modification that the different groups have undergone has to be expressed by ranking them under different groups and subgroups, which we refer to as genera, subfamilies, families, sections, orders, and classes.

We can perhaps better understand this view of classification as a geneological series by considering the case of human languages. If we possessed a perfect pedigree of mankind—showing exactly how each race is related to the other races and the order of their formation— a genealogical arrangement of the human races would afford the best classification of the various languages now spoken throughout the world; and, if all extinct languages, and all intermediate and slowly changing dialects, were to be included, such an arrangement would be the only possible one. Yet it might be that some ancient languages had altered very little over time and had given rise to only a few new languages, while others had become altered a great deal owing to the spreading, isolation, and state of civilization of the several co-descended races and had thus given rise to many new dialects and new languages, all arising from the same initial tongue. The various degrees of difference between the languages of the same initial stock would have to be expressed by groups within groups. But the proper or even the only possible arrangement would still be genealogical, and this would be strictly natural because it would connect together all languages, extinct and recent, by the closest affinities and would give the relationships and origin of each tongue.

In confirmation of this view, let us glance at the classification of varieties that are known or believed to be descended from a single species. These are grouped under the species name, with the subvarieties being grouped under the name of the particular varieties and, in some cases, as with the domestic pigeon, with several other grades of difference. Nearly the same rules are followed as in classifying species. Authors have insisted on the necessity of arranging such varieties on a natural instead of an artificial system. We are cautioned, for instance, not to class two varieties of the pineapple together merely because their fruit, though the most important part, happens to be nearly identical. No one puts the Swedish and common turnip together, even though the esculent and thickened stems are so similar.

Whatever part is found to be most constant is used in classing varieties. The great agriculturist William Marshall, for example, says that the cattle's horns are very useful for this purpose because they are less variable than the shape or color of the body, whereas with sheep the horns are much less useful because they vary more among individuals. In classing varieties, I realize that if we had an actual pedigree, a genealogical classification would be universally preferred; that has, in fact, been attempted in some cases. For we might feel sure, whether there had been more or less modification, that the principle of inheritance would keep together those forms that were allied in the greatest number of points. In tumbler pigeons, though some of the sub-varieties differ in the important character of the length of their beak, yet all are kept together in the same named group from having the common habit of tumbling in the air. But the short-faced breed of tumbler pigeon has nearly or even completely lost this habit of tumbling; even so, without any thought on the subject, these pigeons are kept in the same group. Why? Because they are all clearly related by blood and are alike in some other respects, and thus clearly belong to this tumbler variety.

With species in a state of nature, every naturalist has in fact unintentionally brought descent into his classification, for they include both sexes in the basic category of species. And yet how enormously the two sexes sometimes differ in their most important characteristics is known to every naturalist: scarcely a single fact can be predicated in common for adult males and hermaphroditic individuals of certain barnacle species, and yet no one dreams of categorizing them as separate species. Among orchids, plants that were for a long time placed in three distinct genera (*Monachanthus, Myanthus, and Catasetum*) were immediately considered to simply be varieties of a single species once it was known that all three types of flowers were sometimes produced on the same plant; indeed, I have now shown that they are the male, female, and hermaphrodite forms of the same species.

Similarly, the naturalist includes as one species the various larval stages of the same individual, however much they may differ from each other and from the adult. The same is true with the so-called alternate generations[9] described for some parasitic worms by the Danish biologist Japetus Steenstrup, which can only in a technical sense be considered as the same individual as they look so very different from each other and have such very different lives; he includes monsters and varieties as belonging to the same one species, not from their partial resemblance to the parent form, but because they are all descended from it.

As descent from an original ancestor has universally been used in classing together the individuals of the same species, though the males and females and larvae are sometimes extremely different, and as it has also been used in classing varieties that have undergone a certain (and sometimes a considerable) amount of modification, may not this same element of descent have been unconsciously used in grouping species under genera, and genera under higher groups, all under the so-called natural system? I believe it has been unconsciously thus used. Indeed this is the only way that I can understand the several rules and guides that have been followed by our best

---

[9] Here Darwin refers to an alteration between phases of sexual and asexual reproduction, not the alteration of diploid and haploid stages seen in some plants and algae.

systematists. As we have no written pedigrees, we are forced to trace community of descent by resemblances of any kind. Therefore, we choose those characters that are the least likely to have been modified in relation to the conditions of life to which each species has been recently exposed. Rudimentary structures on this view are as good as, or even sometimes better than other parts of an organism's organization for use in classification. We care not how trifling a character may be—let it be the mere inflection of the angle of the jaw, the manner in which an insect's wing is folded, or whether the skin be covered by hair or by feathers—if it prevail throughout many and different species, especially those having very different habits of life, it assumes high value for showing relationships; for we can account for its presence in so many forms with such different habits only by inheritance from a common parent. We may err in this respect in regard to single points of structure, but when several characters, let them be ever so trifling, concur throughout a large group of beings having different habits we may feel almost sure, on the theory of descent, that these characters have been inherited from an ancient common ancestor; such aggregated characters will clearly have special value in classification.

We can now understand why a species, or a group of species, may depart from its relatives in several of its most important characteristics and yet still be confidently grouped with them. This may be done with confidence—and often is done—as long as a sufficient number of other characters, let them be ever so unimportant, betrays the hidden bond showing community of descent. Even if two forms have not a single character in common, if these extreme forms are nevertheless connected together by a chain of intermediate groups, we may at once infer their community of descent and put them all into the same taxonomic category. As we find organs of high physiological importance—those which serve to preserve life under the most diverse environmental conditions—are generally the most constant, we attach special value to them; but if these same organs, in another group or section of a group, are found to differ much, we at once value them less in our classification. Later in this chapter we shall see why embryological characters are so especially useful in classification. Geographical distribution may sometimes also be brought usefully into play in classifying large genera because all the species of the same genus, inhabiting any distinct and isolated region, are in all probability descended from the same ancestral parents.

Analogical Resemblances

We can understand, on the above views, the very important distinction between real genealogical relationships and analogical resemblances caused simply by independent adaptations to the same environmental selective agents. Lamarck first called attention to this subject, and he has been ably followed by Macleay and others. The resemblance in the shape of the body and in the fin-like anterior limbs between dugongs and whales, for example, and between these two orders of mammals and fishes, are merely analogical (i.e., not indicating an evolutionary relationship). So is the resemblance between a mouse and a shrew-mouse (genus *Sorex*), which belong to different orders, as is the still closer resemblance, insisted on by the English biologist St. George Jackson Mivart,[10] between the mouse and a small marsupial animal (genus *Antechinus*) of Australia. These latter resemblances may be easily accounted for, I believe, by independent adaptation for similarly active movements through thickets and herbage, and with concealment from enemies.

Among insects there are innumerable instances of such misclassification. For example, Linnaeus, misled by external appearances, actually classified a homopterous insect[11] as a moth. We see something of the same kind even with our domestic varieties, as in the strikingly similar body shapes of the body in the improved breeds of the Chinese and common pig, which are descended from what turn out to be entirely different ancestral species, and in the similarly thickened stems of the common and specifically distinct Swedish turnip. The resemblance between the greyhound and the race horse is hardly more fanciful than the analogies that have been drawn by some authors between other widely different animals.

On the view that characters are of real importance for classification only if they reveal descent from a common ancestor, we can clearly understand why analogical or adaptive characters, although of the utmost importance to the welfare of the organisms that possesses them, are almost valueless to the systematist. For animals belonging to two very separate and distinct lines of descent may have gradually become adapted to similar conditions and thus have gradually and independently assumed a close external resemblance; but such resemblances will not reveal—in fact will rather tend to conceal—their actual degree of blood relationship. We can thus also understand the apparent paradox that the very same characters are merely analogous with each other—and thus of no value in assessing relationships—when the members of one group are compared with the members of another group but show true affinities when the members of the same group are compared with each other: thus the shape of the body and the fin-like limbs are merely analogous characteristics when whales are compared with fishes, being independent adaptations for swimming through the water for the members of both classes. But between the several members of the whale family, the shape of the body and the fin-like limbs offer characters exhibiting true genealogical affinity: for as those parts are so nearly similar throughout the whole family, we cannot doubt that they have been inherited from a common ancestor. And so it is with fishes.

Numerous cases could be given of striking resemblances between single parts or organs that have been adapted for the same functions in very distinct beings. A good instance is afforded by the close resemblance of the jaws of the dog and the Tasmanian wolf (*Thylacinus cynocephalus*: Figure 14.2),[12] animals that are placed widely apart

---

[10] Mivart was initially an ardent follower of Darwin's theory of natural selection, but later become one of its fiercest opponents.

[11] The homopterans include aphids, cicadas, leaf hoppers, and scale insects, all of which have sucking mouthparts.

[12] The Tasmanian wolf is no more: it was rare in 1965, and extinct by 1982.

Figure 14.2 Tasmanian wolf.

in the natural classification system. But this resemblance is confined to general appearance, as in the prominence of the canines and in the cutting shape of the molar teeth. Closer inspection reveals that the teeth really differ a great deal in these two groups of animals: thus the dog has on each side of the upper jaw four premolars and only two molars, while the *Tasmanian* wolf has three premolars and four molars. The molars also differ greatly in their relative size and structure for animals in the two groups. The adult dentition is preceded by a widely different milk dentition. Anyone may, of course, deny that the teeth in either case have been adapted for tearing flesh through the natural selection of successive variations; but if this be admitted in the one case, it is unintelligible to me that it should be denied in the other. I am glad to find that so high an authority as Professor William Henry Flower has come to this same conclusion.[13]

The extraordinary cases given in Chapter 6 ("Difficulties with the Theory") of widely different types of fishes possessing electric organs, of widely different insects possessing luminous organs, and of orchids and asclepiads both having pollen-masses with viscid discs, come under this same head of analogical resemblances. These cases are so wonderful that they were introduced by skeptics as difficulties or objections to our theory! However, in all such cases, we can detect some fundamental difference in the growth or development of the parts, and generally in their matured structure, indicating independent origins of the structures in the different groups. **The end gained is the same, but the means, though appearing superficially to be the same, are essentially different.** The principle formerly alluded to under the term of "analogical variation" has probably often come into play in these cases; that is, the members of the same class, although only distantly related, have inherited so much in

---

[13] William Flower was a leading authority on mammals. In 1884, he was appointed as Director of the British Museum of Natural History in London, upon the retirement of Sir Richard Owen.

common in their constitution that they are apt to vary under similar exciting causes in a similar manner, and this would obviously aid in the acquisition through natural selection of parts or organs strikingly like each other, independently of their direct inheritance from a common ancestor.

There is another and especially curious class of cases in which a close external resemblance between organisms that are not closely related to each other depends not on adaptation to similar habits of life but rather to protection from enemies. I allude here to the wonderful manner in which certain butterflies imitate, as first described by Mr. Henry Walter Bates,[14] other and quite distinct butterfly species. This excellent observer has shown that in some districts of South America, where, for instance, a clearwing butterfly species (genus *Ithomia*) abounds in gaudy swarms, another butterfly belonging to a different genus (*Leptalis*) is often found mingled in the same flock; the latter so closely resembles the clearwing butterfly in every shade and stripe of color, and even in the shape of its wings, that Mr. Bates, with his eyes sharpened by 11 years of collecting, was, though always on his guard, continually deceived as to which species an individual belonged. When the mocker (i.e., the imitating) butterfly and the mocked (i.e., the imitated butterfly) are caught and compared, they are found to be very different in their essential structure and to belong not only to distinctly different genera, but also often to distinctly different families.

Had this mimicry occurred in only one or two instances, it might have been passed over as a strange coincidence. But, if we proceed from a region where one *Leptalis* butterfly imitates an *Ithomia* butterfly, another mocking and mocked species belonging to the same two genera, equally close in their resemblance, may be found. Altogether no less than 10 genera of butterflies have been found that include some species that closely imitate other butterfly species. Interestingly, the mockers and the mocked always inhabit the same region; we never find an imitator living far from the form that it imitates. The mockers are almost invariably rare insects, whereas the mocked species in almost every case abounds in swarms. In the same district in which a species of *Leptalis* closely imitates a species of *Ithomia*, there are sometimes other lepidopterans mimicking the same *Ithomia* species: thus, in the same place, butterfly species of three different genera and even a moth are found that all closely resemble a butterfly species that belongs to a fourth genus. It deserves special notice that many of the mimicking forms of the *Leptalis* butterfly, as well as of the mimicked forms, can be shown by a graduated series to be merely varieties of the same species, while others are undoubtedly distinct species.

But why, it may be asked, are certain forms treated as the mimicked and others as the mimickers? Mr. Bates satisfactorily answers this question by showing that the form which is imitated keeps the usual dress characteristic of the group to which it

---

[14] Have you ever heard of *Batesian mimicry*, in which one group of otherwise defenseless organisms comes to possess the physical warning signals that characterize a different, chemically defended species? That's the Bates that Darwin is talking about here, based on work that Bates did in the rainforests of Brazil, mostly in the 1850s.

belongs, while the counterfeiters have changed their dress and no longer resemble their closest relatives.

We are next led to enquire why certain butterflies and moths so often assume the dress of a different and quite distinct form; why, to the perplexity of naturalists, has nature condescended to such tricks of the stage? Mr. Bates has, no doubt, hit on the true explanation. The mocked forms, which always abound in numbers, must habitually escape destruction to a large extent, otherwise they could not exist in such swarms. A large amount of evidence has now been collected showing that these mocked forms are distasteful to birds and other insect-devouring animals. On the other hand, the mocking forms that inhabit the same district are comparatively rare and belong to rare groups; hence, they must suffer habitually from some danger, for otherwise, from the number of eggs laid by all butterflies, they would in three or four generations swarm over the whole country.

Now if a member of one of these persecuted and rare groups were to assume a dress so similar to that of a well-protected, chemically defended species that it continually deceived the practiced eyes of an entomologist, it would often deceive predaceous birds and insects as well, and thus often escape being eaten. Mr. Bates may almost be said to have actually witnessed the process by which the mimickers have come to so closely resemble the mimicked, for he found that some of the forms of the *Leptalis* butterfly that mimic so many other butterflies varied in an extreme degree. In one district several varieties were found, and of these one alone resembled, to a certain extent, the common *Ithomia* butterfly of the same district. In another district there were two or three varieties, one of which was much commoner than the others, and this one closely mocked another form of *Ithomia*. From facts of this nature, Mr. Bates concludes that the *Leptalis* first varies, and when a variety happens to resemble in some degree any common, chemically protected butterfly inhabiting the same district, this variety, from its resemblance to a flourishing and little persecuted kind, has a better chance of escaping destruction from predaceous birds and insects and is consequently more likely to survive and reproduce, "the less perfect degrees of resemblance being generation after generation eliminated, and only the others left to propagate their kind." Here indeed we have an excellent illustration of natural selection in action.

Messrs. Robert Wallace and the British-South African entomologist Roland Trimen have likewise described several equally striking cases of imitation among the lepidopterans of the Malay Archipelago and Africa, and with some other insects as well. Mr. Wallace has also detected one such case with birds, although we have no examples involving the larger quadrupeds.[15] The observation that insects show such imitation with much greater frequency than is seen in other animals is probably a consequence of the insects' small size. For one thing insects cannot defend themselves physically, excepting indeed the kinds furnished with a stinger, and I have never

---

[15] Quadrupeds are four-legged vertebrates, such as cattle, dogs, cats, and lizards.

heard of an instance of such kinds mocking other insects, though they themselves are mocked by other species. Neither can insects easily escape capture from the larger animals that prey on them by simply flying away. Therefore, speaking metaphorically, insects are reduced, like most weak creatures, to trickery and dissimulation: imitate a chemically defended species and by such trickery become less vulnerable yourself.

It should be observed that the process of imitation probably never took place between forms that were widely dissimilar in color. But, starting with species that initially resemble each other to some degree, a closer and closer resemblance, if beneficial, could readily be gained by the above means, and, if the imitated form was subsequently and gradually modified through any agency, the imitating form would be led along the same track and thus be altered to almost any extent so that it might ultimately assume an appearance or coloring wholly unlike that of the other members of the family to which it belonged.

There is, however, some difficulty with this explanation, for it is necessary to suppose in some cases that ancient members belonging to several distinct groups, before they had diverged to their present extent, accidentally resembled a member of another and protected group closely enough to afford some slight protection, this having given the basis for the subsequent, gradual acquisition of the most perfect resemblance.

On the Nature of the Affinities Connecting Organic Beings

As the modified descendants of dominant species belonging to the larger genera tend to inherit the advantages that made the groups to which they belong large and their parents dominant, they are almost sure to spread widely and to seize on more and more places in the economy of nature. The larger and more dominant groups within each taxonomic class thus tend to go on increasing in size over time and, consequently, supplant many smaller and feebler groups. Thus we can account for the fact that all organisms, both recent and extinct, are included within a few great orders and still fewer classes. To show just how few the higher groups are in number and how widely they are spread throughout the world, consider this striking fact: the discovery and exploration of Australia has not added a single insect species belonging to a new taxonomic class. The same is true for the vegetable kingdom: as I have learned from Dr. Hooker, it has added only two or three families of small size, all within existing classes.

In the chapter on geological succession (Chapter 11), I attempted to show, on the principle of each group having generally diverged much in character during the gradual process of modification over a great many generations, how it is that the more ancient forms of life often present characters that are to some degree intermediate between those of existing groups. A few of the old and intermediate forms have transmitted descendants to the present day that are but little modified from

Figure 14.3 Duck-billed platypus.
Credit: Heinrich Harder (1858–1935), Public domain, via Wikimedia Commons.

their ancestors: these constitute our so-called aberrant groups. The more aberrant any form is, the greater must be the number of connecting forms that have been exterminated and utterly lost over time. And we have evidence of aberrant groups having suffered severely from extinction for they are almost always represented by extremely few species, and such species as do occur are generally very distinct from each other, which again implies extinction of the intermediate forms. The genera *Ornithorhynchus* (the duck-billed platypus: Figure 14.3) and *Lepidosiren* (the South-African lungfish), for example, would not have been less aberrant had each been represented by a dozen species instead of by a single species, as they are today, or by two or three. We can, I think, account for this fact only by looking at aberrant groups as forms that have been largely conquered and exterminated by more successful competitors in the past, with just a few members still persisting under unusually favorable conditions.

The English naturalist Mr. George Robert Waterhouse[16] has remarked that when a member belonging to one group of animals exhibits an affinity to a quite distinct group, this affinity in most cases is general and not special: thus, according to Mr. Waterhouse, the rodent known as bizcacha[17] is most nearly related to marsupials, but, in the points in which it approaches this order, its relations are general (i.e., not to any one marsupial species more than to another). As these points of affinity are believed to be real and not merely adaptive and representing convergent evolution,

[16] Years before the first edition of *The Origin of Species* was published, Waterhouse had been invited to accompany Darwin on the voyage of the *Beagle* but had declined to do so. He later took possession of Darwin's collection of mammals and beetles at the British Museum of Natural History and wrote a book about the natural history of mammals.

[17] Today these animals are usually called viscachas. They live in South America (Argentina, Bolivia, Chile, and Peru). Although they resemble rabbits (but have longer tails), the two groups are not closely related.

Figure 14.4 Bizcacha (now known as the viscacha). These animals live in dry, rocky places in the Andes.

they must be due, according to our view, to inheritance from a common ancestor. Therefore, we must suppose either (1) that all rodents, including the bizcacha (Figure 14.4), branched off from some ancient marsupial ancestor, which will naturally have been more or less intermediate in character with respect to all existing marsupials, or (2) that both rodents and marsupials branched off from a common ancestor and that both groups have since undergone much modification in divergent directions. On either view we must suppose that the bizcacha has retained, by inheritance, more of the character of its ancient ancestor than have other rodents; it follows then that it will not be specially related to any one existing marsupial species but will instead be indirectly related to all (or nearly all) marsupials, from having partially retained the character of their common ancestor or of some early member of the group. On the other hand, of all marsupials, as Mr. Waterhouse has remarked, members of the genus *Phascolomys*[18] resemble most nearly not any one species, but rather the general order of rodents. In this case, however, it may be strongly suspected that the resemblance is only analogous, owing to *Phascolomys* having become adapted to habits very similar to those of a rodent. The elder De Candolle has made nearly similar observations on the general nature of the affinities of distinct families of plants.

On the principle of the gradual multiplication and gradual divergence in character of the species descended from a common ancestor, together with their retention by inheritance of some characters in common, we can understand the excessively complex

---

[18] *Phascolomys* is a genus of prehistoric Australian marsupials in the wombat family (Family Phascolomyidae). The largest species, *Phascolomys gigas*, weighed as much as 200 kg (450 lb). *Phascolomys* existed alongside an even larger marsupial, *Diprotodon*, which weighed as much as three tons and was distantly related to wombats. Both disappeared at the end of the Late Pleistocene in a Quaternary extinction event, together with many other large Australian animals.

and radiating affinities by which all the members of the same family or higher group are connected together: the common ancestor of a whole family that is now broken up by extinction into distinct groups and subgroups will have transmitted some of its characters, modified in various ways and to various degrees, to all of the descendant species. These species will therefore all be related to each other by circuitous lines of affinity of various lengths (as may be seen in Figure 4.11), mounting up through many predecessors. As it is difficult to show the blood relationship between the numerous kindred of any ancient and noble human family, even by the aid of a genealogical tree, and almost impossible to do so without this aid, we can understand how extraordinarily difficult it is for naturalists to describe, without the aid of a diagram, the various affinities that they perceive to exist between the many living and extinct members of any one great natural class of organisms.

Extinction, as we have seen in Chapter 4, has played an important part in defining and widening the gaps between the various groups in each class of organisms. We may thus understand why classes of organisms are so distinct from each other—for instance, why birds are so distinctly different from all other vertebrate animals—by the belief that the many ancient forms of life that formerly connected the various early ancestors have been utterly lost. At one point in time, the early ancestors of birds were connected with the early ancestors of the other—and at that time less differentiated—vertebrate classes. Those intermediate forms are now extinct.

There has been much less extinction of the forms of life that once connected fishes with Batrachians. There has been still less extinction within some whole classes, such as the Crustacea: for here the most wonderfully diverse forms are still linked together by a long and only partially broken chain of affinities. Extinction has only defined the groups; it has by no means made them, for if every form that has ever lived on this Earth were suddenly to reappear, a natural classification, or at least a natural arrangement, would be possible. We can easily see this by turning to Figure 4.11: the letters A to L can be thought to represent 11 ancient Silurian genera, some of which have produced large groups of modified descendants, with every link in each branch and sub-branch still alive and the links not greater than those between existing varieties. In this case it would be quite impossible to give definitions distinguishing the several members of the several groups from their more immediate parents and descendants. Yet the arrangement in the diagram would still hold good and would be natural, for, on the principle of inheritance, all the forms that descended, for instance, from ancestor A, would have something in common with that ancestor and with each other. In a tree we can distinguish this or that branch, though at the actual fork the two unite and blend together. Although we could not, as I have said, define the several groups, we could nevertheless pick out types, or forms, representing most of the characters of each group, whether large or small, and thus give a general idea of the value of the differences between them. This is what we should be driven to do if we were ever to succeed in collecting all the forms in any one class that have lived throughout all time and space. Assuredly we shall never succeed in making so perfect a collection. Nevertheless, in certain classes, we are making progress toward this end; Milne Edwards has lately insisted, in an able and recent paper, on the high importance of looking to types, whether or not we can separate and define the groups to which such types belong.

Finally, since natural selection follows from the struggle for existence and almost inevitably leads to extinction and divergence of character in the descendants of any one parent species, we can see why it explains that great and universal feature in the affinities of all organic beings—namely, their subordination in group within group. We use the element of descent in classing the individuals of both sexes and of all ages under one species even though they may have but few characters in common; we use descent in classifying acknowledged varieties as members of a particular species, however different they may be from their parents. *I believe that this element of descent is the hidden bond of connection that naturalists have sought under the term of "the Natural System."* On this idea of "the natural system" being, in so far as it has been perfected, genealogical in its arrangement, with the grades of difference expressed by the terms genera, families, orders, etc., we can understand the source of the rules that we are compelled to follow in our system of classification. We can understand (1) why we value certain resemblances far more than others; (2) why rudimentary and useless organs, or others of trifling physiological importance, are so useful in deducing affiliations; (3) why, in finding the relations between one group and another, we summarily reject analogous or adaptive characters; and yet (4) why we can use these same characters to determine relationships within the limits of the same group. We can also clearly see how it is that all living and extinct forms can be grouped together within a few great taxonomic classes, and how the several members of each class are connected together by the most complex and radiating lines of affinities. We shall probably never disentangle the inextricable web of the affinities between the members of any one class, but when we have a distinct object in view, and do not simply look to some unknown "plan of creation," we may hope to make sure but slow progress toward this end.[19]

Professor Haeckel in his *Generelle Morphologie* and other works, has recently brought his great knowledge and abilities to bear on what he calls "phylogeny," or the lines of descent of all organic beings from ancient ancestors. In drawing up the several series, he trusts chiefly to embryological characters but receives aid from homologous and rudimentary organs, as well as from the successive periods at which the various forms of life are believed to have first appeared in our geological formations. He has thus boldly made a great beginning, and shows us how classification will surely be treated in the future.

---

[19] The current availability of sophisticated molecular techniques and data analysis is now finally allowing us to do just that.

Morphology

We have seen that the members of any one class of organisms, independently of their habits of life, resemble each other in the general plan of their organization. This resemblance is often expressed by the term "unity of type," or by saying that the several parts and organs in the different species of the class are "homologous."[20] The whole subject is included under the general term of "morphology." This is one of the most interesting departments of natural history, and may almost be said to be its very soul. **What can be more curious than that the hand of a man, formed for grasping, that of a mole for digging, the leg of the horse, the paddle of the porpoise, and the wing of the bat, should all be constructed on the same pattern, and should include similar bones, in the same relative positions to each other?** And how curious it is, that the hind feet of the kangaroo, which are so well fitted for bounding over the open plains; and those of the climbing, leaf-eating koala, equally well fitted for grasping the branches of trees; and those of the ground-dwelling, insect or root-eating, bandicoots; and those of some other Australian marsupials, should all be constructed on the same extraordinary plan, namely with the bones of the second and third digits extremely slender and enveloped within the same skin, so that they appear as a single toe furnished with two claws. Notwithstanding this similarity of pattern, it is obvious that the hind feet of these several animals are used for as widely different purposes as it is possible to conceive. The case is rendered all the more striking by the American opossums, which follow nearly the same habits of life as some of their Australian relatives and yet have their feet constructed on the ordinary plan. Professor Flower, from whom these statements are taken, remarks in conclusion, "We may call this 'conformity to type,' without getting much nearer to an explanation of the phenomenon;" and he then adds "but is it not powerfully suggestive of true relationship, of inheritance from a common ancestor?"

Geoffroy St. Hilaire has strongly insisted on the high importance of relative position or connection in homologous parts; they may differ to almost any extent in form and size and yet remain connected together in the same invariable order. We never find, for instance, the bones of the arm and forearm, or of the thigh and leg, transposed in any individual. That explains why the same names can be given to the homologous bones in widely different animals. We see the same great law in the construction of the mouths of insects: What can be more different than the immensely long spiral proboscis of a sphinx-moth, the curious folded proboscis of a bee or bug, and the great jaws of a beetle? Yet all these organs, although serving for such widely different purposes, are formed by infinitely numerous modifications of an upper lip, mandibles, and two pairs of maxillae. The same law governs the construction of the

---

[20] This term, "homologous," is still used today and refers to traits with a common evolutionary origin (i.e., traits that have evolved from a common ancestor).

mouths and limbs of all the different crustaceans. And so it is with the many different flowers of plants.

Nothing can be more hopeless than to attempt to explain this similarity of pattern in members of the same class by utility or by the doctrine of final causes. The hopelessness of the attempt has been expressly admitted by Richard Owen in his most interesting work on the "Nature of Limbs." On the ordinary view of the independent creation of each being, we can only say that so it is: that it has pleased the Creator to construct all the animals and plants in each great class on a uniform plan. But this is not a scientific explanation.

On the other hand, the explanation is pretty simple on the theory of the selection of successive slight modifications in successive generations, each being profitable in some way to the modified form but often also affecting, by correlation, other parts of the organism's organization. In changes of this nature, there will be little or no tendency to alter the original pattern or to transpose any of the parts. The bones of a limb might become shortened and flattened to any extent while at the same time becoming enveloped in thick membrane so as to serve as a fin; or a webbed hand might have all its bones, or certain bones, lengthened to any extent, with the membrane connecting them increased, so as to serve as a wing; yet none of these modifications would alter the framework of the bones or the relative connection of the parts.

If we suppose that an early ancestor—the archetype, as it may be called—of all mammals, birds, and reptiles had its limbs constructed on the existing general pattern, for whatever purpose they served, we can at once understand the plain signification of the homologous construction of the limbs in all members of the class. And so it is with the mouths of insects: we have only to suppose that their common ancient ancestor had an upper lip, mandibles, and two pairs of maxillae, these parts having been perhaps very simple in form; natural selection, over enormous periods of time, will then account for the infinite diversity in structure and function of the mouths of insects that we see today. Nevertheless, it is conceivable that the general pattern of an organ might become so much obscured as to be finally lost through the reduction and ultimately by the complete abortion of certain parts, or by the fusion of other parts, or by the doubling or multiplication of others, variations that we know to be within the limits of possibility. In the paddles of the gigantic extinct sea-lizards, for example, and in the mouths of certain suctorial crustaceans, the general pattern seems thus to have become partially obscured.

Here is another and equally curious branch of our subject: namely, "serial homologies," which involve comparing the different parts or organs in the same individual rather than of the same parts or organs in different members of the same class. Most physiologists believe that the bones of the skull are homologous—that is, corresponding in number and in relative connection with the elemental parts of a certain number of vertebrae. The anterior and posterior limbs in all members of the higher vertebrate classes are plainly homologous. So it is with the wonderfully complex jaws and legs of crustaceans: they are built on the same basic plan in the members of all species. Similarly, the relative position of the sepals, petals, stamens, and pistils of a flower, as well as their intimate structure, makes sense only on the view that they consist of modified leaves arranged in a spire. In monstrous plants, we often get direct evidence of the possibility of one organ being transformed into another, and we can actually see, during the early or embryonic stages in the development of flowers, as well as in crustaceans and many other animals, that organs which become extremely different when mature are exactly alike earlier in development.

How inexplicable are such cases of serial homologies on the ordinary view of special creation! Why should the vertebrate brain be enclosed in a box composed of such numerous and such extraordinarily shaped pieces of bone apparently representing vertebrae? As Owen has remarked, the benefit derived from the yielding of the separate pieces of the mammalian skull during childbirth[21] will by no means explain why the skulls of birds and reptiles are constructed in the same way. And why should similar bones have been created to form both the wing and the leg of a bat, structures that are used for totally different purposes, namely flying and walking? And why should one crustacean species that has an extremely complex mouth formed of many parts consequently always have fewer legs; or conversely, why should those crustacean species with many legs have simpler mouths? And with flowers, why should the sepals, petals, stamens, and pistils in each flower, though fitted for such distinct purposes, all be constructed on the same basic pattern?

On the theory of natural selection, we can, to a certain extent, answer all of those questions. We need not consider here how the bodies of some animals first became divided into a series of segments, or how they became divided into right and left sides, with corresponding organs, for such questions are almost beyond investigation. Some serial structures probably result from cells multiplying by division, entailing the multiplication of the parts developed from such cells. It must suffice for our purpose to bear in mind that an indefinite repetition of the same part or organ is common, as Professor Owen has remarked, in all of the less specialized forms: therefore the unknown ancestor of the Vertebrata probably possessed many vertebrae; the unknown ancestor of the Articulata[22], many segments; and the unknown ancestor of flowering plants, many leaves arranged in one or more spires. We have also formerly seen that parts that are repeated many times in an organism are eminently liable to vary, not only in number, but also in form. Consequently, such parts, being already present in considerable numbers and being highly variable, would naturally provide the materials needed for adaptation to widely different purposes, yet they would generally retain, through inheritance, plain traces of their original or fundamental resemblance. They would retain this resemblance all the more as the variations that afforded the basis for their subsequent modification through natural selection would tend from the first to be similar, as the parts are alike in the early stages of growth, and would be

---

[21] This "yielding" is what allows the baby's skull to be compressed during childbirth, ultimately allowing us to have larger brains.

[22] The Articulata is an older taxonomic grouping that related the annelids and arthropods, both of which have bodies with many distinct segments.

subjected to nearly the same environmental conditions. Such parts, whether more or less modified, unless their common origin became wholly obscured, would be "serially homologous."

Among the molluscs (members of the phylum Mollusca), although the parts in distinct species can be shown to be homologous, only a few serial homologies can be indicated, such as the eight shell plates of chitons; that is, we are seldom able to say whether one part is homologous with another part in the same individual. And we can understand this fact quite easily, as among the molluscs, even in the least complex members of the group, we do not find nearly so much indefinite repetition of any one part as we find in the other great classes of the animal and vegetable kingdoms.[23]

But morphology is a much more complex subject than it at first appears, as has been shown in a recent and remarkable paper by Mr. E. Ray Lankester. Mr. Lankester has drawn an important distinction between certain classes of cases that have all been equally ranked by naturalists as homologous. He proposes to call those structures that resemble each other in different animals, owing to their descent with modification from a common ancestor "homogenous"; resemblances that cannot thus be accounted for in that way, he proposes to call "homoplastic." For instance, he believes that the hearts of birds and mammals are as a whole homogenous—that is, they have all been derived over an enormous number of generations from a common ancestor. On the other hand, he believes that the four cavities of the heart in those two groups are homoplastic—that is, that they have been independently developed. Mr. Lankester also gives as evidence the close resemblance of the parts on the right and left sides of the body, as in humans, and in the successive segments of the same individual animal, as in earthworms; here we have parts commonly called "homologous" but which bear no relation to the descent of distinct species from a common ancestor. Homoplastic structures are the same as those that I have referred to as analogous modifications or resemblances. Their formation may be attributed in part to distinct organisms—or to distinct parts of the same organism—having varied in an analogous manner and in part to similar modifications having been preserved for the same general purpose or function, of which many instances have been given.

Naturalists frequently speak of the vertebrate skull as being formed of modified vertebrae, and of the jaws of crabs as being modified legs, and of the stamens and pistils of flowers as modified leaves. But it would in most cases be more correct, as Professor Huxley has remarked, to speak of both skull and vertebrae, jaws and legs, and so forth as having been modified, not one from the other as they now exist, but rather from some common and simpler structure. Most naturalists, however, use such language only in a metaphorical sense: they are far from meaning that during many generations of descent, primordial organs of any kind—vertebrae in the one case and legs in the other, for example—have actually been converted into skulls or jaws. Yet

---

[23] The diversity of form and function within the phylum Mollusca is truly astounding. The phylum includes members as diverse as oysters and clams, which lack a central nervous system, and cephalopods, such as the octopus and squid, which are remarkably complex both anatomically and behaviorally.

so strong is the appearance of this having occurred that naturalists can hardly avoid employing language that has this very clear implication. But according to the views I have presented here, such language may in fact be used literally! The wonderful fact of the jaws of a crab retaining numerous characters of true, but extremely simple legs, for instance, is at least in part explained by recognizing that crabs probably did indeed retain those characteristics through inheritance, over the course of many, many generations.

Development and Embryology

This is one of the most important subjects in the whole round of natural history. The metamorphosis of insects, with which everyone is familiar, is generally abrupt, with distinct transitions from one stage to the next; but the transformations are actually numerous and gradual, although concealed. A certain ephemerous insect[24] (the "mayfly," genus *Chloeön*) molts more than 20 times during its development, as shown by Sir J. Lubbock, and undergoes a certain amount of change each time; in this case we see the act of metamorphosis performed in a very gradual manner. Many insects, and especially certain crustaceans, show us what wonderful changes of structure can be effected during development. Such changes, however, reach their acme in the so-called alternate generations of some of the lower animals. It is, for instance, an astonishing fact that a delicate branching hydrozoan, studded with polyps that resemble small sea anemones and attached to a submerged rock, should produce, first by budding and then by transverse division, a host of huge floating jellyfish![25] These jellyfish then go on to produce sperm and eggs; the fertilized eggs develop into swimming microscopic larvae that eventually attach themselves to rocks and soon develop into branching polyps again—and so on in an endless cycle.

The idea that the processes of alternating generations and of ordinary metamorphosis are essentially identical has been greatly strengthened by Wagner's discovery that the larva or maggot of a certain small fly (*Cecidomyia*) goes on to produce other larvae asexually, and these larvae eventually develop into mature males and females and go on to propagate their kind in the ordinary manner by eggs and sperm. It may be worth notice that when Wagner's remarkable discovery was first announced, I was asked how was it possible to account for the larvae of this fly having acquired the power of a sexual reproduction. As long as the case remained unique, no answer could be given. But now Grimm has shown that another fly, a species in the nonbiting midge genus *Chironomus*, reproduces itself in nearly the same manner, and he believes that this occurs frequently among other species in the order. The main

---

[24] Members of the insect order Ephemoptera; the larval stages may live for years, but the adults rarely live for more than a day.

[25] Jellyfish larvae eventually settle to the sea bottom and form an anemone-like "polyp" individual that grows and eventually splits into a series of layers, each of which eventually swims away and develops into a separate but genetically identical jellyfish.

difference here is that it is the pupa, and not the larva, of *Chironomus* that has this power; and Grimm further shows that this case, to a certain extent, "unites that of the Cecidomyia with the parthenogenesis of the Coccidae," with the term "parthenogenesis" implying here that the mature females of the Coccidae are capable of producing fertile eggs without any contribution from the male. Certain animals belonging to several classes are now known to have the power of ordinary reproduction at an unusually early age, and we have only to accelerate parthenogenetic reproduction by gradual steps to an earlier and earlier age—*Chironomus* showing us an almost exactly intermediate stage, viz., that of the pupa—and we can perhaps account for the marvelous case of the *Cecidomyia*.

It has already been stated that various parts may be exactly alike in the same individual during early embryonic development, but become widely different and serve for widely different purposes in the adult. So again it has been shown that although the embryos of the most distinct species belonging to the same class are generally closely similar, they become, when fully developed, widely dissimilar. A better proof of this latter fact cannot be given than by the embryologist Karl von Baer's statement that "the embryos of mammalia, of birds, lizards and snakes, probably also of turtles, are in the earliest states exceedingly like one another, both as a whole and in how their parts develop; so much so, in fact, that we can often distinguish the embryos of these widely different groups only by their size. In my possession are two little preserved embryos whose names I didn't attach to the vials, and at present I am quite unable to say to what class of vertebrates they belong. They may be lizards or small birds, or even very young mammals, so complete is the similarity in the mode of formation of the head and trunk in these different animals. The extremities, however, are still absent in these embryos. But even if they had existed in the earliest stage of their development we should learn nothing, for the feet of lizards and mammals, the wings and feet of birds, no less than the hands and feet of man, all arise from the same fundamental form."

The larvae of most crustaceans, at corresponding stages of development, closely resemble each other even though the adults may look very different; and so it is with very many other animals. A trace of the law of embryonic resemblance occasionally lasts until a rather late age: thus birds of the same genus, and of allied genera, often resemble each other in their immature plumage; we see this, for example, in the spotted feathers of young thrushes. In the cat tribe, the adults of most species are striped or spotted in lines; stripes or spots can also be plainly distinguished in the whelp of the lion and the puma. We occasionally, though rarely, see something of the same kind in plants; thus the first leaves of the gorse plant (also known as the "furze," genus *Ulex*) and the first leaves of the phyllodineous acacias[26] are produced on both sides of the stem or divided like the ordinary leaves of the pea and bean family, Leguminosae.

---

[26] These are members of the pea family, Fabaceae.

The points of structure in which the embryos of widely different animals within the same class resemble each other are often unrelated to their environmental conditions. We cannot, for instance, suppose that the peculiar loop-like courses of the arteries near the branchial slits during the embryonic development of fish are related to similar conditions seen in the young mammal that is nourished in the womb of its mother, in the egg of the bird that is hatched in a nest, and in the spawn of a frog under water. We have no more reason to believe in such a relation than we have to believe that the similar bones in the hand of a man, wing of a bat, and fin of a porpoise are related to similar conditions of life. And no one supposes that the stripes on a young lion or the spots on the young blackbird are of any use to these animals.

The case, however, is different when an animal, during any part of its pre-juvenile development, is active and free-living and has to provide for itself. This period of activity may come on earlier or later in life, but whenever it comes on, the adaptation of the larva to its conditions of life is just as perfect and as beautiful as it is in the adult animal. Sir J. Lubbock has recently shown in just how important a manner this has acted in his remarks on the close similarity of the larvae of some insects belonging to very different orders, and on the dissimilarity of the larvae of other insects within the same order, according to their habits of life. Owing to such adaptations, the similarity of the larvae of allied animals is sometimes greatly obscured, especially when there is a division of labor during the different stages of development, as when the same larva has to search for food during one stage of development and has to search for a place to attach during another stage. Cases can even be given of the larvae of allied species, or groups of species, differing more from each other than do the adults. In most cases, however, the larvae, though active, still obey, more or less closely, the law of common embryonic resemblance. Barnacles, as intriguing members of the arthropod superclass Crustacea, afford a good example of this: even the illustrious Cuvier did not perceive that a barnacle was in fact a crustacean. But even a glance at the larval stage shows this in an unmistakable manner (Figure 14.5). So again the two main divisions of barnacles—the stalked and directly attached barnacles—though differing widely in external appearance, have larvae that are, in all stages of their development, barely distinguishable from each other.

Embryos generally increase in organization as they develop. I use this expression even though I am aware that it is hardly possible to define clearly what is meant by an organism's organization being "higher" or "lower." But probably no one will dispute that a butterfly is higher (i.e., more complex in organization) than a caterpillar. In some cases, however, the mature animal must be considered as lower in the scale than the larva, as with certain parasitic crustaceans. Let us consider barnacles once again: the larvae in the first stage have three pairs of locomotive appendages, a simple single eye, and a prosociformed mouth, with which they must feed a great deal for they increase much in size as they develop. In the second larval stage of barnacle development, similar to the chrysalis stage of butterflies, the "cyprid" larva (Figure 14.5b) now has six pairs of beautifully constructed swimming legs, a pair of magnificent compound eyes, and extremely complex antennae. And yet they have a closed

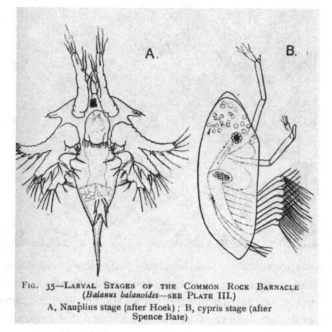

FIG. 35—LARVAL STAGES OF THE COMMON ROCK BARNACLE
(*Balanus balanoides*—SEE PLATE III.)
A, Nauplius stage (after Hoek) ; B, cypris stage (after
Spence Bate)

Figure 14.5 The planktonic larval stages of the common barnacle. The larva on the left is a nauplius, and the larva on the right is the final larval stage, the cyprid, which attaches to a rock, shell, or other solid surface and metamorphoses into a juvenile barnacle. Most marine (and many freshwater) crustaceans, including copepods and many decapods, include nauplius larvae in their development; the nauplius larva is unique to crustaceans.

and imperfect mouth in that stage and cannot feed at all: the larval cyprid's sole function is simply to explore solid surfaces, using their well-developed organs of sense, and to reach by their active powers of swimming a proper place upon which to attach and to undergo their final metamorphosis to adult form. When this is completed they are fixed in that place for the rest of their lives: their legs are now converted into prehensile organs for feeding, and they indeed again obtain a well-constructed mouth. But they have no antennae, and their two complex larval eyes are now reconverted into a minute, single, simple, light-sensitive eye-spot. In this last and complete state, adult barnacles may be considered as either more highly or more lowly organized than they were in the larval condition. But in some barnacle genera, some larvae metamorphose and become simultaneous hermaphrodites[27] as adults, having the ordinary structure, while others develop into what I have called "complemental males," in which the male is now a mere sack that lives attached to a female for only a short time

---

[27] Each individual has both male and female reproductive organs.

Figure 14.6 A hooded cuttlefish (*Sepia prashadi*). Cuttlefish are cephalopod molluscs and thus are close relatives of squid, octopus, and the chambered nautilus. They have a lightweight, brittle, internal shell that they use for flotation. Humans commonly use their shells as a calcium source for caged birds and a variety of other pets. Cephalopods are members of the phylum Mollusca, and thus—remarkably—have a common evolutionary origin with snails and bivalves.

and has neither mouth, stomach, or any other organ of importance, excepting those for reproduction.

We are so used to seeing a difference in structure between the embryo and the adult that we are tempted to look at this difference as in some necessary manner contingent on growth. But there is no reason why, for instance, the wing of a bat or the fin of a porpoise should not have been sketched out with all their parts in proper proportion as soon as any part became visible in embryonic development. In some whole groups of animals and in certain members of other groups this is indeed the case, and the embryo does not differ widely from the adult at any stage of development. Owen, for example, has remarked in regard to cuttlefish[28] (Figure 14.6), "there is no meta-morphosis; the cephalopodic character of the animal is manifested long before the parts of the embryo are completed." Terrestrial snails and freshwater crustaceans are born having their proper forms, while the marine members of these same two great classes pass through considerable and often great changes during their development. Spiders, again, barely undergo any metamorphosis during their development. In con-trast, the larvae of most insects pass through a worm-like "caterpillar" stage, whether they are active and adapted to diversified lifestyles and habitats or are inactive from being placed in the midst of proper nutriment or from being fed by their parents;

---

[28] Cuttlefish are cephalopod molluscs, closely related to octopus and squid.

but in some few cases, as in that of aphids, if we look to Professor Huxley's admirable drawings, we see hardly any trace of the worm-like stage in its development.

Sometimes it is only the earlier developmental stages that fail to appear. Thus, Fritz Müller has made the remarkable discovery that the larvae of certain shrimp-like crustaceans first develop into a simple nauplius larva (Figure 14.5) and, after passing through two or more zoeal stages (Figure 14.7), then through a "mysis" stage before finally metamorphosing one last time to acquire their mature structure. Now in the whole great malacostracan order to which these crustaceans belong, no other member is as yet known to first develop as a nauplius larva, although many appear as zoeas; nevertheless, Müller argues that if there had been no suppression of development in other malacostracans, all crustaceans would have first developed as nauplii.

How, then, can we explain these several facts in embryology—namely, (1) the very general, though not universal, difference in structure between the embryo and the adult; (2) the various parts in the same individual embryo, which ultimately become very unlike and serve for diverse purposes, being very similar in early development; (3) the common, but not invariable, resemblance between the embryos or larvae of species belonging to the same class that look very different from each other as adults; (4) the embryo often retaining, while within the egg or womb, structures that are of no use to it, either at that or at a later period of life; (5) larvae that have to provide for their own wants being perfectly adapted to the conditions surrounding them; and last but not least, (6) the fact that certain larvae are more structurally complex than the mature animal into which they develop? I believe that all these facts can be explained as described in the following paragraph.

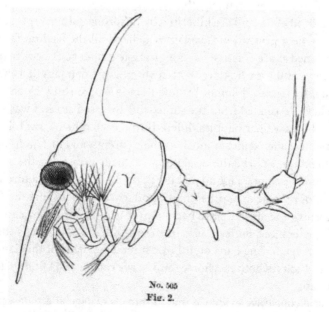

No. 505
Fig. 2.

Figure 14.7 A crustacean zoea larva.

It is commonly assumed that slight variations or individual differences seen among adults of any particular group will necessarily appear early in development. We have little evidence on this point, but what we have certainly points the other way; for it is notorious that breeders of cattle, horses, and various fancy animals cannot positively tell, until sometime after birth, what the strengths and weaknesses of their young animals will be. We see this plainly in our own children: we cannot tell early on whether a child will be tall or short, or what his or her precise features will be as they age. The question is not at what period of life any variation may have been *caused*, but rather at what period the effects are *displayed*. The cause may have acted, and I believe often has acted, on one or both parents before the act of generation. It is worth noticing that it is of no importance to a very young animal, as long as it is nourished and protected by its parent, whether most of its characters are acquired a little earlier or later in life. It would not matter, for instance, whether or not a bird that obtained its food by having a much-curved beak as an adult possessed a beak of this shape when young as long as it was fed by its parents during that time.

As I stated in Chapter 1, at whatever age any variation first appears in the development of the parent, it tends to reappear at a corresponding age in their offspring. Certain variations can *only* appear at corresponding ages; for instance, peculiarities in the caterpillar, cocoon, or imago states of the silk-moth can't appear at other stages of development; or, again, in the full-grown horns of cattle. But variations which, for all that we can see, might have appeared equally well either earlier or later in life than we see them appear likewise tend to reappear at a corresponding age in the offspring and parent. I am far from meaning that this is invariably the case, but it is a worthy generalization.

These two principles—namely, that slight variations generally appear at a not very early period of life and are inherited by offspring at a corresponding not early period in the next generation—explain, I believe, all the leading facts in embryology mentioned above. But first let us look at a few analogous cases in our domestic varieties. Some authors who have written about dogs maintain that the greyhound and bulldog, though so different in how they look, are really closely related varieties that have descended from the same wild ancestral stock. I was therefore curious to see how far their puppies differed from each other. I was told by breeders that the puppies differed just as much as their parents, and this, judging by the eye, did seem to be almost to be the case. But on actually measuring the adult dogs and their six-day-old puppies, I found that the puppies had not yet acquired nearly their full amount of proportional difference. Similarly, I was told that the foals of cart- and racehorses—breeds that have been almost wholly formed by careful selection over many generations under domestication—differed as much as the full-grown animals; however, having had careful measurements made of the dams and of the three-day-old colts of both racehorses and heavy carthorses, I find that this is by no means the case.

As we have conclusive evidence that all breeds of pigeon are descended from a single wild species, I compared the young pigeons within 12 hours after being hatched with characteristics of their parents. I carefully measured the proportions (but will not here give the details) of the beak, width of mouth, length of nostril and of eyelid, and size of feet and length of leg in the wild parent species and in pouters, fantails, runts, barbs, dragons, carriers, and tumblers. Now some of these birds, when mature, differ in so extraordinary a manner in the length and form of the beak and in other characters, that they would certainly have been ranked as belonging to distinctly different genera if they had been found in nature. But when the nestling birds of these several breeds were placed in a row, though most of them could just be distinguished, the proportional differences in the above specified points were incomparably less than in the full-grown birds. Some characteristic points of difference among the breeds—for instance, differences in the width of mouth—could hardly be detected at all in the young. I did, however, find one remarkable exception to the rule: the young of the short-faced tumbler differed from the young of the wild rock-pigeon and of the other breeds in almost exactly the same proportions as in the adult stage.

These facts are explained by the above two principles. **Breeders select their dogs, horses, pigeons, etc., for breeding when their subjects are nearly grown up. They care not at all whether the desired qualities are acquired earlier or later in life, as long as the full-grown animal possesses them.** And the cases just given, particularly that of the pigeons, show that the characteristic differences that have been accumulated by man's selection and that give value to his breeds do not generally appear at a very early period of life and are inherited at a correspondingly not early period in development. But the case of the short-faced tumbler, which already possessed its proper characters when only 12 hours old, proves that this is not the universal rule; for here the characteristic differences must either have appeared at an earlier period than usual or, if not so, the differences must have been inherited, not at a corresponding age, but at an earlier age.

Now, let us apply these two principles to species in nature. Let us take a group of birds, descended from some ancient form and modified through natural selection for different lifestyles. Then, from the many slight successive variations having supervened in the several species at a not very early age and having been inherited by offspring at a corresponding age, the young will have become but little modified, and they will still resemble each other much more closely than do the adults, just as we have seen with the breeds of the pigeon. We may extend this view to widely distinct structures and to whole classes of organisms. The forelimbs, for instance, which once served as legs to a remote ancestor, may have become, through a long course of modification, adapted in one descendant to act as hands, in another as paddles, in another as wings; but, on the above two principles, the forelimbs will not have been much modified in the embryos of these several forms despite the great differences in forelimb structure in adults. Whatever influence long continued use or disuse may have had in modifying the limbs or other parts of any species, this will chiefly or solely have affected it when the individual was nearly mature, when it was compelled to use its full powers to gain its own living; the effects thus produced will subsequently have been transmitted to the offspring at a corresponding nearly mature age, as explained in a previous chapter. Thus the young will not be modified, or will be modified only in a slight degree, through the effects of the increased use or disuse of parts.[29] With some animals the successive variations may have supervened at a very early period of life, or the steps may have been inherited by offspring at an earlier age than that at which they first occurred. In either of these cases the young or embryo will closely resemble the mature parent form, as indeed we have seen with the short-faced tumbler. And this is the rule of development in certain whole groups, or in certain subgroups alone, as with cuttlefish, land snails, freshwater crustaceans, spiders, and some insects. With respect to the final cause of the young in such groups not passing through any metamorphosis, we can see that this would follow from the following contingencies: namely, from the young having to provide at a very early age for their own wants and from their following the life-style of their parents; for in this case it would be indispensable for their existence that they should be modified in the same manner as their parents. Again, with respect to the singular fact that many terrestrial and freshwater animals do not undergo any metamorphosis in their development while marine members of the same groups typically do pass through such transformations, Fritz Müller has suggested that the process of slowly modifying and adapting an animal to live on the land or in fresh water, instead of in the sea, would be greatly simplified by its not passing through any larval stage; for it is not probable that places well-adapted for both the larval and mature stages, under such new and greatly changed habits of life, would commonly be found unoccupied or ill-occupied by other organisms. In this case the gradual acquisition of the adult structure at an earlier and earlier age would be favored by natural selection—and all traces of former metamorphoses would finally be lost from the life history.

If, on the other hand, the young of some animal benefitted by following habits of life slightly different from those of the parent form, and consequently were to be constructed on a slightly different plan, or if it benefitted a larva that already looked different from its parent to change still further, then, on the principle of inheritance at corresponding ages, the young or the larvae might be rendered by natural selection more and more different from their parents to any conceivable extent, generation after generation. Differences in the larva might also become correlated with successive stages of its development so that the larva, in the first stage of its development, might come to differ greatly from the larva in the second stage of its development, as is in the case with many animals. The adult might also become adapted for sites or habits in which organs of locomotion or of the senses, etc., would be useless; in this case, the metamorphosis would be retrograde, involving the loss of such organs and capabilities.

From the remarks just made we can see how by changes of structure in the young—in conformity with changes in lifestyles—together with inheritance at corresponding ages, animals might come to pass through stages of development that are perfectly

---

[29] As noted previously, there is still no evidence for this Lamarckian idea that structural changes gained through the use or disuse of parts are transmitted to offspring.

distinct from the ancient condition of their adult ancestors. Most of our best authorities are now convinced that the various larval and pupal stages of insects have thus been acquired through adaptation and not through simple inheritance from some ancient form. The curious case of *Sitaris*—a blister beetle that passes through certain unusual stages of development—will illustrate how this might occur. The first larval form is described by the French entomologist M. Jean-Henri Fabre as an active, minute insect that is furnished with six legs, two long antennae, and four eyes. These larvae hatch in the nests of bees. In the spring, when the male bees emerge from their burrows—which they do before the females—the blister beetle larvae spring on them and afterwards crawl onto the females while those females are still paired with the males for mating. As soon as the female bee deposits her eggs on the surface of the honey stored in the cells, the *Sitaris* larvae leap onto the eggs and devour them. The larvae then undergo a complete change: their eyes disappear, their legs and antennae become rudimentary, and they feed on honey. Now they more closely resemble ordinary insect larvae; ultimately, they undergo a further transformation and finally emerge as the perfect, normal beetle.

Now if an insect undergoing transformations like those exhibited by *Sitaris* were to become the ancestor of a whole new class of insects, the course of development of this new class would be widely different from that of any of our existing insects; certainly, the first larval stage would not represent the former condition of any adult and ancient form.

On the other hand it is highly probable that with many animals the embryonic or larval stages give us a very good idea of what the ancestor of the whole group looked like as an adult. Among crustaceans, forms wonderfully distinct from each other, namely sectorial parasites, barnacles, entomostracans,[30] and even members of the Malacostraca all appear at first as minute, swimming nauplius larvae; as these larvae live and feed in the open sea and are not adapted for any peculiar habits of life, it is probable that an independent adult animal resembling a nauplius once existed, a very long time ago, and subsequently produced, along several divergent lines of descent, the above-named great crustacean groups. Similarly, the embryos of mammals, birds, fishes, and reptiles seem very likely to be the modified descendants of some ancient ancestor that as an adult possessed gills, a swim bladder, four fin-like limbs, and a long tail, all fitted for an aquatic life.

As all organic beings—both extinct and recent—which have ever lived can be arranged within a few great classes, and as all species within each class have, according to our theory, been connected together by fine gradations, the best, and—if our collections were nearly perfect—the only possible arrangement, would be genealogical, descent from a common ancestor being the hidden bond of connection that naturalists have been seeking under the term of the "Natural System." On this view we can understand how it is that, in the eyes of most naturalists, the structure of the embryo

---

[30] Entomastracans were at the time one of two major groups of crustaceans, including copepods, ostracods, and branchiopods. Other crustaceans were included in the other major subgroup, the Malacostraca.

is even more important for classification than that of the adult. In two or more groups of animals, however much the adults may differ from each other in structure and habits if they pass through closely similar embryonic stages, we may feel assured that they are all descended from one and the same ancestor and are therefore closely related. Thus, similarities in embryonic structure reveal community of descent.

On the other hand, dissimilarities in embryonic development do *not* disprove community of descent: the developmental stages may have been suppressed in one of any two groups or may have been so greatly modified through adaptation to new habits of life as to be no longer recognizable. Even in groups in which the adults have been modified to an extreme degree, community of origin is often revealed by the structure of the larvae. We have seen, for instance, that barnacles, though externally so like oysters and other shellfish, are at once known by their larvae to be crustaceans. As the embryo often shows us more or less plainly the structure of the less modified and ancient ancestor of the group, we can see why the adults of ancient and extinct forms so often resemble the developmental stages of existing species of the same class. Agassiz believes this to be a universal law of nature; we may hope hereafter to see that law proved true. It can, however, be proven true only in those cases in which the ancient state of the group's ancestor has not been wholly obliterated, either by successive variations having supervened at a very early period of growth or by such variations having been inherited at an earlier age than that at which they first appeared. It should also be borne in mind that although the law may very well be true, it may remain for a long period—or forever—incapable of being convincingly proven owing to the geological record not extending far enough back in time. The law will not strictly hold good in those cases in which an ancient form became adapted in its larval state to some special line of life and transmitted the same larval state to a whole group of descendants; for such larvae will not resemble any still more ancient form in its adult state.

Thus, it seems to me that the leading facts in embryology, which are second to none in importance, are explained very simply as follows: variations in the many descendants from some particular ancient ancestor, having appeared at a not very early period of life, have been inherited at a corresponding period. The study of embryology rises greatly in interest when we look at the embryo as a picture, more or less obscured, of the ancestor—either in its adult or larval state—of all the members of the same great class.

Rudimentary, Atrophied, and Aborted Organs

Organs or parts in this strange condition, bearing the plain stamp of uselessness, are extremely common, or even general, throughout nature. It would be impossible to name any one of the higher animals in which some part or other is not now in a rudimentary condition. Among the mammals, for instance, the males possess rudimentary mammary glands; in snakes, one lobe of the lungs is rudimentary; in birds, the small, freely moving alula (also called the "bastard-wing") on the leading edge of the wings of modern (and some extinct) birds may safely be considered as a rudimentary digit (and in some species the whole wing is so far rudimentary that it cannot be used for flight). And what can be more curious than the presence of teeth in fetal whales, which when grown up have not a single tooth in their mouths, or the teeth that never cut through the gums in the upper jaws of unborn calves?

Rudimentary organs plainly declare their origin and meaning in various ways. There are beetles belonging to closely related species—or even to the same identical species—that have either full-sized and perfect wings, or mere rudiments of membrane that often lie under normal wing-covers firmly soldered together; in these cases it is impossible to doubt that the rudiments represent wings. Rudimentary organs sometimes retain their potentiality: this occasionally occurs with the mammaries of some male mammals, which have been known to become well-developed and to secrete milk. So again, there are normally four developed and two rudimentary teats in the udders of the cow genus *Bos,* but the rudimentary teats in our domestic cows sometimes become well-developed and yield milk. In regard to plants, the petals are sometimes rudimentary and sometimes well developed in different individuals of the same species. In certain plants that have separate sexes, Kölreuter found that by crossing one species in which the male flowers included a rudiment of a seed-producing pistil with a hermaphrodite species having of course a well-developed pistil, the pistil rudiment in the hybrid offspring was much larger; this clearly shows that the rudimentary and perfect pistils are essentially alike in nature.

An animal may also possess some parts in a perfect state—perfectly developed—and yet they may be useless and thus in this sense still be considered to be "rudimentary." For example, the tadpole of the common salamander or water-newt, as Mr. G. H. Lewes remarks, "has gills, and passes its existence in the water; but the *Salamandra atra,* which lives high up among the mountains, brings forth its young full-formed, with no specialized free-living larval stage.[31] This animal never lives in the water. Yet if we open a gravid female, we find tadpoles inside her with exquisitely feathered gills; and when placed in water they swim about like the tadpoles of the water-newt. Obviously this aquatic organization of the tadpole has no reference to the future life of the animal, nor has it any adaptation to its embryonic condition; it has solely reference to ancestral adaptations, it repeats a phase in the development of its progenitors."[32]

An organ that serves two different purposes may become rudimentary or utterly aborted for one—even the more important—purpose and yet remain perfectly efficient for the other. Thus, in plants, the role of the female pistil is to allow the pollen tubes to reach the ovules within the ovary. The pistil consists of a stigma

---

[31] This is now referred to as "direct development": the embryo develops directly into the adult morphology without passing through a free-living larval stage.

[32] This is a very good example of something that makes sense only in the context of evolution. The adults are fully terrestrial, and their offspring are born as miniatures of the adults. And yet there is a well-formed tadpole larva in the life history, which basically metamorphoses before birth.

supported on the style. But in some members of the widespread angiosperm family Compositae,[33] the male florets, which of course cannot be fertilized, have a rudimentary pistil, for it is not crowned with a stigma, and yet the style remains well developed and is clothed in the usual manner with hairs that serve to brush the pollen out of the surrounding and conjoined anthers. Again, an organ may become rudimentary for its proper purpose and be used instead for a distinctly different one: in certain fishes, for example, the swim bladder seems to be rudimentary for its normal function of achieving buoyancy but has become converted into a nascent breathing organ or lung. Many similar instances could be given.

Useful organs, however little they may be developed, ought not to be considered as rudimentary unless we have reason to suppose that they were formerly more highly developed. They may be in an early stage of development progressing toward further development. **On the other hand, truly rudimentary organs are either quite useless, such as teeth that never cut through the gums, or almost useless, such as the wings of an ostrich, which serve merely as sails.** As organs in this condition would formerly, when even less well-developed, have been of even less use than at present, they cannot formerly have been produced through variation and natural selection, which acts solely by the preservation of useful modifications. They have instead been partially retained by the power of inheritance and now relate to a former state of things when they did perform a useful function. It is, however, often difficult to distinguish between rudimentary organs and nascent organs—organs at an early stage of evolutionary development—for we can judge only by analogy whether a part is capable of further development, in which case it deserves to be called "nascent." Organs in this condition will always be somewhat rare, for beings thus provided will commonly have been replaced by their descendants with the same organ in a more perfect state and consequently will have become extinct long ago. The penguin's wing, for example, is of high service to the animal, acting as a fin when the penguin is swimming; it could, therefore, represent the nascent state of the wing. Not that I believe this to be the case: it is more probably a reduced organ, modified for a new function. But a nascent function is still an interesting possibility. The wings of the flightless kiwi bird in New Zealand (genus *Apteryx*), on the other hand, is quite useless and is truly rudimentary. Richard Owen considers the simple filamentary limbs of the South American lungfish, *Lepidosiren paradoxa*, as the "beginnings of organs which attain full functional development in higher vertebrates." However, according to the view lately advocated by Dr. Gunther, they are probably the remnants of a formerly better-developed condition, consisting of the persistent axis of a fin with the lateral rays or branches aborted. The mammary glands of the duck-billed platypus of Australia (*Ornithorhynchus anatinus*) may be considered, in comparison with the udders of a cow, as in a nascent condition. Similarly, the egg-retaining "ovigerous frena" of certain barnacles that no longer provide attachment to the developing eggs ("ova") and are feebly developed, are probably nascent branchiae.

---

[33] This family includes daisies, asters, and sunflowers.

Rudimentary organs found in individuals of a single species are very liable to vary in their degree of development and in other respects in closely related species; the extent to which the same organ has been reduced occasionally differs a good deal. This latter fact is well exemplified by the state of development in the wings of female moths belonging to the same family. Rudimentary organs may be utterly aborted; this implies that, in certain animals or plants, parts are entirely absent, where analogy would lead us to expect to find them and which are occasionally found in monstrous individuals. Although in most species found within the plant family Scrophulariaceae, the fifth stamen is utterly aborted, yet we may conclude that a fifth stamen once existed, for a rudiment of it is found in many other species within the family; in fact, this rudiment occasionally becomes perfectly developed, as is sometimes seen in the common snapdragon. In tracing the homologies of any part in different members of the same class of organisms, nothing is more common or more useful in understanding the relations of the parts than the discovery of rudiments. This is well shown in the drawings given by Richard Owen of the leg bones of the horse, ox, and rhinoceros, which show very clearly that all are related.

Rudimentary organs, such as teeth in the upper jaws of whales and in cows, sheep, and other ruminants, can often be detected in the embryo but wholly disappear as development continues. It is also, I believe, a universal rule that a rudimentary part is larger in the embryo, relative to the adjoining parts, than in the adult; thus the organ at this early age is less rudimentary or cannot even be said to be rudimentary in any degree. Hence rudimentary organs in the adult are often said to have retained their embryonic condition.

I have now given the leading facts with respect to rudimentary organs. In reflecting on them, all readers must be struck with astonishment: for the same reasoning power which tells us that most body parts and organs are exquisitely adapted for certain purposes tells us with equal plainness that rudimentary or atrophied organs are imperfect and useless. In works on natural history, rudimentary organs are generally said to have been created "for the sake of symmetry," or in order "to complete the scheme of nature." But this is merely a restatement of the fact, not an explanation. Nor is it consistent with itself. The boa constrictor, for example, has rudiments of hind limbs and of a pelvis; if it be said that these bones have been retained "to complete the scheme of nature," why, as Professor Weismann asks, have they not been retained by other snakes, which do not possess even a vestige of these same bones?

What would we think about an astronomer who maintained that satellites revolve in elliptic courses round their planets "for the sake of symmetry" because the planets thus revolve round the sun? An eminent physiologist (who I will not name) accounts for the presence of rudimentary organs by supposing that they serve to excrete matter in excess, or matter injurious to the an organism's physiology; but can we suppose that the minute papilla, which often represents the pistil in male flowers and which is formed of mere cellular tissue, can act in that same way and for the same reasons? Can we suppose that the rudimentary teeth that are subsequently absorbed as development proceeds are beneficial to the rapidly growing embryonic calf by removing matter so precious as phosphate of lime? When a man's fingers have been amputated, imperfect nails have been known to appear on the stumps; I could as soon believe that these vestiges of nails are developed in order to excrete horny matter as that the rudimentary nails on the fin of the manatee have been developed for this same purpose!

On the view of descent with modification, however, the origin of rudimentary organs is comparatively simple, and we can understand to a large extent the laws governing their imperfect development. We have plenty of cases of rudimentary organs in our domestic productions, as for example, the stump of a tail in our otherwise tailless breeds; the vestige of an ear in earless breeds of sheep; the reappearance of minute dangling horns in otherwise hornless breeds of cattle, particularly (according to William Youatt[34]) in young animals; and the state of the whole flower in the cauliflower. We often see rudiments of various parts in monsters, but I doubt whether any of these cases throw light on the origin of rudimentary organs in a state of nature, other than by simply showing that rudiments can indeed be produced; for the balance of evidence clearly indicates that species under nature do not undergo great and abrupt changes. But we learn from the study of our domestic productions that the disuse of parts leads to their reduced size, and that that result is inherited in future generations.

It seems likely that disuse of organs has been the main agent in rendering them rudimentary. It would at first lead by slow steps to the more and more complete reduction of a part until at last it became rudimentary, as in the case of the eyes of animals that inhabit dark caverns and of the wings of birds inhabiting oceanic islands, which have seldom been forced by beasts of prey to take flight and have thus ultimately lost the power of flying. Again, an organ that is useful under certain conditions might become injurious under others, as with the wings of beetles living on small and wind-exposed islands; in this case natural selection will have aided in reducing the organ, until it was rendered harmless and rudimentary.

Any change in structure and function that can be brought about by small stages is within the power of natural selection to accomplish, so that an organ rendered useless or injurious for one purpose through changed habits of life might be subsequently modified and used for a very different purpose. An organ might also be retained for just one of its several former functions. Organs that were originally formed by the aid of natural selection may well be variable when rendered useless, for their variations can no longer be checked by natural selection. All of this agrees well with what we see in nature. Moreover, at whatever period of life either disuse or selection reduces an organ—and this will generally be when the being has come to maturity and begins to exert its full powers of action—the principle of inheritance at corresponding ages will tend to reproduce the organ in its reduced state at the same mature age; but it will seldom affect it in the embryo. Thus we can understand the greater size of rudimentary organs in the embryo in comparison with the adjoining parts and their lesser

[34] William Youatt (1776–1847) was a British veterinarian who wrote a series of books on the management and diseases of farm animals, including cattle.

relative size in the adult. If, for instance, the digit of an adult animal was used less and less during many generations owing to some change of habits, or if an organ or gland was less and less functionally exercised, we may infer that it would gradually become reduced in size in the adult descendants of this animal but would likely retain nearly its original standard of development in the embryo.

There remains, however, the following difficulty. After an organ has ceased being used and has become in consequence much reduced, how can it be still further reduced in size until the merest vestige is left, and how can it be finally quite obliterated? It is scarcely possible that disuse can go on producing any further effect after the organ has once been rendered functionless. Some additional explanation is required here, but unfortunately I cannot give it. If, for instance, we could be prove that every part of the organization tends to vary in a greater degree toward diminution than toward augmentation of size, then we should be able to understand how an organ that has become useless would be rendered, independently of the effects of disuse, rudimentary, and would at last be wholly suppressed, for the variations toward diminished size would no longer be checked by natural selection. The principle of the economy of growth, explained in a former chapter, by which the materials forming any part that is not useful to the possessor are saved as far as is possible, will perhaps come into play in rendering a useless part rudimentary. But this principle will almost necessarily be confined to the earlier stages in the process of reduction; for we cannot suppose that a minute papilla, for instance, representing in a male flower the pistil of the female flower and formed merely of cellular tissue, could be further reduced or absorbed for the sake of economizing nutriment.

Finally, as rudimentary organs—by whatever steps they may have been degraded into their present useless condition—are a record of a former state of things and have been retained solely through the power of inheritance, we can understand (on the genealogical view of classification) how it is that systematists, in trying to place organisms in their proper places in the natural system, have often found rudimentary parts to be as useful as parts of high physiological importance—or even sometimes more useful. Rudimentary organs may be compared with letters that are still retained in the spelling of a word, even though those letters are no longer pronounced; they nevertheless serve as a clue to the word's derivation. On the view of descent with modification, we may conclude that the existence of organs in a rudimentary, imperfect, and useless condition, or quite aborted, far from presenting a strange difficulty—as they assuredly do on the old doctrine of special creation—might even have been anticipated in accordance with the views explained here.

Summary

In this chapter I have attempted to show all of the following: (1) the rules followed and the difficulties encountered by naturalists in their classifications; (2) that all organic beings throughout all time are arranged in groups within groups; (3) that all living and extinct organisms are united into a few grand classes by complex, radiating, and circuitous lines of affinities; (4) that the characters of greatest value are those that are constant and prevalent, whether of high or of the most trifling importance, or, as with rudimentary organs, of no functional importance at all; and (5) the wide opposition in value between analogical or adaptive characters, and characters based on true affinity. All naturally follow if we admit the common parentage of allied forms together with their modification through variation and natural selection, with the contingencies of extinction and divergence of character. In considering this view of classification, it should be borne in mind that the element of descent has long been universally used in ranking together the sexes, ages, dimorphic forms, and acknowledged varieties of the same species, however much they may differ from each other in structure. If we extend the use of this element of descent—the one certainly known cause of similarity between and among organic beings—we shall understand what is meant by the Natural System: it is genealogical in its attempted arrangement, with the levels of acquired difference marked by the terms varieties, species, genera, families, orders, and classes.

On this same view of descent with modification, most of the great facts in Morphology also become intelligible—whether we look to the same pattern displayed by the different species within a single class of organisms in their homologous organs, to whatever purpose applied, or to the serial and lateral homologies in each individual animal and plant.

On the principle of successive slight variations being inherited by offspring at a corresponding period in development, we can understand the leading facts in embryology: the close resemblance in the individual embryo of the parts that are homologous, but which when matured become widely different in structure and function; and the resemblance of the homologous parts or organs in allied—though distinct—species, though fitted in the adult state for habits as different as is possible. Larvae are active developmental stages that have become specially modified to a greater or less degree in relation to their habits of life, with their modifications inherited at a corresponding early age. On these same principles, and bearing in mind that when organs are reduced in size, either from disuse or through natural selection, it will generally be at that period of life when the being has to provide for its own wants, and, bearing in mind how strong is the force of inheritance, we might even have anticipated the occurrence of rudimentary organs. The importance of embryological characters and of rudimentary organs in classification makes perfectly good sense on the view that a natural arrangement must be genealogical.

Finally, the several classes of facts that have been considered in this chapter seem to me to proclaim very plainly that the innumerable species, genera, and families with which our world is inhabited are all descended, each within its own class or group, from common parents and have all been gradually modified over the long course of descent to the present day. I should without hesitation adopt this view, even if it was not supported by any other facts or arguments.

Key Issues to Talk and Write About

1. What does Darwin mean when he talks about "incipient species"?
2. Why does Darwin believe that embryological characteristics should play a very important role in determining species relationships?
3. According to Darwin, how does the distribution of human languages around the world relate to our system of taxonomic classification?
4. What are "analogical" characteristics? How do they differ from characteristics that are "homologous"?
5. What is Batesian mimicry and, according to Darwin, what selects for it? How does Batesian mimicry complicate the use of morphological similarities in classifying species?
6. List the evidence that Darwin gives in support of evolution by natural selection in this chapter.

15

Recapitulation and Conclusion

In this final chapter, Darwin begins by reminding us of the main objections that had been advanced against his and Wallace's theory of descent with modification since it was first proposed in 1858, and he summarizes his responses to those objections. He then summarizes his main argument in favor of the theory, reminding us of the importance of natural variability in traits of all sorts, the inheritance of those variations by offspring, the impact of variation on survival and reproductive success and the evidence in favor of those factors having contributed, over many millions of years, to the variety of plant and animal species we now have on our planet. In some of the most marvelous and inspiring paragraphs in the book, he then writes about the changes that this understanding of the process of natural selection—and of its role in creating the diversity of animal and plant species we see around us—will bring about in the sorts of research that will be undertaken in the future and in our understanding of the world around us—and our place in that world.

As this entire book has been one long argument, let me briefly recapitulate the leading facts and inferences.

I do not deny that many serious objections may be advanced against my theory of descent with modification through variation and natural selection, and I have endeavored to present them all openly and fairly. Nothing can at first appear more difficult to believe than that the most complex organs and instincts that we now find among living organisms have been perfected, not through means superior to human reason, but rather by the very gradual accumulation of innumerable slight variations, each good in some way for its individual possessor. The difficulty would seem to be insuperably great. Nevertheless, this difficulty cannot be considered real if we admit the following propositions: (1) that we clearly see at least some individual differences in all aspects of an organism's organization and instincts; (2) that there is, and has always been, a struggle for existence, and that this struggle must lead to the preservation of those deviations of structures or instincts that increase the likelihood of survival and successful mating; and (3) that each of an organism's organs may have existed in the past in different degrees of perfection, with each stage having provided some advantage to its owner. The truth of these propositions cannot, I think, be disputed.

The Readable Darwin. Second Edition. Jan A. Pechenik, Oxford University Press. © Oxford University Press 2023.
DOI: 10.1093/oso/9780197575260.003.0016

It is, no doubt, extremely difficult even to conjecture by what small steps many structures have been gradually perfected, particularly among broken and failing groups of organisms that have suffered much extinction; but since we now see so many strange gradations in nature, we ought to be extremely cautious in saying that any organ or instinct, or any whole structure, could not have arrived at its present state by many graduated steps. There are, it must be admitted, cases of special difficulty opposed to the theory of natural selection. One of the most curious of these is the existence in the same community of two or three well-defined castes of sterile female worker ants, which, because of this sterility, cannot pass individual differences on to their offspring: there are no offspring for these workers! I have, however, attempted to show how these difficulties can be mastered.

With respect to the almost universal sterility of males and females of different species when first crossed—forming such a remarkable contrast with the almost universal fertility of varieties when crossed—I must refer readers to the recapitulation of the facts given at the end of Chapter 9; those fact seem to me to show conclusively that this sterility is no more a special endowment than is the incapacity of two distinct kinds of trees to be grafted together; rather, it is incidental on differences confined to the reproductive systems of the intercrossed species. We see the truth of this conclusion in considering the vast difference in results when one species is crossed reciprocally with another—that is, when one species is first used as the father in a cross with another species and then as the mother. Analogy from the consideration of dimorphic and trimorphic plants of a single species clearly leads to the same conclusion, for when the forms are illegitimately united by us, they yield few seeds or no seed and their offspring—if they have any—are more or less sterile; and yet these forms belong to the same undoubted species and differ from each other only in their reproductive organs and functions.

Although the fertility of varieties when intercrossed, and of their mongrel offspring, has been claimed as universal by so many authors, there is growing evidence against this claim, particularly considering the facts given on the high authority of the German botanists Karl Friedrich von Gartner and Joseph Kolreuter.

A double and parallel series of facts seems to throw much light on the sterility of species when first crossed and of their hybrid offspring. On the one side, there is good reason to believe that slight changes in the conditions of life give vigor and fertility to all organic beings. We know also that a cross between the distinct individuals of the same variety, and between distinct varieties, increases the number of their offspring, and certainly gives them increased size and vigor. This is largely due to the forms being crossed having been exposed to somewhat different conditions of life; for I have ascertained by a laborious series of experiments that if all the individuals of the same variety are subjected during several generations to the same conditions, the good derived from crossing is often much diminished or wholly disappears. This is one side of the case. On the other side, we know that species that have long been exposed to nearly uniform conditions either perish when they are subjected under confinement to new and greatly changed conditions or, if they survive, are rendered sterile, though retaining perfect health. This does not occur—or it occurs only in a very slight degree—with our domesticated productions, which have long been exposed to fluctuating conditions. Hence when we find that hybrids produced by a cross between two distinct species are few in number, owing to their perishing soon after conception or at a very early age, or if surviving that they are rendered more or less sterile, it seems highly probable that this is due to their having been in fact subjected to a great change in their conditions of life, from being compounded of two distinct organizations. He who will explain in a definite manner why, for instance, an elephant or a fox will not breed under confinement in its native country, while the domestic pig or dog will breed freely under the most diversified conditions, will at the same time be able to explain why two distinct species, when crossed, as well as their hybrid offspring, are generally rendered more or less sterile, while two domesticated varieties when crossed, and their mongrel offspring, are perfectly fertile.

Turning to geographical distribution, the difficulties encountered on the theory of descent with modification are serious enough: Why do we sometimes find what seem to be identical or closely related species living in such widely separated parts of the world? The answer is that all the individuals of the same species and all the species of the same genus, or an even higher group (e.g., families and orders), are descended from common parents; therefore, in however distant and isolated parts of the world they may now be found, they must in the course of successive generations have traveled from some one point to all the others where we now find them. We are often wholly unable even to conjecture how this could have been accomplished.[1] Yet, as we have reason to believe that some species have retained the same specific form for very long periods of time—immensely long as measured by years—too much stress ought not to be laid on the occasional wide diffusion of the same species; during very long periods there will always have been a good chance for wide migration by many means. A broken or interrupted range may often be accounted for by the extinction of the species in the intermediate regions.

It cannot be denied that we are as yet very ignorant as to the full extent of the various climatic and geographical changes that have affected the Earth during modern periods. Such changes will often have facilitated migration. As an example, I have attempted to show how potent the recent glacial period[2] has been on the distribution of the same and of allied species throughout the world. We are as yet profoundly ignorant of the many occasional means of transport. With respect to distinct species of the same genus that inhabit distant and isolated regions, as the process of modification has necessarily been slow, all the means of migration will have been possible during a very long period; the difficulty of the wide diffusion of the species of the same genus is, in consequence, in some degree lessened.

According to the theory of natural selection, a vast number of intermediate forms must have linked together all the species in each group by gradations as fine as those

---

[1] As I noted in Chapters 11 and 12, we now know that in many cases these distributions are explained by continental drift, an idea that was first developed by Alfred Wegener in 1912.

[2] This period ended about 15,000 years ago, after having lasted for nearly 100,000 years.

shown by our present varieties. It may be asked, then, "Why do we not see these linking forms all around us? Why are not all organic beings blended together in an inextricable chaos?" Well, with respect to existing forms, we should remember that we have no right to expect (excepting in rare cases) to discover *directly* connecting links between them; rather we should expect to see links between each existing form and some extinct and subsequently supplanted form. Even on a wide area that has remained continuous for a long time and within which the climatic and other conditions of life change insensibly as we move from a region occupied by one species into another district occupied by a closely allied species, we have no reason to expect to find intermediate varieties in the intermediate zones very often. This is so because it seems that only a few species within any particular genus ever undergo change, the other species becoming utterly extinct and leaving no modified descendants. Of the species that do change over time, only a few within the same region will change at the same time, and all modifications are effected only slowly. I have also shown that the intermediate varieties that probably at first did exist in the intermediate zones would likely be supplanted by the related forms on either side for the latter, from existing in greater numbers, would generally be modified and improved at a quicker rate than the intermediate varieties, which existed in fewer numbers; the intermediate varieties, then, would eventually be supplanted and exterminated.

This raises yet another difficulty. If we accept this doctrine of the inevitable extermination of an infinite number of connecting links between the living and the extinct inhabitants of the world, and at each successive period between the extinct and still older species, then why do we not find such connecting links in every geological formation? Why does not every collection of fossil remains afford plain evidence of the slow gradation and mutation of the forms of life? Although geological research has undoubtedly revealed the former existence of many links, bringing numerous forms of life much closer together, it does not yield the infinitely many fine gradations between past and present species required by my theory; this is in fact the most obvious of the many objections that may be urged against it. Why, again, do whole groups of allied species seem to appear so suddenly (though this impression is often false) on the successive geological stages? Although we now know that organic beings appeared on this globe at a period incalculably remote, long before the lowest bed of the Cambrian system was deposited, why do we not find beneath this system great piles of strata stored with the remains of the ancestors of the Cambrian fossils?[3] For, on my theory of evolution by natural selection, such strata must somewhere have been deposited at these ancient and utterly unknown epochs of the world's history.

I can answer these questions and objections only on the supposition that the geological record is far more imperfect than most geologists believe. The number of specimens in all our museums is absolutely as nothing compared with the countless generations of countless species that have certainly existed over time. The

---

[3] We have now found multicellular life forms in pre-Cambrian formations, from more than 600 million years ago.

characteristics of the parent form of any two or more presently existing species would not all be directly intermediate between those of its modified offspring, any more than the crop and tail of today's rock-pigeon are directly intermediate between those of its descendants, the pouter and fantail pigeons. Without possessing most of the intermediate links, we should not be able to recognize a species as the parent of another and modified species even if we were to examine the two ever so closely; and, of course, owing to the imperfection of the geological record, we have no right to expect to find so many links. If two or three, or even more linking forms were discovered, they would simply be ranked by many naturalists as so many new species, more especially if found in different geological substages, even if their differences were ever so slight. Numerous existing doubtful forms could be named that are probably varieties: But who will pretend that, in future ages, so many fossil links will be discovered that naturalists will be able to decide whether or not these doubtful forms ought to be called varieties of a single species?

It should also be noted that only a small portion of the world has been geologically explored. And only organic beings of certain classes can be preserved as fossils, at least in any great number. Many species when once formed never undergo any further change but become extinct without leaving modified descendants, and the periods during which species have undergone modification, though long as measured by years, have probably been short in comparison with the periods during which they retained the same form. Because the dominant and widely ranging species vary most frequently and vary the most, and because varieties are often at first only local, the likelihood of discovering intermediate links in any one geological formation is reduced. Local varieties will not spread into other and distant regions until they are considerably modified and improved; when they have spread, and are now discovered in a geological formation, they appear as if suddenly created there and will be simply classed as new species. Most geological formations have been intermittent in their accumulation, and their duration has probably been shorter than the average duration of particular species. Successive geological formations are in most cases separated from each other by blank intervals representing long periods of time, for fossiliferous formations thick enough to resist future degradation can, as a general rule, be accumulated only where much sediment is deposited on the subsiding bed of the sea. During the alternate periods when the land is being elevated above the water or remains stationary, the geological record will generally be blank. During these latter periods there will probably be more variability in the forms of life, while during periods of subsidence there should be more extinction.

With respect to the absence of strata rich in fossils beneath the Cambrian formation, I can recur only to the hypothesis given in Chapter 10: namely, that although our continents and oceans have endured for an enormous period in nearly their present relative positions, we have no reason to assume that this has always been the case.[4]

---

[4] Indeed, as noted earlier, we now know about plate tectonics and its impact on continental drift.

Consequently, formations much older than any that we now know about may lie buried beneath the great oceans.

With respect to the lapse of time not having been sufficient since our planet was consolidated for the assumed amount of organic change that my theory proposes to have occurred, a difficulty strongly urged by Sir William Thompson, well, this is probably one of the gravest objections as yet advanced. In response, I can only say, first, that we do not know at what rate species change, as measured by years, and second, that many philosophers are not yet willing to admit that we know enough of the constitution of the universe and of the interior of our globe to speculate with safety on its past duration.[5]

That the geological record is imperfect all will admit, but few will be inclined to admit that it is imperfect to the degree required by our theory. If we look to long enough intervals of time, geology plainly declares that species have all changed and that they have in fact changed in the manner required by the theory, for they have changed slowly and in a graduated manner. We clearly see this in the fossil remains from consecutive geological formations invariably being much more closely related to each other than are the fossils found in widely separated formations.

Such is the sum of the several chief objections and difficulties which may justly be brought against the theory of evolution by natural selection; and I have now briefly recapitulated the answers and explanations which, as far as I can see, may be given. I have felt these difficulties far too heavily during many years to doubt their weight. But it deserves special notice that the more important objections relate to questions on which we are confessedly ignorant—nor do we even know how ignorant we are! We do not know all the possible transitional gradations between the simplest and the most perfect organs. And it cannot be pretended that we know all the varied means by which species have become more widely distributed during the long lapse of years, or that we know how imperfect the geological record is at present. As serious as these several objections are, in my judgment they are by no means sufficient to overthrow the theory of descent with subsequent modification.

Now let us turn to the other side of the argument and consider the observations and logical arguments that support the theory. Under domestication we see much variability in physical characteristics—variability that is governed by many complex but poorly understood laws including correlated growth, compensation, the increased use and disuse of parts, and the definite action of the surrounding conditions.[6] There is much difficulty in ascertaining just how largely our domestic productions have been modified over time, but we may safely infer that the amount has been

[5] We now know that the planet Earth is approximately 4.5 billion years old; the first multicellular life that we know of (cyanobacteria-like organisms) lived about 3 billion years ago, and the oldest animal fossils found so far, discovered in glacial deposits in southern Australia, are about 635 million years old. Darwin would have been very excited to have learned about this in his lifetime.

[6] Remember, Darwin knew nothing about the genetic basis of variation or inheritance.

large and that the modifications can be passed along to future generations for long periods. As long as the conditions of life remain the same, we have reason to believe that any modification that has already been inherited for many generations may continue to be inherited for an almost infinite number of additional generations. On the other hand, we have evidence that variability, when it has once come into play, does not cease under domestication for a very long period; nor do we know in fact that it ever ceases, for new varieties are still occasionally produced by our oldest domesticated productions.

Variability is not actually caused by people. We only unintentionally expose organic beings to new conditions of life; nature then acts on the organization and somehow causes it to vary. **But we can (and in fact do) select the variations given to us by nature, and thus we can accumulate them in any desired manner over a great many generations.** We have thus gradually adapted particular animals and plants for our own benefit or pleasure. Breeders may do this methodically, or they may do it unconsciously by preserving the individuals most useful or pleasing to them without any intention of altering the breed. It is certain that breeders can largely influence the character of a breed by selecting, in each successive generation, individual differences so slight as to be inappreciable except by an educated eye. This unconscious process of selection has been the great agency in forming the most distinct and useful of our domestic breeds. That many breeds produced by us have to a large extent the character of natural species, is shown by the inextricable doubts whether many of them are simply varieties or are aboriginally distinct species.

There is no reason why the principles of selection that have acted so efficiently under domestication should not have also acted in nature. In the survival of favored individuals and races during the constantly recurrent struggle for existence, we see a powerful and ever-acting form of selection. The struggle for existence inevitably follows from the high geometrical ratio (2 individuals, 4 individuals, 8, 16, 32, 64, 128 individuals, etc.) of increase that is common to all organisms, both plant and animal. This high rate of increase is proven by calculation and by observation: by the rapid increase of many animals and plants during a succession of peculiar seasons and when firmly established in new countries. **Thus more individuals are born than can possibly survive.** A grain in the balance may determine which individuals shall live and which shall die, and which variety or species shall increase in number and which shall decrease—or finally become extinct. The struggle for existence will generally be most severe among individuals of the same species, as those individuals will come in all respects into the closest competition with each other; it will be almost equally severe between varieties of the same species, and next in severity between members of different species within the same genus.

The struggle can also often be severe between beings remote in the scale of nature. The slightest advantage in certain individuals, at any age or during any season, over those with which they come into competition, or a better adaptation in however slight a degree to surrounding physical conditions, will, in the long run, turn the balance in favor of some individuals over others.

With animals having separated sexes, there will in most cases be a struggle between the males for the possession of the females. The most vigorous males, or those which have most successfully struggled with their conditions of life, will generally leave the most offspring. But success will often depend on the males having special weapons or other means of defense, or particularly compelling charms to attract and subdue females; even a slight advantage will lead to victory for the bearer.

As geology plainly proclaims that each area of land has undergone great physical changes over long periods of time, we might have expected to find that plants and animals have varied under nature in the same way as they have varied under domestication[7]. And if there *has* been any variability under nature, it would be an unaccountable fact if natural selection had not come into play in dealing with that variation. It has often been asserted—although the assertion is incapable of being proven—that the amount of variation under nature is a strictly limited quantity. Man, however, though acting on external characters alone and often doing so capriciously, can produce a great result within a short period by adding up mere individual differences in his domestic productions over the generations. And everyone admits that all species present clear differences among individuals. But, besides such differences, all naturalists admit that *natural* varieties also exist within species, varieties that are considered sufficiently distinct to be worth recording in systematic works. No one has drawn any clear distinction between the differences seen among individuals and slight varieties within a species and those seen between more plainly marked varieties and subspecies and species. On separate continents, and on different parts of the same continent, when divided by barriers of any kind, and on outlying islands, what a multitude of forms exist! Some experienced naturalists rank them as varieties, others as geographical races or subspecies, and others as distinct, though closely related, species! The **distinction between individuals, varieties, and species is thus unclear, something that is easily explained only by our theory of gradual evolution from a common ancestor.**

If, then, animals and plants do vary, let it be ever so slightly or slowly, why should not those variations or individual differences that are in any way beneficial be preserved and accumulate through natural selection (i.e., the survival of the fittest)? If we can by patience select variations that are useful to us, then why, under changing and complex conditions of life, should not variations useful to nature's living productions often arise and be preserved or selected? What limit can be put to this power, acting during long ages and rigidly scrutinizing the whole constitution, structure, and habits of each creature, favoring the good and rejecting the bad? I can see no limit to this power in slowly and beautifully adapting each form to the most complex relations of life. The theory of natural selection, even if we look no further than this, seems to be in the highest degree probable. **I have already recapitulated, as fairly as I could, the**

---

[7] As noted earlier, it was believed at the time that variation was somehow caused by changes in climate and other physical conditions. The genetic basis of variation was not yet understood.

opposed difficulties and objections and my responses to those objections: now let us turn to the special facts and arguments in favor of the theory.

On the view that species are only strongly marked and permanent varieties and that each species first existed as a variety of an earlier ancestral species, we can see why it is that no clear line of demarcation can be drawn between species (which are commonly supposed to have been produced by special acts of creation) and varieties (which are acknowledged to have been produced by secondary laws). On this same view we can now understand how it is that in a region where many species of a genus have been produced—and where they now flourish—these same species should present many varieties: for where the *formation* of new species has been active, we might expect, as a general rule, to find it *still* active, and this is the case if varieties are indeed incipient species. Moreover, the species of the larger genera—which afford the greater number of varieties (e.g., incipient species)—themselves retain to a certain degree the character of varieties, differing from each other less than do the species of smaller genera. The closely allied species within a larger genus apparently have restricted ranges, and in their affinities they are clustered in little groups around other species—in both respects resembling varieties. These are strange relations if we believe that each species was independently and specially created, but they are fully intelligible if each existed first as a variety of some ancestral species.

As each species tends by its geometrical rate of reproduction to increase inordinately in numbers over time, and as the modified descendants of each species will be enabled to increase by as much as they become more diversified in habits and structure so as to be able to seize on many and widely different environmental niches, there will be a constant tendency for natural selection to preserve the most adaptively divergent offspring of any one species. **Hence during a long-continued course of modification, the slight differences that characterize varieties of any particular species tend to become gradually augmented into the greater differences that characterize species within the same genus.** Over time, new and improved varieties will inevitably supplant and exterminate the older, less improved and intermediate varieties; in this way, species eventually become to a large extent well-defined and distinct objects.

Dominant species belonging to the larger groups within each class tend to give rise to new and dominant forms; each large group therefore tends to become still larger over time and more divergent in character. But as all groups cannot thus go on increasing in size, for the world would not hold them, the more dominant groups eventually beat the less dominant. This tendency for the large groups to go on increasing in size and diverging in character, together with the inevitable contingency of much extinction, explains very nicely the arrangement of the classification system that we now have, with all the forms of life being placed in groups that are contained within other groups, all within a few great classes, which has prevailed throughout all time. **This grand fact of the grouping of all organic beings under what is called the Natural System of Classification, is utterly inexplicable on the theory of "special creation."**

As natural selection acts solely by accumulating slight, successive, favorable variations over long periods of time, it can produce no great or sudden modifications: it can act only by small and slow steps. Hence, the canon of "*Natura non facit saltum*,"[8] which every fresh addition to our knowledge tends to confirm, is completely intelligible on this theory. And we can see why, throughout nature, the same general end is gained by an almost infinite diversity of means—for once acquired, every useful peculiarity is long inherited by generation after generation of offspring, and structures already modified in many different ways have to become adapted for the same general purpose. We can, in short, see why nature is prodigal in variety though stingy with innovation. But why this would be a law of nature if each species had been independently created by a "creator," no one can explain.

Many other facts are, it seems to me, explicable on this theory. How strange it is that any bird with the form of a woodpecker should prey on insects on the ground; that upland geese, which rarely or never swim, would possess webbed feet; that a thrushlike bird should dive and feed on subaquatic insects; and that a petrel should have the habits and structure fitting it for the life of an auk! And so on, in endless other cases. But on the view that each species is constantly trying to increase in number, with natural selection always ready to adapt the slowly varying descendants of each to any unoccupied or ill-occupied niche in nature, these facts cease to be strange; they might even have been anticipated.

We can to a certain extent understand how it is that there is so much beauty throughout nature: this may be largely attributed to the agency of selection. Sexual selection has given the most brilliant colors, elegant patterns, and other ornaments to the males and sometimes to both sexes of many birds, butterflies, and other animals. With birds it has often rendered the voice of the male musical to the female as well as to our ears. Flowers and fruit have been rendered conspicuous by brilliant colors in contrast with the green foliage in order that the flowers may be easily seen, visited, and fertilized by insects and so that the seeds will be disseminated by birds. How it comes that certain colors, sounds, and forms should also give pleasure to man and the lower animals—that is, how the sense of beauty in its simplest form was first acquired—we do not know any more than we know how certain odors and flavors were first rendered agreeable. To us, however, beauty is not universal—something that must be admitted by everyone who will look at some venomous snakes, at some fishes, and at certain hideous bats with a distorted resemblance to the human face.

Because natural selection acts through competition, it adapts and improves the inhabitants of each country only in relation to their co-inhabitants. Thus we need feel no surprise at the species of any one country—species supposed on the ordinary view to have been created and specially adapted for that country—being beaten and supplanted by the naturalized productions from another land. Nor should we marvel if all the contrivances in nature be not, as far as we can judge, absolutely perfect, as in the case even of the human eye, or if some of them be abhorrent to our ideas of fitness.

---

[8] "Nature does not make leaps."

We need not marvel that the sting of the bee, when used against its enemy, causes the bee's own death; at drones being produced in such great numbers for one single act of mating and then being slaughtered by their sterile sisters; at the astonishing waste of pollen by our fir trees; at the instinctive hatred of the queen bee for her own fertile daughters; at ichneumonid wasps[9] feeding within the living bodies of caterpillars; and at other such cases. On the theory of natural selection, the wonder, indeed, is that more cases lacking in absolute perfection have not been detected.

The complex and little-known laws that govern the production of varieties are the same, as far as we can judge, as those which have governed the production of distinct species. Correlated variation seems to have played an important part in shaping both varieties and species, so that when one part has been modified, other parts have been necessarily modified as well. With both varieties and species, reversions to long-lost characters occasionally occur. How inexplicable on the theory of creation is the occasional appearance of stripes on the shoulders and legs of the various species in the horse genus *Equus* and of their hybrids! But how simply is this fact explained if we believe that these species are all descended from a striped ancestor, in the same manner as the several domestic breeds of the pigeon are descended from the blue and barred rock-pigeon.

On the common view that each species has been independently created as we see them today, why should specific characters, or those by which the species of the same genus differ from each other, be more variable than the generic characters in which they all agree? Why, for instance, should a flower's color be more likely to vary within any one species of a genus if the other species possess differently colored flowers than if all possessed the same colored flowers? But if species are only well-marked varieties, of which the characters have become to a high degree permanent, we can understand this fact; for they have already varied since they branched off from a common ancestor in certain characters, by which they have come to be specifically distinct from each other. These same characters, then, would be more likely again to vary than the generic characters that have been inherited without change for an immense period.

Similarly, the theory of special creation cannot explain why a part developed in a very unusual manner in just one species within a large genus, and that therefore, as we may naturally infer, is of great importance to that species, should be eminently liable to variation. On our view, this part has undergone an unusual amount of variability and modification since the several species branched off from a common ancestor. We might therefore expect the part generally to be still variable. But a part may be developed in the most unusual manner—like the wing of a bat—and yet not be more variable than any other structure if that part is common to many subordinate forms (i.e., if it has been inherited for a very long period). For in this case it will have been rendered constant by long-continued natural selection.

---

[9] The family Ichneumonidae is found within the insect order Hymenoptera.

Instincts, marvelous as some are, offer no greater difficulty than do physical characteristics on the theory of the natural selection of successive, slight, but profitable modifications. We can thus understand why nature moves by graduated steps in endowing different animals of the same class with their several instincts. I have attempted to show how much light the principle of gradation throws on the admirable architectural powers of the hive-bee. Habit no doubt often comes into play in modifying instincts, but it certainly is not indispensable, as we see in the case of neuter insects, which leave no offspring to inherit the effects of long-continued habit. If we accept that all the species of the same genus have descended from a common parent and have inherited much in common, we can understand how it is that related species, when placed under widely different conditions of life, follow nearly the same instincts. We can understand, for example, why the thrushes of tropical and temperate South America line their nests with mud just as our British species do. In addition, on the view of instincts having been slowly acquired through natural selection, we need not marvel at some instincts not being perfect and being liable to mistakes, and at many instincts causing other animals to suffer.

If species are only well-marked and permanent varieties, we can at once see why their crossed offspring should follow the same complex laws in their degrees and kinds of resemblance to their parents as we see in the crossed offspring of acknowledge varieties, in being absorbed into each other by successive crosses and in other such points. This similarity would be a strange fact if species had been independently created and varieties had been produced through secondary laws.

If we admit that the geological record is imperfect to an extreme degree, then those facts that the record does give strongly support the theory of descent with modification. The record also shows that new species have come onto the stage only slowly and at successive intervals, and that the amount of change after equal intervals of time differs widely in different groups. The extinction of species and of whole groups of species, which has played so conspicuous a part in the history of the living world, almost inevitably follows from the principle of natural selection: old forms are inevitably supplanted by new and improved forms. Moreover, neither single species nor groups of species ever reappear when the chain of ordinary generation is once broken. The gradual spread of dominant forms into new areas, with the slow modification of their descendants, causes the forms of life, after long intervals of time, to appear as if they had changed simultaneously throughout the world. The fact that the fossil remains found within each geological formation are in some degree intermediate in character between the fossils found in the formations above and below them is simply explained by their intermediate position in the chain of descent. And the grand fact that all extinct beings can be classed with all recent beings naturally follows from the idea that the living and the extinct are both the offspring of common parents. And as species have generally diverged in character during their long course of descent and modification, we can understand why it is that the more ancient forms, or early ancestors of each group, so often occupy a position in some degree intermediate between existing groups.

Recent forms are generally looked upon as being, on the whole, higher in the scale of organization than ancient forms. Indeed, they must be higher, in so far as the later and more improved forms have conquered the older and less improved forms in the struggle for life; they have also generally had their organs more specialized for different functions. This fact is perfectly compatible with numerous beings still retaining simple and but little improved structures, fitted for simple conditions of life; it is likewise compatible with some forms having regressed in organization by having become at each stage of descent better fitted for new and degraded habits of life.

Last, the wonderful law of the long endurance of related forms on the same continent—of marsupials in Australia; of anteaters, tree sloths, and armadillos in America; and other such cases—is completely intelligible and well-explained by descent with modification: within the same country, the existing and the extinct will be closely allied by descent from a common ancestor.

Looking to geographical distribution, if we admit that there has been during the long course of ages much migration from one part of the world to another, owing to former changes in climate and geography and to the many occasional and unknown means of dispersal, then we can understand, on the theory of descent with modification, most of the great leading facts in the worldwide distribution of organisms on our planet. We can see why there should be so striking a parallelism between the distribution of organisms throughout space and their geological succession throughout time for, in both cases, the organisms have been connected by the bond of ordinary generation, and the means of modification have been the same. We see the full meaning of the wonderful fact that has so clearly struck every observant traveler: namely, that on the same continent, under the most diverse conditions—under heat and under cold, on mountain and lowland, on deserts and marshes—most of the inhabitants within each great class of organisms are plainly related to each other; they are related because they are the descendants of the same ancestors and early colonists.

On this same principle of former migration—combined in most cases with modification—we can understand, by the aid of the glacial period, the great similarity of some few plants (and the close alliance of many others) on the most distant mountains and in the northern and southern temperate zones and, likewise, the close relationship between some of the inhabitants of the sea in the northern and southern temperate latitudes even though they are separated by the whole intertropical ocean. Although two countries may present physical conditions as closely similar as the same species ever require, we need feel no surprise at finding their inhabitants to be widely different if they have been for a long period completely sundered from each other for, as the relation of organism to organism is the most important of all relations, and as the two countries will have received colonists at various periods and in different proportions from some other country or from each other, the course of gradual modification in the two areas will inevitably have been different.

On this view of migration with subsequent gradual modification, we see why oceanic islands are inhabited by only a few species, but, of these, why many are peculiar or endemic forms—found there and nowhere else. We clearly see why species belonging to those groups of animals that cannot cross wide spaces of the ocean (e.g., frogs and terrestrial mammals, for example) do not inhabit oceanic islands, and why, on the other hand, new and peculiar species of bats—animals that *can* traverse the ocean—are often found on islands far distant from any continent. Such cases as the presence of peculiar species of bats on oceanic islands and the absence of all other terrestrial mammals are utterly inexplicable on the theory of independent acts of creation.

The existence of closely related representative species living in any two separated areas implies, on the theory of descent with modification, that the same parent forms formerly inhabited both areas. Indeed, we almost always find that wherever many closely related species are found inhabiting two distinct areas, some identical species are still found in both. Wherever many closely related yet distinct species occur together in one place, we also find doubtful forms and varieties belonging to the same groups. Indeed, it is a very general rule that the inhabitants of each area are clearly related to the inhabitants of the nearest source from which immigrants might have come. We see this in the strikingly close relationship between nearly all the plants and animals of the Galapagos Archipelago, of the Juan Fernández Islands[10] in the Eastern Pacific Ocean, and of the other American islands to the plants and animals of the neighboring American mainland. Similarly close relationships are seen between species found in the Cape de Verde Archipelago and the other African islands to those living on the African mainland. These facts cannot be explained by the theory of special creation—but they make perfect sense on our theory of descent with modification.

We have seen that all past and present living beings can be arranged within a few great classes, in groups within groups and with the extinct groups often falling in between the recent groups. This fact makes perfect sense assuming the theory of natural selection to be true, with its contingencies of extinction and divergence of character. On these same principles we see how it is that the mutual relationships between the forms within each class are so complex and circuitous. We also see why certain characters are far more useful than others for the purpose of classification; why adaptive characters, though of paramount importance to the beings that own them, are of hardly any importance in classification; why characters derived from rudimentary parts, though of no service to their owners, are often of high value for classification; and why embryological characters are often the most valuable of all. The real relationships of all organic beings, in contradistinction to their adaptive resemblances, are due to inheritance: to their "community of descent." Our "Natural System of Classification" is a genealogical arrangement, with the acquired grades of difference marked by the terms varieties, species, genera, families, etc.; we have to discover the lines of descent by examining the most permanent characters, whatever they may be, without regard to their functional importance in the lives of the organisms under study.

---

[10] Juan Fernández was a sixteenth-century Spanish explorer.

The similar framework that we see of bones in the human hand, the wing of a bat, the fin of the porpoise, and the leg of the horse; the fact that we see the same number of vertebrae forming the neck of the giraffe and that of the elephant; along with innumerable other such facts, at once explain themselves on the theory of descent with slow and slight successive modifications over time. The similar pattern of bones that we find in the wing and in the leg of a bat, even though those limbs are used for such different purposes; the similar patterns in the jaws and legs of a crab; the similar patterns in the petals, stamens, and pistils of a flower—all are likewise largely intelligible on the view of the gradual modification of parts or organs that were originally the same in an early ancestor in each of these groups.

On the principle of successive variations not always supervening at an early age and being inherited at a corresponding not early period of life, we clearly see why the embryos of mammals, birds, reptiles, and fishes should be so very similar in appearance and be so unlike the adult forms. And we may cease marveling at the embryo of an air-breathing mammal or bird having branchial slits (i.e., gills) and arteries running in loops, like those of a fish that has to obtain oxygen dissolved in water using well-developed branchiae: those surprising features reflect the morphological characteristics of an ancient ancestor.

Disuse of parts, aided sometimes by natural selection, will often have reduced their size and development so that they have become useless under changed habits or environmental conditions: this fully explains the existence of rudimentary organs. But disuse and subsequent selection will generally act on a creature only after it has come to maturity and has to play its full part in the struggle for existence and will thus have little power on an organ during early life; hence the organ will not be reduced or become rudimentary at this early age. A calf, for instance, has inherited teeth that never cut through the gums of the upper jaw; they must have inherited that trait from an early ancestor that had well-developed teeth. The teeth in the mature animal were possibly reduced long ago through disuse, owing to the tongue and palate, or the lips, having become fully capable, through natural selection, of browsing without their aid. In the calf, however, the teeth have been left unaffected by selection or disuse, and, on the principle of inheritance at corresponding ages, those useless teeth have simply been inherited from a remote period to the present day. On the view of each organism with all its separate parts having been specially created, how utterly inexplicable is it that organs that are plainly useless—such as the teeth in the embryonic calf or the shriveled wings that we find under the soldered wing-covers of many beetles—should be found so frequently. Nature may be said to have taken pains to reveal her scheme of modification to us by means of such rudimentary organs and of embryological and homologous structures; it's just that we have been too blind to understand her meaning.

I have now recapitulated the facts and considerations that have thoroughly convinced me that species have been modified during a long course of descent. This has been effected chiefly through the natural selection of numerous successive, slight, favorable variations over very long periods of time, aided in an important manner by the inherited effects of the use and disuse of parts and in a relatively unimportant manner by the direct action of external conditions and by variations which seem to us in our ignorance to arise spontaneously. It appears that I formerly underrated the frequency and value of these latter forms of variation as leading to permanent modifications of structure independently of natural selection. But as my conclusions have lately been much misrepresented, and as it has been stated that I attribute the modification of species exclusively to natural selection, I may be permitted to remark that in the first edition of this work, and subsequently, I placed in a most conspicuous position—namely, at the close of the Introduction—the following words: "I am convinced that natural selection has been the main but not the exclusive means of modification." This has been of no avail. Great is the power of steady misrepresentation. But the history of science shows that, fortunately, this power does not long endure.

It can hardly be supposed that a false theory would explain, in so satisfactory a manner as does the theory of natural selection, the several large classes of facts that I have detailed above. It has recently been objected that this is an unreliable method of arguing; but it is a method commonly used in judging the common events of life and one that has often been used by the greatest natural philosophers. The undulatory theory of light has thus been arrived at, for example, and the belief that the Earth revolves on its own axis was until lately supported by hardly any direct evidence. It is no valid objection that science has so far thrown no light on the far higher problem of the essence of or the origin of life. Who can explain what is "the essence" of the attraction of gravity? No one now objects to following out the results that stem from this unknown element of attraction, even though Leibnitz formerly accused Newton of introducing "occult qualities and miracles into philosophy."

I see no good reasons why the views given in this volume should shock the religious feelings of anyone. It is satisfactory, as showing how transient such impressions are, to remember that the greatest discovery that humans have ever made, namely, the law of the attraction of gravity, was also attacked by Leibnitz, "as subversive of natural, and inferentially of revealed, religion." A celebrated author and divine has written to me that "he has gradually learned to see that it is just as noble a conception of the Deity to believe that He created a few original forms capable of self-development into other and needful forms, as to believe that He required a fresh act of creation to supply the voids caused by the action of His laws."

Why, it may be asked, until recently did nearly all the most eminent living naturalists and geologists not believe that species could change over time? It certainly cannot be claimed that organisms in nature do not show variation; it cannot be proven that there has been only a limited amount of variation in the course of long ages; and no clear distinction has been drawn—or can be drawn—between species and well-marked varieties.[11] As noted earlier, it cannot be maintained that species

---

[11] We now generally define species by their reproductive isolation from other species. But hybridization between the members of recognized species does sometimes occur in nature, fitting in very nicely with Darwin's arguments.

when intercrossed are invariably sterile or that varieties when crossed are invariably fertile, or that sterility is a special endowment and sign of special creation. The belief that species were unchangeable was almost unavoidable as long as the history of the world was thought to be of short duration. And now that we have acquired some idea of the remarkably large lapse of time, we are too apt to assume, without proof, that the geological record is so perfect that it would have afforded us plain evidence of the changeability of species if they had undergone such changes.

But the chief cause of our natural unwillingness to admit that one species has been transformed into different and distinct species is that we are always slow in admitting any great changes of which we do not see the steps. The difficulty is the same as that felt by so many geologists when Lyell first insisted that long lines of inland cliffs had been formed and great valleys excavated by such agencies as erosion and uplift that we still see at work. The mind cannot possibly even grasp the full meaning of what "one million years" really means; it cannot add up and perceive the full effects of many slight variations that have been accumulated during an almost infinite number of generations over millions and millions of years.

Although I am fully convinced of the truth of the views given in this abstract of my ideas, I by no means expect to convince experienced naturalists whose minds are stocked with a multitude of facts all viewed, during a long course of years, from a point of view directly opposite to mine. It is so easy to hide our ignorance through expressions such as the "plan of creation" and "unity of design." But these are merely restatements of fact, not explanations of anything. Anyone whose disposition leads him to attach more weight to unexplained difficulties than to the explanation of a certain number of facts will certainly reject the theory. A few naturalists, endowed with much flexibility of mind, and who have already begun to doubt the immutability of species, may already be influenced by this book. But I look with confidence to the future, to young and rising naturalists who will be able to view both sides of the question with impartiality. Those who are led to believe that species are indeed changeable over time will do good service by conscientiously expressing their conviction; only in that way can the load of prejudice by which this subject is overwhelmed be removed.

Several eminent naturalists have recently published their belief that a multitude of reputed species in each genus are not actually separate, legitimate species after all, but merely variations of a species. But they also write that other species are real and have been independently created. This seems to me a strange conclusion to arrive at. They admit that a multitude of forms, which until lately they themselves thought were special creations and which are still thus looked at by most naturalists, and which consequently have all the external characteristic features of true species—they admit that these have been produced by variation, but they refuse to extend the same view to other and slightly different forms. Nevertheless, these naturalists do not pretend that they can define, or even conjecture, which are the created forms of life and which are those produced by secondary laws. They admit variation as a *vera causa* (a "true cause") in one case, and yet they arbitrarily reject it in another, without assigning any distinction in the two cases. The day will come when this will be given as a curious illustration of the blindness of preconceived opinion.

These authors seem no more startled at a miraculous act of creation than at an ordinary birth. But do they really believe that at innumerable periods in the Earth's history certain elemental atoms have been commanded suddenly to flash into living tissues? Do they really believe that, at each supposed act of creation, one individual or many were produced? Were all the infinitely numerous kinds of animals and plants created as eggs or seed, or were they brought into existence fully grown? And, in the case of mammals, were they created bearing the false marks of nourishment from the mother's womb?

Undoubtedly some of these same questions cannot be answered by those who believe in the appearance or creation of only a few forms of life or of some one form alone. Several authors have maintained that it is as easy to believe in the divine creation of a million beings as in the creation of but one. However, Pierre de Maupertuis's philosophical axiom "of least action" leads the mind more willingly to admit the smaller number, and certainly we ought not to believe that innumerable beings within each great class have been deliberately created with plain, but deceptive and misleading, marks of descent from a single parent.

As a record of a former state of things, I have retained in the foregoing paragraphs and elsewhere several sentences implying that naturalists believe in the separate creation of each species, and I have been much censured for having thus expressed myself. But undoubtedly this was the general belief when the first edition of the present work appeared, in 1859. I formerly spoke to very many naturalists on the subject of evolution, and never once met with any sympathetic agreement. It is probable that some did then believe in evolution, but they were either silent or expressed themselves so ambiguously that it was not easy to understand their meaning. Things are now wholly changed, and almost every naturalist admits the great principle of evolution. There are, however, some who still think that species have suddenly given birth, through quite unexplained means, to new and totally different forms. But, as I have attempted to show, weighty evidence can be opposed to the admission of great and abrupt modifications. Under a scientific point of view, little advantage is gained by believing that new forms are suddenly developed in an inexplicable manner from old and widely different forms, over the old belief in the creation of species from the dust of the earth. And the scientific view leads to further investigation.

It may be asked how far I extend the doctrine of the modification of species. The question is difficult to answer: the more distinct the forms are that we consider, the more the arguments in favor of community of descent become fewer in number and less in force. But some arguments of the greatest weight extend very far. All the members contained within whole classes of organisms are connected together by a chain of affinities, and all can be classed on the same principle, in groups within groups. Moreover, fossil remains sometimes fill up very wide intervals of forms between existing orders.

Organs in a rudimentary condition plainly show that an early ancestor must have had the organ in a fully developed condition, and this in some cases implies an enormous amount of modification in the descendants. Throughout whole classes of organisms, various structures are formed on the same pattern, and, at a very early age, the embryos of those organisms closely resemble each other. Therefore I cannot doubt that the theory of descent with modification embraces all members belonging to the same great class or kingdom. I believe that all animals are descended from at most only four or five ancestors, and all plants from an equal or lesser number.

Analogy would lead me one step further: namely, to the belief that all animals and plants are descended from some one prototype. But analogy may be a deceitful guide. Nevertheless, all living things have much in common: in their chemical composition, their cellular structure, their laws of growth, and their liability to injurious influences. We see this even in so trifling a fact as that the same poison often similarly affects plants and animals; or that the poison secreted by the gall-fly produces monstrous growths on the wild rose or oak tree.

With all organic beings—except perhaps some of the very lowest—sexual reproduction seems to be essentially similar. With all, as far as we presently known, the germinal vesicle is the same, suggesting that all organisms start from a common origin. If we look even to the two main divisions of living organisms—namely, to the animal and vegetable kingdoms—certain low forms are so far intermediate in character that naturalists have disputed to which of the two kingdoms they belong. As Harvard University's Professor Asa Gray has remarked, "the spores and other reproductive bodies of many of the lower algae may claim to have first a characteristically animal, and then an unequivocally vegetable existence." Therefore, on the principle of natural selection with slow divergence of character, it does not seem incredible that, from some such low and intermediate form, both animals and plants may have been gradually developed; and, if we admit that, then we must likewise admit that all of the organisms that have ever lived on this earth may be descended from some one primordial form.

But this inference is chiefly grounded on analogy, and it is immaterial whether or not it be accepted. No doubt it is possible, as Mr. G. H. Lewes has urged, that at the first commencement of life many different forms were evolved; but if so, we may conclude that only a very few have left modified descendants. For, as I have recently remarked in regard to the members of each great kingdom, including our own group, the Vertebrata, we have distinct evidence in their embryological, homologous, and rudimentary structures that, within each kingdom, all the members are descended from a single ancestor.

When the views advanced here by me and by Mr. Alfred Wallace are accepted, or when analogous views on the origin of species are generally admitted, we can dimly foresee that there will be a considerable revolution in the study and understanding of natural history. Systematists will be able to pursue their work as they do at present, but they will not be incessantly haunted by the shadowy doubt whether this or that form is or is not a true species. This, I feel sure—and I speak from experience—will be no slight relief. The endless disputes about whether or not some 50 species of British blackberry plant are good species will cease. Systematists will have only to decide (not that this will be easy) whether or not any particular form is sufficiently constant and distinct from other forms to be capable of definition; and, if definable, whether the differences are sufficiently important to deserve a specific name. This latter point will become a far more essential consideration than it is at present, for differences between any two forms, however slight, if not blended by intermediate gradations, are now looked at by most naturalists as sufficient to raise both forms to the rank of species.

Hereafter we shall be compelled to acknowledge that the only valid distinction between species and well-marked varieties is that varieties are known to be—or at least are believed to be—presently connected by intermediate gradations, whereas species were only formerly thus connected. Thus, although continuing to consider the present existence of intermediate gradations between any two forms, we shall be led to weigh more carefully and to value more highly the actual amount of difference between them. It is quite possible that forms now generally acknowledged to be merely varieties may hereafter be thought worthy of status as separate species; and, in this case, scientific and common language will come into accordance. In short, we shall have to treat species in the same manner as those naturalists treat genera: by admitting that genera are merely artificial combinations of species made for convenience. This may not be a cheering prospect, but we shall at least be freed from the vain search for the undiscovered and undiscoverable essence of the term "species."

The other and more general departments of natural history will rise greatly in interest. The terms used by naturalists—such as affinity, relationship, community of type, paternity, morphology, adaptive characters, rudimentary and aborted organs—will cease to be metaphorical and will have a plain significance. When we no longer look at an organism as a savage looks at a naval ship, as something wholly beyond his comprehension; when we regard every naturally occurring plant or animal as one that has had an extremely long history; when we contemplate every complex structure and instinct as the summing up of many contrivances, each useful to the possessor, in the same way as any great mechanical invention is the summing up of the labor, the experience, the reason, and even the blunders of numerous workmen; when we thus view each organic being, how far more interesting—and I speak here from experience—does the study of natural history become!

A grand and almost untrodden field of inquiry will be opened: on the causes and laws of variation, on correlation, on the effects of use and disuse, on the direct action of external conditions, and so forth. The study of our domestic productions will rise immensely in value. A new variety raised by humans will be a far more important and interesting subject for study than one more species added to the infinitude of already recorded species. Our classifications will come to be, as far as they can be so made, genealogies, and will then truly give what may be called "the plan of creation." The rules for classifying will no doubt become simpler when we have a definite object in view. We possess no pedigree or heraldic coat of arms showing definitive ancestries and relationships for the organic beings on our planet; rather, we have to discover and trace the many diverging lines of descent in our natural genealogies using characters of any kind that have long been inherited. Rudimentary organs will speak infallibly with respect to the nature of long-lost structures. Species and groups of species that are now called aberrant and that may fancifully be called living fossils will aid us in forming a picture of the ancient forms of life. Embryology will often reveal to us the structure, in some degree obscured, of the prototypes of each great class.

When we can feel assured that all the individuals of a given species, and all the closely allied species of most genera, have, over many generations, descended from one ancient, ancestral parent and have migrated to their present locations from some one ancient birthplace and when we better know the many means of migration, then, by the light that geology now throws—and will continue to throw—on former changes of climate and of the level of the land, we shall surely be enabled to trace in an admirable manner the former migrations of the inhabitants of the whole world. Even at present, by comparing the differences between marine organisms found on the opposite sides of a continent and by comparing the nature of the various inhabitants of that continent in relation to their apparent means of immigration, some light can be thrown on ancient geography.

The noble science of geology loses glory from the extreme imperfection of the fossil record. The crust of the Earth, with its embedded remains, must not be looked at as a well-filled museum, but rather as a poor collection made at random and at rare intervals. The accumulation of each great fossiliferous formation will be recognized as having been made possible by an unusual occurrence of favorable circumstances and the blank intervals between the successive stages that we see before us will be seen as having been of vast duration. But we shall be able to gauge with some confidence the durations of these blank intervals by a comparison of the preceding and succeeding preserved forms.

We must be cautious in attempting to correlate as strictly contemporaneous two geological formations that do not include many identical species, by the general succession of the forms of life. As species are produced and are exterminated by slowly acting and still existing causes, and not by miraculous acts of creation, and as the most important of all causes of organic change is one that is almost independent of altered and perhaps suddenly altered physical conditions—namely, the mutual interactions between organisms, with the improvement of one organism entailing the improvement or the extermination of others—it follows that the amount of organic change in the fossils of consecutive geological formations probably serves as a fair measure of the relative, though not actual lapse of time. A number of species, however, remaining together in their ancestral home might remain for a long period unchanged, while, within the same period, several of these species, by migrating into new geographical areas and coming into competition with foreign associates, might become modified; thus we must not overrate the accuracy of organic change as a measure of time.

In the future I see open fields for far more important researches. Psychology will be securely based on the foundation already well laid by Mr. Herbert Spencer: that of the necessary acquirement of each mental power and capacity by gradual change. And much light will be thrown on the origin of man and his history.

Authors of the highest eminence seem to be fully satisfied with the view that each species has been independently created by a divine Creator. To my mind it agrees better with what we know of the laws impressed on matter by this Creator: that the production and extinction of the past and present inhabitants of the world should have been due to secondary causes, like those determining the birth and death of the individual. When I view all beings not as special creations, but rather as the lineal descendants of some few beings that lived long before the first geological bed of the Cambrian system was deposited, they seem to me to become ennobled. Judging from the past, we may safely infer that not one living species will transmit its unaltered likeness to a distinct future. And of the species now living, very few will transmit offspring of any kind to a far distant future. For the manner in which organisms are grouped shows that the greater number of species in each genus, and all the species in many genera, have left no descendants but have instead become utterly extinct. We can so far take a prophetic glance into the future as to foretell that it will be the common and widely spread species belonging to the larger and dominant groups within each class that will ultimately prevail and that will procreate new and dominant species. As all the living forms of life are the lineal descendants of those that lived long before the Cambrian epoch, we may feel certain that the ordinary succession by generation has never once been broken and that no cataclysm has ever completely desolated the whole world. Hence, we may look with some confidence to a secure future of great length. And, as natural selection works solely by and for the good of each being, all corporeal and mental endowments will tend to progress toward perfection.

It is interesting to contemplate a tangled bank, clothed with many plants of many kinds, with birds singing on the bushes, with various insects flitting about, and with worms crawling through the damp earth, and to reflect that these elaborately constructed forms, so different from each other and yet so dependent upon each other in such a complex manner, have all been produced by laws still acting around us. These laws, taken in the largest sense, are Growth with reproduction; Inheritance, which is almost implied by reproduction; Variability from the indirect and direct action of the conditions of life, and from use and disuse; and a Ratio of Population Increase so high as to lead inevitably to a Struggle for Life and, as a consequence, to Natural Selection, entailing Divergence of Character and the Extinction of less improved forms. Thus, from the war of nature, from famine and death, the most exalted object that we are capable of conceiving, namely, the production of the higher animals, directly follows.

There is grandeur in this view of life, with its several powers having been originally breathed by the Creator into a few forms or into one; and that, while this planet has gone circling on according to the fixed law of gravity, from so simple a beginning, endless forms most beautiful and most wonderful have been, and are being, evolved.

Key Issues to Talk and Write About

1. Now that you have finished reading the entire book, make a list of all the studies that Darwin mentions that he conducted himself.
2. In one sentence each, what do you think are the three most interesting or important points that Darwin makes in this concluding chapter?

Other Books by Charles Darwin

(Drawing by Ardea Thurston-Shane)

1839

The Voyage of the Beagle
This is a lively record of Darwin's travels and scientific observations aboard *HMS Beagle*, one of the most important scientific voyages of the nineteenth century. He left England at the age of 22, in December 1831, for what was intended to be a two-year journey. He returned nearly 5 years later, in October 1836, having spent considerable time adventuring on land in such fascinating places as South America, the Galápagos Islands, the Falkland Islands, South Africa, Australia, and New Zealand.

1842

The Structure and Distribution of Coral Reefs
This is Darwin's original (and largely correct) thinking about how coral reefs and atolls come to be formed through the gradual sinking of islands. This was the first in his series of geology books. His theory wasn't confirmed until 1952.

1844

Geological Observations on the Volcanic Island Visited during the Voyage of the H.M.S. **Beagle**
This is the second of Darwin's geology books, and it includes geological observations that he made while visiting the volcanic island of Saint Jago in Cape Verde, Ascension Island, the island of Saint Helena, the Galápa-gos Islands, New Zealand, Australia, Van Diemen's Land, and the Cape of Good Hope.

1846

Geological Observations of South America
This is the third of Darwin's geology books, based on his explorations of South America on his journey aboard *HMS Beagle*.

1862

On the Various Contrivances by Which British and Foreign Orchids are Fertilised by Insects
Among a great many other things, Darwin was also a gifted and knowledgeable botanist, and this book proves it! Here he discusses the remarkable co-adaptations between plants and insects that result in cross-pollination and fertilization in plants. He introduces the idea of co-evolution and shows the many advantages of cross-fertilization. The book is based on many of his own observations, detailed dissections, and experiments.

1865

On the Movements and Habits of Climbing Plants
This book is again based on many of Darwin's own experiments, as he documents the climbing movements in a variety of different plant species and shows how these fascinating behaviors can be explained by natural selection acting on small inherited variations.

1868

The Variation of Animals and Plants Under Domestication
This book was originally a small part of what Darwin intended to be his monumental work on evolution by natural selection and the only part of that book to actually be published during his lifetime. It includes a fascinatingly detailed (but, we now know, misguided) discussion of his ideas ("pangenesis") about how variations might be caused and inherited.

1871

The Descent of Man and Selection in Relation to Sex
In this book Darwin considers how natural selection applies to human evolution, and he spends considerable time talking about sexual selection, which he discussed only briefly in *The Origin*.

1872

The Expression of the Emotions in Man and Animals
The title sums up the content very nicely. Here is my favorite quote from the book: "Blushing is the most peculiar and the most human of all expressions."

1875

Insectivorous Plants
In this book Darwin gives detailed information about the feeding activities of carnivorous plants, largely based on his own experiments and observations.

1876

The Various Contrivances by Which Orchids Are Fertilised by Insects
This is a remarkably detailed book about how the structures and behaviors of many different orchid species promote transfer of pollen among flowers to achieve cross-fertilization. Darwin argues that these details make sense only if they have arisen through natural selection.

1876

Recollections of the Development of My Mind and Character
This is an autobiography, written for his family. One of his sons later deleted certain sections before publication, but one of his grandchildren restored that material and republished the book in 1958 as *The Autobiography of Charles Darwin 1809–1882.*

1876

The Effects of Cross and Self Fertilisation in the Vegetable Kingdom
In this book Darwin reports the results of studies in which he monitored the growth of offspring from both cross-fertilizations and self-fertilizations in more than 60 plant species, clearly demonstrating the negative effects of inbreeding.

1877

The Different Forms of Flowers on Plants of the Same Species
Here is yet another book based on Darwin's detailed experiments with plants. In this case, he documents the outcomes of crosses between different flower types found within a number of individual plant species, showing that the various flower types take advantage of insect visitations in remarkable ways to maximize seed production and seedling vigor.

1881

The Formation of Vegetable Mould, Through the Actions of Worms, with Observations on Their Habits
This was Darwin's final book, based on about 40 years of detailed observations that he made at his home in Downe, of the feeding, physiology, burrowing behavior, and reproduction of earthworms and their remarkable roles in gradually shaping the landscape.

Darwin also wrote a number of monographs for specialists:

- 1851 *A Monograph of the Sub-class Cirripedia, with Figures of All the Species*
- 1851 *A Monograph on the Fossil Lepadida, or, Pedunculated Cirripedes of Great Britain*
- 1854 *A Monograph on the Fossil Balanidæ and Verrucidæ of Great Britain*
- 1854 *Living Cirripedia, The Balanidæ (or sessile cirripedes); the Verrucidæ*

People Referred to in These Chapters

Darwin was a remarkably active reader and correspondent. He wrote and received more than 15,000 letters during his life and corresponded with about 2,000 people all over the world. It's not surprising, then, that he refers to a great many authors and correspondents in *The Origin of Species*.

Name (year)	Description	Chapter reference
Agassiz, Jean Louis Rodolphe [Louis] (1807–1873)	American zoologist, paleontologist, and geologist; Harvard University	5, 7, 10, 12, 14
Aucapitaine, Baron Henri (1832–1867)	An associate member of the French Society of Natural Sciences (admitted at the age of 17!), and a member of the French Archaeological Society	13
Akbar the Great (1542–1605)	Emperor of India (ruled 1556– 1605)	1, 6, 8
Audubon, John James (1785–1851)	French-American ornithologist and illustrator	13
Babington, Charles Cardale (1808–1895)	English botanist and archaeologist	2
Baker, Samuel White, Sir (1821–1893)	English explorer and naturalist	7
Bakewell, Robert (1725–1795)	English sheep and cattle breeder, a pioneer in the field of artificial selection	1
Barrande, Joachim (1799–1883)	French geologist and paleontologist	10, 11
Bartlett, Edward (1836–1908)	English ornithologist and museum curator	7
Bates, Henry Walter (1825–1892)	English entomologist	8, 13
Bentham, George (1800–1884)	English botanist	2
Birch, Samuel (1813–1885)	English Egyptologist and archaeologist at the British Museum	1
Blyth, Edward (1810–1873)	English pharmacist, zoologist, and ornithologist, became a curator at the museum of the Royal Asiatic Society of Bengal in India	1, 5

Name (year)	Description	Chapter reference
Borrow, George Henry (1803–1881)	English traveler and writer	1
Boue, Ami (1794–1881)	Austrian geologist, born and educated in France	10
Brent, Bernard Peirce (1822–1867)	English bird enthusiast, especially pigeons	8
Broca, Pierre Paul (1824–1880)	French surgeon and anthropologist	7
Bronn, Heinrich Georg (1800–1862)	German geologist and paleontologist, translated *The Origin* into German	7, 11
Brown, Robert (1773–1858)	Scottish botanist and paleobotanist, made especially major contributions to his field with his microscope work	14
Brown-Séquard, Charles Édouard (1817–1894)	French physiologist and neurobiologist	5
de Buzareingues, Girou (1773–1856)	French plant physiologist and agronomist	9
Buckley, John (d. ca. 1787)	English sheep breeder	1
Burgess, Joseph (d. ca. 1807)	English sheep breeder	1, 10
Busk, George (1807–1886)	Russian-born naval surgeon and naturalist	1, 7
Clarke, William Branwhite (1798–1878)	English geologist and paleontologist, active in Australia	12
Claparède, Jean Louis René Antoine Edouard (1832–1871)	Swiss naturalist and invertebrate zoologist	6
Cope, Edward Drinker (1840–1897)	American paleontologist and comparative anatomist who believed in evolution but thought it was driven by changes in the timing of events in embryological development	6
Croll, James (1821–1890)	Scottish scientist who predicted multiple ice ages and developed a theory of climate change based on changes in the Earth's orbit	12
Crüger, Hermann (1818–1864)	German pharmacist and botanist	6
Cunningham, Robert Oliver (1841–1918)	Scottish naturalist	5
Cuvier, Frédéric (1773–1838)	French zoologist and anatomist, younger brother of Georges Cuvier	8, 11

Name (year)	Description	Chapter reference
Cuvier, Georges (1769–1832)	French zoologist, naturalist, historian of science, and politician who did not believe that organisms evolved	6, 8, 10, 11
Dana, James Dwight (1813–1895)	American geologist, Yale University	5
de Azara, Félix (1746–1821)	Spanish army officer, explorer, geographer, and naturalist	3, 6
Dawson, John (1820–1899)	Canadian geologist, a founder of the science of paleobotany	10
de Candolle, Alphonse (1806–1893)	Swiss botanist and politician, son of Augustin Pyramus de Candolle	2, 3, 4, 5, 6, 12, 13
de Candolle, Augustin Pyramus (1778–1841)	Swiss botanist and politician, father of Alphonse de Candolle	3, 5, 7
de Jussieu, Antoine Laurent (1748–1836)	French botanist	7, 14
de Lacépède, Bernard Germain (1756–1825)	French naturalist	7
de Quatrefages, Jean Louis Armand (1810–1892)	French doctor and professor of natural history	9
de Saussure, Henri Louis Frédéric (1829–1905)	Swiss naturalist, zoologist, geologist, physicist, and explorer	6
D'Orbigny, Alcide (1802–1857)	French zoologist, paleontologist, geologist, archaeologist, and anthropologist	10
Downing, Andrew Jackson (1815–1852)	American horticulturist, landscape designer, and author	4
Earle, Windsor (unknown)	English navigator who published (in 1845) an influential pamphlet documenting the physical geography and animal life of Australia and South-Eastern Asia	13
Edwards, W. W. (unknown)	Probably a horse-racing expert	5
Edwards, Henri Milne (1800–1885)	French zoologist, wrote books about insects, arachnids, mammals, and marine invertebrates	14
Eyton, Thomas Campbell (1809–1880)	English naturalist, working with cattle, fish, and birds	9
Fabre, Jean-Henri (1823–1915)	French entomologist and author	4, 8
Falconer, Hugh (1808–1865)	Scottish botanist and paleontologist	10
Flower, William Henry, Sir (1831–1899)	English comparative anatomist and zoologist	7, 14

Name (year)	Description	Chapter reference
Forbes, Edward (1815–1854)	English naturalist and geologist, professor of botany at King's College in London	6, 12, 13
Fries, Elias Magnus (1794–1878)	Swedish botanist and mycologist, director of the Uppsala University botanical garden	2
Ghuret, Gustave (1817–1875)	Well-known French botanist	9
Gosse, Philip Henry (1810–1888)	English zoologist, traveler, and writer	5, 8
Gould, John (1804–1881)	English ornithologist, artist, and taxidermist	8
Gray, Asa (1810–1888)	American botanist at Harvard University	4, 5, 6, 12
Günther, Albert (1830–1914)	German-born zoologist	7
Hearne, Samuel (1745–1792)	English explorer, author, and naturalist	6
Hensen, Christian Victor (1835–1924)	German physiologist and marine biologist	6
Heron, Robert, Sir (1765–1854)	English politician and breeder of unusual animals	4
Heusinger von Waldegg, Karl Friedrich (1792–1883)	Also known as Johann Friedrich Christian Karl Heusinger von Waldegg; German physician and pioneer in comparative pathology	1
Hicks, Henry (1837–1899)	Welsh physician and geologist	10
Hilgendorf, Franz (1839–1904)	German zoologist and paleontologist, with important studies on fossil	10
Hildebrand, Friedrich	German botanist, Bonn	4, 9
Hofmeister, Wilhelm Friedrich Benedikt (1824–1877)	German biologist and botanist, discovered alternation of generations in plants, and studied plant embryology	7
Hopkins, William (1793–1866)	English geologist	10
Godwin-Austen, Robert Albert Cloyne (1808–1884)	German botanist	13
Hooker, Joseph (1814–1879)	English botanist and explorer, and one of Darwin's close friends	2, 4, 5, 7, 12, 13
Huber, François (1750–1831)	Swiss entomologist, pioneered modern beekeeping, father of Pierre Huber	8

Name (year)	Description	Chapter reference
Huber, Pierre (1777–1840)	Swiss entomologist, son of François Huber	8
Hudson, William Henry (1841–1922)	South American (Buenas Aires) naturalist, ornithologist	6, 8
Hunter, John (1728–1793)	Scottish surgeon and anatomist	5
Hutton, Thomas (1807–1874)	English captain in the Bengal Army, wrote papers on the natural history and geology of India	8
Huxley, Thomas Henry (1825–1895)	English comparative anatomist, a leading promoter and defender of Darwin's ideas about evolution	4, 11, 14
James Hector (1834–1907)	Scottish—New Zealand geologist	12
Kirby, William (1759–1850)	English entomologist and Church of England clergyman	5
Knight, Thomas Andrew (1759–1838)	English horticulturist, bee keeper, and plant physiologist	4, 8
Kölreuter, Joseph Gottlieb (1733–1806)	German botanist, an authority of hybridization and pollination	4, 9, 14, 15
Lamarck, Jean-Baptiste de Monet (1744–1829)	French naturalist, one of the first people to propose evolution as a natural force that has shaped present diversity	1, 4, 14
Lambert, Edward (b. 1717)	Englishman known as the "porcupine man," afflicted with Ichthyosis hystrix	7
Landois, Hermann (1835–1905)	German zoologist	6
Lankester, Edwin Ray (1847–1929)	English zoologist, University College	7
Le Roy, Charles (1723–1789)	French naturalist, animal behaviorist	8
Lepsius, Karl Richard (1810–1884)	German Egyptologist, pioneer in the developing field of modern archaeology	1
Lewes, George Henry (1817–1878)	English writer and literary critic, published on physiology and the nervous system	7
Linnaeus, Carl (1707–1778)	Swedish naturalist; created the classification categories for animals and plants and the binomial system of naming species.	3, 13
Lockwood, Samuel, Rev. (1819–1894)	American naturalist	7
Louis François Ramond (1755–1827)	French geologist, botanist, and politician	12

Name (year)	Description	Chapter reference
Lubbock, John Avebury, Sir (1834–1913)	English banker, politician, anthropologist, botanist, entomologist, a close friend of Darwin	2, 5, 6, 8
Lucas, Prosper (1808–1888)	French medical doctor and specialist in the study of heredity	3, 4, 7 9
Lyell, Charles, Sir (1797–1875)	Scottish-born geologist	10, 11, 12, 13
Malm, August Wilhelm (1821–1882)	Swedish zoologist	7
Malthus, Thomas (1766–1834)	English political economist and priest of the Church of England	3
Marshall, William (1745–1818)	English agriculturist, horticulturist, and writer; wrote about the origins of British cattle breeds	1
Martin, William Charles Linneaus (1798–1864)	English author of natural history books and museum curator	5, 13
Matteucci, Carlo (1811–1868)	Italian physicist and animal physiologist	6
McDonnell, Robert (1828–1889)	Irish surgeon and anatomist	6
Mendel, Gregor (1822–1884)	German-speaking scientist and friar who established many rules of plant heredity	5
Merrell, S. A. (b. 1828)	American homeopathic doctor	8
Miller, William Hallowes (1801–1880)	Welsh-born mineralogist and crystallographer, professor at Cambridge University	8
Milne-Edwards, Henri (1800–1885)	French zoologist and physiologist and the 27th child in his family!	4, 6
Mivart, St. George Jackson (1827–1900)	English comparative anatomist, accepted evolution but not natural selection	6, 7, 14
Morren, Charles François Antoine (1807–1858)	Belgian botanist and horticulturist	7
Müller, Adolf (unknown)	German naturalist	8
Müller, Johan Friedrich Theodor (Fritz) (1822–1897)	German naturalist, worked in the forests of Brazil	4, 6, 7, 8
Müller, Johannes Peter (1801–1858)	German physiologist and comparative anatomist	7, 14
Murchison, Roderick (1792–1871)	Scottish geologist who first defined the Silurian period (about 444 to 419 mya) of the Paleozoic era	10
Murie, James (1832–1925)	Scottish pathologist and naturalist	6
Murray, Andrew (1812–1878)	English lawyer, entomologist, and botanist	5

Name (year)	Description	Chapter reference
Naudin, Charles Victor (1815–1899)	French botanist	5
Newman, Henry Wenman (1788–1865)	English army officer	3
Nitsche, Hinrich (1845–1902)	German zoologist	7
Owen, Richard, Sir (1804–1892)	English comparative anatomist and paleontologist, a very vocal opponent of Darwin's theory of evolution by natural selection	5, 6, 8, 10, 11, 14
Pacini, Filippo (1812–1883)	Italian anatomist	6
Paley, William, Rev. (1743–1805)	English theologian and priest of the Church of England, famous for his watchmaker analogy supporting the concept of intelligent design	6
Pallas, Peter Simon (1741–1811)	German zoologist, botanist, and geographer, led many research expeditions in Russia	5
Perrier, Jean Octave Edmond (1844–1921)	French zoologist	7
Pictet, François Jules (1809–1892)	Swiss zoologist and paleontologist	10, 11
Pierce, James (unknown)	American explorer, geographer, and geologist	4
Pliny the Elder (23–79 AD)1	Roman scholar, naturalist, and encyclopedist	1
Poole, Skeffington, Col. (b. 1803)	English army officer and authority on the horses of India	5
Pouchet, Charles Henri Georges (1833–1894)	French naturalist	7
Radcliffe, Charles Bland (1822–1889)	English physician	6
Ramsey, Andrew (1814–1891)	Scottish geologist	10
Ramsay, Edward Pierson (1842–1916)	Australian amateur entomologist and ornithologist	8
Rengger, Johann Rudolph (1795–1832)	Swiss physician and naturalist	3
Richardson, John, Sir (1787–1865)	Scottish navy surgeon, Arctic explorer, and naturalist	6
Saint-Hilaire, Augustin François César Prouvençal de (1779–1853)	French botanist and naturalist, also known as Auguste de Saint-Hilaire	7, 14
Saint-Hilaire, Étienne Geoffroy (1772–1844)	French vertebrate biologist, specializing in embryology and comparative anatomy, father of Isidore Geoffroy Saint-Hilaire	5, 6, 14

Name (year)	Description	Chapter reference
Saint-Hilaire, Isidore Geoffroy (1805–1861)1	French zoologist, particularly interested in developmental abnormalities, son of Étienne Geoffroy Saint-Hilaire	1, 5
Salvin, Osbert (1835–1898)	English ornithologist and entomologist	7
Sedgwick, Adam (1785–1873)	British geologist and priest, one of the founder of modern geology and one of Darwin's early teachers	10
Schiødte, Jørgen Matthias Christian (1815–1884)	Danish entomologist and naturalist	5, 7
Schlegel, Hermann (1804–1884)	German naturalist and ornithologist	5
Schöbl, Joseph (unknown)	Czech anatomist (Prague) English	7
Scoresby, William (1789–1857)	explorer of the Arctic	7
Silliman, Benjamin (1779–1864)	American chemist, Yale University	5
Smith, Charles Hamilton (1776–1859)	English army officer, naturalist, and artist	5
Smith, Frederick (1805–1879)	English entomologist and authority on ants	8
Smitt, Fredrik Adam (1839–1904)	Swedish biologist	7
Somerville, John Southey (1765–1819)	English farmer and agriculturist; 15th Lord Somerville	1
Spencer, Herbert (1820–1903)	English philosopher and journalist	1, 3, 15
Spencer, John Charles (1782–1845)	English politician and agriculturist; Viscount Althorp, 3rd Earl Spencer	1
Sprengel, Christian Konrad (1750–1816)	German botanist and Lutheran priest, particularly known for his pioneering work on the pollination of flowers by insects	4, 5
Steenstrup, Japetus (1813–1897)	Danish zoologist, biology, and professor	14
St. John, Charles William George (1809–1856)	English naturalist and sportsman	8
Tegetmeier, William Bernhard (1816–1912)	English naturalist, journalist, pigeon fancier, and poultry expert, also known for his work on bees	8
Thwaites, George Henry Kendrick (1811–1882)	English botanist, entomologist, and government official	5

Name (year)	Description	Chapter reference
Traquair, Ramsay (1840–1912)	Scottish naturalist, paleontologist, and authority on flatfish and fish fossils	7
Van Mons, Jean Baptiste (1765–1842)	Belgian chemist, horticulturist, and breeder of pears, producing more than 40 varieties in his lifetime	1
Verlot, Bernard (1836–1897)	French botanist	8
Virchow, Rudolf (1821–1902)	German doctor, writer, and biologist	6
von Baer, Karl Ernst (1792–1876)	Estonian zoologist, geologist, and embryologist, Königsberg University	4
von Eschwege, Wilhelm Ludwig (1775–1855)	German geologist, geographer, mineralogist and engineer of German mines; did much of his work in Portugal and Brazil	10
von Gärtner, Karl Friedrich (1772–1850)	German botanist with particular expertise in plant hybridization	9
von Goethe, Johann Wolfgang (1749–1832)	German poet and naturalist	5
von Heer, Oswald (1809–1883)	Swiss paleobotanist, entomologist, and geologist	4
von Helmholtz, Hermann (1821–1894)	German physician and physicist	6
von Humboldt, Friedrich Alexander (1769–1859)	German geographer, naturalist, explorer, philosopher and scientist	10
von Nägeli, Carl Wilhelm (1817–1891)	Swiss botanist, maintained a teleological view of evolution	7
von Nathusius, Hermann (1809–1879)	German animal breeder	6
Wagner, Moritz Friedrich (1813–1887)	German explorer, naturalist, and a leading proponent of the idea that geographical isolation can promote speciation	4
Wallace, Alfred Russel (1823–1913)	Welsh born naturalist, explorer, and anthropologist; co-discoverer of the theory of evolution by natural selection	1, 2, 4, 6, 7, 8, 10, 12, 13, 14, 15
Walsh, Benjamin (1808–1869)	American entomologist, and the first State Entomologist of Illinois; strong supporter of Darwin's theory	5

Name (year)	Description	Chapter reference
Waterhouse, George Robert (1810–1888)	English entomologist, zoologist, and geologist	5, 8
Watson, Hewett Cottrell (1804–1881)	English botanist and phrenologist	2, 4, 5, 6
Westwood, John Obadiah (1805–1893)	English entomologist and paleographer	2, 5
Wollaston, Thomas Vernon (1822–1878)	English entomologist and conchologist	2, 5, 6
Woodward, Henry (1832–1921)	English geologist and paleontologist, especially known for his work with fossilized arthropods	10, 11
Wright, Chauncey (1830–1875)	American mathematician and philosopher	7
Wyman, Jeffries (1814–1874)1	American anatomist and ethnologist, professor at Harvard University	1
Yarrell, William (1784–1856)	English zoologist	7
Youatt, William (1776–1847)	English veterinary surgeon and author of a series of handbooks on domesticated animals	1, 4

Illustration Credits

Chapter 1

Figure 1.1   © PanuRuangjan/istock.

Figure 1.2   Drawing by Ardea Thurston-Shaine.

Figure 1.3   Pigeons bred by Aycan Seckin (A), Nathanael Medley (B), and Anthony Allel (C); photographs by David McIntyre. D: Illustration by Chinami Michaels.

Figure 1.4   Pigeons bred by James Ashton (A), Annette de Bruycker (B), Eli Stottman (C), and Frank J. D'Alessandro (D); photographs by David McIntyre.

Figure 1.5   Original figure created by Casey Diederich.

Figure 1.6   A: © russwitherington1/istock. B: Courtesy of David Hillis, Double Helix Ranch.

Figure 1.7   A: David McIntyre. B: © brackish_nz/istock.

Figure 1.8   A: © pigphoto/istock. B: © Kim Nguyen/ Shutterstock.

Chapter 2

Figure 2.1   Courtesy of Jeffrey W. Lotz, Florida Department of Agriculture and Consumer Services, Bugwood.org.

Figure 2.2   A: Courtesy of Dezidor/Wikipedia, under a Creative Commons Attribution 3.0 Unported License, creativecommons.org/licenses/by/3.0/deed.en. Used unmodified. B: © JonnyJim/istock.

Chapter 3

Figure 3.1   © StevenRussellSmithPhotos/Shutterstock.

Figure 3.2   © Suljo/istock.

Figure 3.3   Figure created by Casey Diederich.

Figure 3.4   © Elliotte Rusty Harold/Shutterstock.

Figure 3.5   © AndreAnita/ Shutterstock.

Figure 3.6   Courtesy of Jan Pechenik.

Figure 3.7	A: Courtesy of Dr. Thomas G. Barnes, University of Kentucky/US Fish and Wildlife Service. B: David McIntyre. C: © godrick/ istock.
Figure 3.8	A: © Alfredo Maiquez/Shutterstock. B: © Alexander Erdbeer/ Shutterstock.
Figure 3.9	© Ruud Morijn/Shutterstock.
Figure 3.10	A: © Dirk Freder/istock. B: Courtesy of James Gath-any/Centers for Disease Control.
Figure 3.11	A: © Vitalii Hulai/istock. B: © wojciech_gajda/ istock.

Chapter 4

Figure 4.1	A:© Dr. Morley Read/Shutterstock. B: © Henrik_L/istock.
Figure 4.2	A: © feath-ercollector/Shutterstock. B: © neil hardwick/Shutterstock.
Figure 4.3	Courtesy of Clemson University, USDA Cooperative Extension Slide Series/Bugwood.org.
Figure 4.4	© alexomelko/istock.
Figure 4.5	Courtesy of Timothy Knepp/U.S. Fish and Wildlife Service.
Figure 4.6	© szefei/istock.
Figure 4.7	Illustration by Elizabeth Card.
Figure 4.8	A: © IgorGorelchenkov/istock. B: © helovi/istock.
Figure 4.9	A: © worldswildlifewonders/Shutterstock. B: From Castelnau, Francis, comte de, 1859. *Expédition dans les parties centrales de l'Amérique du Sud, de Rio de Janeiro à Lima, et de Lima au Para.*
Figure 4.10	A: © Alexia Khrushcheva/istock. B: © Makarova Viktoria/ Shutterstock.
Figure 4.11	Rendered from Darwin's original by Casey Diederich.
Figure 4.12	© QueenTut/istock.
Figure 4.13	© Lebendkulturen.de/Shutterstock.
Figure 4.14	© cbpix/istock.
Figures 4.15, 4.16	David McIntyre.

Chapter 5

Figure 5.1	A:© Stubblefield Photography/Shutterstock. B: David McIntyre.
Figure 5.2	© Larsek/Shutterstock.
Figure 5.3	© Cosmin Manci/Shutterstock.
Figure 5.4	A: © CreativeNature.nl/Shut-terstock. B: From Alcide Dessalines d'Orbigny, 1847. *Voyage dans l'Amérique méridionale.*
Figure 5.5	© Dchauy/Shutterstock.
Figure 5.6	David McIntyre.

Figure 5.7   © Shawn Hempel/Shutterstock.
Figure 5.8   A: © NNehring/istock. B: Operation Deep Scope 2005 Expedition: NOAA Office of Ocean Exploration. C: © Lebendkulturen. de/Shutterstock.
Figure 5.9   © Peter Zijlstra/istock.
Figure 5.10   © Robert Kyllo/ Shutterstock.
Figure 5.11   Pigeons bred by James Ashton (A) and Kelsea Reid (B); photographs by David McIntyre.
Figure 5.12   A: © Sergei25/Shutterstock. B: Watercolor by Nicolas Marechal, 1793.

Chapter 6

Figure 6.1   A: © Erni/Shutterstock. B: © Joe McDonald/Steve Bloom Images/ Alamy. C: © Ryan M. Bolton/ Shutterstock.
Figure 6.2   Illustration by John Gerrard Keulemans, 1876.
Figure 6.3   Courtesy of the National Oceanic and Atmospheric Administration.
Figure 6.4   © Joab Souza/ Shutterstock.
Figure 6.5   © David Dohnal/ Shutterstock.
Figure 6.6   © MikeLane45/istock.
Figure 6.7   © visceralimage/Shutterstock.
Figure 6.8   Courtesy of Wagner Machado Carlos Lemes, under a Creative Commons BY 2.0 license, creativecommons.org/licenses/by/2.0/. Used unmodified.
Figure 6.9   A: Courtesy of Lieutenant Elizabeth Crapo, NOAA Corps. B: Illustration by John Gould (1804–1881). C: © SteveOehlenschlager/ istock.
Figure 6.10   Courtesy of Andrew2606 at en.wikipedia, under a Creative Commons Attribution BY 3.0 license, creativecommons.org/licenses/by/3.0/ deed.en. Cropped from original.
Figure 6.11   A: Courtesy of Andrew D. B: © pcnorth/Shutterstock. C: © PhotonCatcher/Shutterstock. D: Illustration by Johann Friedrich Naumann, 1899.
Figure 6.12   David McIntyre.
Figure 6.13   © micro_photo/istock.
Figure 6.15   © ankh-fire/istock.
Figure 6.16   A: Illustration by Paul Louis Oudart, 1847. B: Courtesy of Roberto Pillon, under a Creative Commons Attribution BY 3.0 license, creativecommons. org/licenses/by/3.0/deed.en. Cropped from original.
Figure 6.17   Courtesy of David Sim, under a Creative Commons BY 2.0 license, creativecommons.org/licenses/by/2.0/. Cropped from original.
Figure 6.18   Courtesy of Christopher Pooley and Eric Erbe/ USDA ARS, EMU.
Figure 6.19   © humbak/Shutterstock.

Figure 6.20	A: David McIntyre. B: Illustration from Darwin, C., 1862. *On the three remarkable sexual forms of* Catasetum tridentatum. J. Proc. Linnean Society 6: 151.
Figure 6.21	A: © valeriopardi/istock. B: © villy_yovcheva/istock.
Figure 6.22	A: David McIntyre. B: © MarcelClemens/Shutterstock.
Figure 6.23	Illustration by Ernst Haeckel, 1904.
Figure 6.24	© marlee/ Shutterstock.
Figure 6.25	© Cosmin Manci/Shutterstock.
Figure 6.26	David McIntyre.

Chapter 7

Figure 7.2	A: © imigra/istock. B: Illustration by Robert Bruce Horsfall, 1913.
Figure 7.3	© Chantal de Bruijne/Shutterstock.
Figure 7.4	© Johan Larson/Shutterstock.
Figure 7.5	A: © PushishDon-hongsa/istock. B: Courtesy of Randall Wade (Rand) Grant, under a Creative Commons BY 2.0 license, creativecommons. org/licenses/by/2.0/. Cropped from original.
Figure 7.6	A: © Brian Lasenby/Shutterstock. B: © Angela Arenal/istock. C: © KarenMassier/istock. D: © o2beat/istock. E: © Wayne Lynch/All Canada Photos/Corbis.
Figure 7.7	From Warne, F., 1893. *The Royal Natural History.*
Figure 7.8	A: © Chris Moody/Shutterstock. B: From Goode, G. B., and T. H. Bean, 1896. *Oceanic Ichthyology.*
Figure 7.9	© lofilolo/istock.
Figure 7.10	A: From Brusca, R., and Brusca, G. 2003. *Invertebrates.* Associates, Sunderland, MA. B: © Gary C. Togno-ni/Shutterstock.
Figure 7.11	From British Museum, 1901. *A Guide to the shell and starfish galleries of the British Museum.* Insets from Brusca, R., and Brusca, G. 2003. *Invertebrates.* Associates, Sunderland, MA.
Figure 7.12	© Martin Fowler/Shutterstock.
Figure 7.13	David McIntyre.
Figure 7.14	From Edwards, S., 1827. Botanical Register vol. 13: 1108.
Figure 7.15	© Milosz_G/Shutterstock.
Figure 7.16	Painting by Heinrich Harder (1858–1935).

Chapter 8

Figure 8.1	© Mark William Penny/Shutterstock.
Figure 8.2	A: © Koo/Shutterstock. B: © Jemini Joseph/ Shutterstock.
Figure 8.3	© Alexander Wild.

Figure 8.4   A: David McIntyre. B: © Inven-tori/istock.

Figure 8.5   A: © red2000/istock. B: Courtesy of Wildfeuer, under a Creative Commons BY 2.5 license, creativecommons.org/licenses/by/2.5/. Cropped from original.

Figure 8.6   © Redmond O. Durrell/Alamy.

Chapter 9

Figure 9.1   Drawn by Ardea Thurston-Shane

Figure 9.2   https://search.creativecommons.org/photos/5d7e8460-b487-463f-8de1-59d9d11eb227 "Scarlet Pimpernel (Anagallis arvensis)" by jkirkhart35 is licensed under CC BY 2.0

Figure 9.3   https://search.creativecommons.org/photos/d6e1f0fe-45a1-4d4e-8f7c-35f5c2b8dcc2 "Male Chinese goose head" by howcheng is licensed under CC BY-SA 2.0

Figure 9.4   https://search.creativecommons.org/photos/dacc5f2e-8f5a-46f9-a578-f4ba86e1c96b "Hippeastrum aulicum" is licensed under CC BY-SA 4.0

Figure 9.5   https://www.researchgate.net/figure/Pictures-of-representative-equids-A-Donkey-Pega-B-horse-Mangalarga-Marchador-C_fig1_308673939

Chapter 10

Figure 10.1   Stick trying to decide

Figure 10.2   https://search.creativecommons.org/photos/7f652060-69fe-4ed5-abf6-8676debbf2ea

Figure 10.3   https://commons.wikimedia.org/w/index.php?curid=85146689

Figure 10.4   https://commons.wikimedia.org/wiki/File:Bellacartwrightia_species_trilobite.jpg

Figure 10.5   A. https://search.creativecommons.org/photos/bf932379-80d8-419c-b027-96b88e91df89; B. https://search.creativecommons.org/photos/06f9f696-f9d2-48e0-95e3-2139ae7547ea

Chapter 11

Figure 11.1   https://search.creativecommons.org/photos/03089a4f-8904-474e-bb4c-d5ee4345b71e

Figure 11.2A   https://commons.wikimedia.org/wiki/File:Megatherium_american
um_Marcus_Burkhardt.jpg

Figure 11.2B   https://search.creativecommons.org/photos/f55653be-f16f-46c9-
9dae-7d974524a89a

Chapter 12

Figure 12.1   https://search.creativecommons.org/photos/b3df5c57-9d06-4924-
aece-ba3a6a465859

Figure 12.2   https://search.creativecommons.org/photos/6152f762-7bf6-4319-
90ee-4aa2c1779597

Chapter 13

Figure 13.1   https://commons.wikimedia.org/wiki/File:Dyticus_submargina
tus.jpg

Figure 13.2   https://search.creativecommons.org/photos/b968e64f-8274-4d69-
a941-b4ddc0c2b5c8

Chapter 14

Figure 14.1A   https://commons.wikimedia.org/wiki/File:Xanthoceras_radicle.jpg ;

Figure 14.1B   https://commons.wikimedia.org/wiki/File:Dycotyledon_seed_diag
ram-en.svg

Figure 14.2   https://commons.wikimedia.org/wiki/File:Tasmanian_wolf.jpg

Figure 14.3   https://commons.wikimedia.org/wiki/File:Duck_billed_platypus_s
chnabeltier.jpg

Figure 14.4   https://commons.wikimedia.org/wiki/File:Viscacha_at_Villamar.jpg

Figure 14.5   https://commons.wikimedia.org/w/index.php?sort=relevance&-
search=barnacle+nauplius&title=Special:Search&profile=advan
ced&fulltext=1&advancedSearch-current=%7B%7D&ns0=1&ns6=
1&ns12=1&ns14=1&ns100=1&ns106=1#/media/File:FMIB_46414_
Larval_stages_of_the_common_rock_barnacle_(Balanus_balanoi
des).jpeg

Figure 14.6   https://search.creativecommons.org/photos/f47a5807-7859-4b9f-
9f71-6b27c3b6e656

Figure 14.7   https://digitalcollections.lib.washington.edu/digital/collection/fis
himages/id/51193

Index

For the benefit of digital users, indexed terms that span two pages (e.g., 52–53) may, on occasion, appear on only one of those pages.

Tables, figures, and boxes are indicated by *t*, *f*, and *b* following the page number. Page numbers followed by an "n" and a number indicate that the information is in a numbered footnote.